Smart Innovation, Systems and Technologies

Volume 374

The Smart Innovation, Systems and Technologies book series encompasses the topics of knowledge, intelligence, innovation and sustainability. The aim of the series is to make available a platform for the publication of books on all aspects of single and multi-disciplinary research on these themes in order to make the latest results available in a readily-accessible form. Volumes on interdisciplinary research combining two or more of these areas is particularly sought.

The series covers systems and paradigms that employ knowledge and intelligence in a broad sense. Its scope is systems having embedded knowledge and intelligence, which may be applied to the solution of world problems in industry, the environment and the community. It also focusses on the knowledge-transfer methodologies and innovation strategies employed to make this happen effectively. The combination of intelligent systems tools and a broad range of applications introduces a need for a synergy of disciplines from science, technology, business and the humanities. The series will include conference proceedings, edited collections, monographs, handbooks, reference books, and other relevant types of book in areas of science and technology where smart systems and technologies can offer innovative solutions.

High quality content is an essential feature for all book proposals accepted for the series. It is expected that editors of all accepted volumes will ensure that contributions are subjected to an appropriate level of reviewing process and adhere to KES quality principles.

Indexed by SCOPUS, EI Compendex, INSPEC, WTI Frankfurt eG, zbMATH, Japanese Science and Technology Agency (JST), SCImago, DBLP.

All books published in the series are submitted for consideration in Web of Science.

Roumen Kountchev · Srikanta Patnaik ·
Wenfeng Wang · Roumiana Kountcheva
Editors

Multidimensional Signals, Augmented Reality and Information Technologies

Proceedings of 3DWCAI 2023

Springer

Editors
Roumen Kountchev
Technical University of Sofia
Sofia, Bulgaria

Wenfeng Wang
Shanghai Institute of Technology
Shanghai, China

Srikanta Patnaik
Interscience Institute of Management
and Technology
Bhubaneswar, Odisha, India

Roumiana Kountcheva
TK Engineering
Sofia, Bulgaria

ISSN 2190-3018 ISSN 2190-3026 (electronic)
Smart Innovation, Systems and Technologies
ISBN 978-981-99-7010-0 ISBN 978-981-99-7011-7 (eBook)
https://doi.org/10.1007/978-981-99-7011-7

This Springer imprint is published by the registered company Springer Nature Singapore Pte Ltd.
The registered company address is: 152 Beach Road, #21-01/04 Gateway East, Singapore 189721, Singapore

Paper in this product is recyclable.

Preface

This book contains part of the papers approved for publication in the proceedings of the Second World Conference on Intelligent and 3D Technologies (WCI3DT2023), which comprises two books, entitled:

AI Methods and Applications in 3D Technologies and *Multidimensional Signals, Augmented Reality and Information Technologies.*

The conference took part on May 26–28, 2023, in Shanghai, China.

The aim of WCI3DT 2023 was to provide a wide forum to academicians and practitioners where they to present latest scientific results and to exchange ideas in the area of Artificial intelligence and Deep learning and their applications in augmented reality and 3D technologies. The conference organizers were focused at the establishment of an effective platform for all participants to introduce their work to scientists, engineers and students from all over the world.

After reviewing, 61 papers were accepted for presentation and publication in the conference proceedings, from which 31 are in this volume. The selected works present contemporary research works based on the multidimensional signal processing, such as a special system for FSAC racing car; a new approach for soil moisture prediction, a method for corn grain crushing rate, a fault detection method for seismic images, new implementation ideas in augmented reality and 3D animation, creation of intelligent VR systems based on Deep Learning (DL), implementation of digital twins technology, human pose estimation and head anatomy modeling based on self-attention mechanism and DL, feature extraction based on time-series images, intelligent robots control in complicated environment, and many other interesting ideas. The chapters are arranged in three groups, which cover different parts of the related scientific areas:

- Multidimensional Signal Processing (9 chapters);
- Augmented Reality and Deep learning (14 chapters);
- IT-based Applications (8 chapters).

In memory of Prof. Dr. Roumen Kountchev, Dr. Sc., *Best Paper Award* was established by the Organizing Committee. Due to high number of very good works, the selection was extremely difficult. Two papers, selected as *winner paper*, were published in the volume *AI Methods and Applications in 3D Technologies*. The winner papers are:

"*Extended Prototypical Network for Few-shot Learning*", with authors: Jingjing Zhang, Lujie Cui, Wenfeng Wang, and Lalit Mohan Patnaik;

and:

"*Low Carbon Scheduling of Thermal Power Unit Thermal Storage Capacity Based on Particle* Swarm *Optimization*", with authors: Wenbin Cao, Mingkai Wang, Bo Han, Qi Chai, Hongtao Li, and Ying Liu.

The book editors express their special thanks to IRNet International Academic Communication Center who organized this conference in correspondence with their dedication to build platforms for scholars and researchers for better knowledge sharing, together with providing full-scale conference services that meet the standards of the renowned publishing organizations.

We also thank Prof. Lakhmi Jain (Honorary chair), Prof. Wenfeng Wang, Prof. Srikanta Patnaik, Prof. Gang Sun, and Prof. Xudong Jiang (General chairs), Dr. Yonghang Tai, Dr. Shoulin Yin, and Prof. Hang Li (Organizing chairs), and S.R. Dr. Roumiana Kountcheva (International advisory chair).

The editors express their warmest thanks to the excellent Springer team for making this book possible.

Sofia, Bulgaria
Bhubaneswar, India
Shanghai, China
Sofia, Bulgaria
August 2023

Prof. Dr. Roumen Kountchev
Prof. Dr. Srikanta Patnaik
Prof. Dr. Wenfeng Wang
Dr. Roumiana Kountcheva

Contents

About the Editors

Roumen Kountchev Ph.D., D. Sc. is a professor at the Faculty of Telecommunications, Department of Radio Communications and Video Technologies at the Technical University of Sofia, Bulgaria. His scientific areas of interest are: digital signal and image processing, image compression, multimedia watermarking, video communications, pattern recognition and neural networks. Professor Kountchev has 420 papers published in magazines and conference proceedings; 20 books; 48 book chapters; and 20 patents. At present, he is a member of Euro Mediterranean Academy of Arts and Sciences and the president of the Bulgarian Association for Pattern Recognition (member of IAPR). He has been a plenary speaker at more than 30 international scientific conferences and symposia and edited several books published in Springer SIST series and was a co-editor of Special issues for Symmetry MDPI.

Dr. Srikanta Patnaik is a Professor in the Department of Information Technology & Systems Management at I.I.M.T., Bhubaneswar, India. He has received his Ph.D. (Engineering) in Computational Intelligence from Jadavpur University, India in 1999 and supervised more than 25 Ph.D. and 30 M. Tech theses in the area of Machine Intelligence, Soft Computing Applications and Re-engineering. Dr. Patnaik has published more than 100 research papers in international journals and conference proceedings. He is the author of 3 textbooks and edited more than 100 books and few invited book chapters, published by leading international publishers like Springer-Verlag, Kluwer Academic, SPIE, IEEE, ACM and IOS Press etc. Dr. Patnaik was the principal investigator of TAPTEC project "Building Cognition for Intelligent Robot", sponsored by All India Council for Technical Education, New Delhi and Major Research Project "Machine Learning and Perception Using Cognition Methods", sponsored by University Grant Commission. He is the Editor-in-Chief of International Journal of Information and Communication Technology and International Journal of Computational Vision and Robotics published from Inderscience Publishing House, England and also Editor-in-Chief of the book series on "Modeling and Optimization in Science and Technology", published from Springer, Germany.

Prof. Wenfeng Wang is the editor-in-chief of IJANS and IJEEE; EBM of Nature—Scientific Reports; a general chair of the World Conference on Intelligent and 3D Technologies; and a reviewer of many SCI journals, including some top ones—Nature Computational Science, Expert System with Applications, Water Research, Science China-Information Sciences, Science of the Total Environment, Environmental Pollution, IEEE Transactions on Automation Science and Engineering, etc.; Prof. Wenfeng Wang is the plenary/keynote speaker of many international conferences—AMICR2019, IACICE2020, NAMSP2021, AMMCS2021, ICCEAI2021, AICS2021, 3DIT-MSP&DL2020, CSAMCS 2021, AMMCS2021, ICPEM2021, ATAMI2021, ICRDET2022 etc.

Roumiana Kountcheva got her M.Sc. and Ph.D. at the Technical University of Sofia, Bulgaria, and in 1992, she got the title Senior Researcher. At present, she is the vice president of TK Engineering, Sofia. She had postgraduate trainings in Fujitsu and Fanuc, Japan. Her main scientific interests are in image processing, image compression, digital watermarking, pattern recognition, image tensor representation, neural networks, CNC and programmable controllers. She has more than 200 publications and 5 patents (EU and USA). R. Kountcheva was the plenary speaker at 23 international scientific conferences and scientific events. She edited several books published in Springer SIST series and was the co-editor of Special issues for Symmetry MDPI. She is a honorary member of the Honorable Editorial Board of the nonprofit peer-reviewed open-access IJBST Journal Group.

Part I
Multidimensional Signal Processing

Chapter 1
Design and Optimization of Steering-by-Wire System for FSAC Racing Car

Guogang Liu and Yingjie Fu

Abstract The Formula Student Autonomous China is the most prestigious innovation competition among the series of popular science events initiated by the China Society of Automotive Engineers. Compared to traditional race cars, Autonomous formula vehicles emphasize the autonomous driving capability. The steering-by-wire system is a crucial module in the lateral control of the car, and its performance directly affects the obstacle avoidance and turning ability of the vehicle. The steering system should possess sensitive, reliable, and stable control performance. In the design of the steering system for the new season, based on the mechanical technology inspection and installation testing from the previous year, we have identified a series of issues with the steering system, such as excessive free play in the steering wheel, looseness in the universal joint, and inadequate optimization of the Ackermann steering trapezoid. To address these problems, this paper will focus on four aspects: steering system parameter design according to the competition rules, steering trapezoid optimization using MATLAB, finite element analysis using ANSYS, and simulation optimization of the steering system using Adams car. The aim is to optimize the design of the steering system for the WUTA team in the 2023 season, improve the vehicle's maneuverability, and achieve better control accuracy. This paper aims to provide design references for the WUTA team's steering system in the 2023 season and share optimization design experience for other Autonomous formula teams.

G. Liu (✉)
School of Automotive Engineering, Wuhan University of Technology, Wuhan, China
e-mail: guog_liu1123@whut.edu.cn

Y. Fu
International Education College, Wuhan University of Technology, Wuhan, China
e-mail: fu17362984739@whut.edu.cn

R. Kountchev et al. (eds.), *Multidimensional Signals, Augmented Reality and Information Technologies*, Smart Innovation, Systems and Technologies 374,
https://doi.org/10.1007/978-981-99-7011-7_1

1.1 Introduction

1.1.1 Status of Foreign Research

In the 1980s, some Western countries began to explore autonomous driving [1] technology. However, due to limitations in key technologies such as computer technology and image processing, the research progress was not smooth. It was not until the late 1990s that Japan successfully conducted tests on autonomous vehicles, taking a leading position. This successful test promoted the development of autonomous driving technology and set an example for other countries.

With the continuous advancement of technology, in 2001, the Defense Advanced Research Projects Agency (DARPA) [2] in the United States held the first DARPA Grand Challenge for autonomous driving, which attracted widespread attention. This challenge prompted autonomous driving technology to become a hot research and development area, with many companies, universities, and research institutions investing in research and development of autonomous driving technology.

Over time, autonomous driving technology has made more breakthroughs. In 2010, Google started investing in autonomous driving technology and conducted the first public road test [3]. This test marked the entry of autonomous driving technology into the application phase in the real world and attracted global attention and discussion.

The development of autonomous driving technology is not only driven by technological advancements but also benefits from the support and investment of governments and private sectors worldwide. Many countries have recognized the immense potential of autonomous driving technology and have formulated policies and plans to encourage and promote related research and innovation. Autonomous driving technology has become a competitive field globally, and countries hope to take a leading position in this field and enjoy the economic and social benefits brought by autonomous driving technology.

1.1.2 Current Research Status in China

In recent years, China's attention to autonomous driving technology has gradually increased. Although relatively lagging behind, Chinese universities have achieved a series of accomplishments in the field of autonomous driving through collaborative efforts from various backbone institutions. The National University of Defense Technology is one of China's important research institutions in the field of autonomous driving [4]. They independently developed the first unmanned vehicle and achieved a speed of 72 km/h during testing, which was the highest level in the domestic field of high-speed control for autonomous driving at that time. This achievement marked an important breakthrough for China in autonomous driving technology. Jilin University has conducted in-depth research in artificial intelligence technology, information fusion technology, and has made significant progress in the research of modified actual vehicles. They are committed to applying advanced technology to the field of autonomous driving, promoting the development of related technologies [5]. Chang an University has also achieved certain accomplishments in the field of autonomous driving. They have carried out hardware modification and research on actual vehicles and have participated in autonomous driving challenges multiple times, achieving good results in the competition [6]. These efforts have made positive contributions to the development of autonomous driving technology in China. Additionally, Tongji University has also made certain contributions in the field of autonomous driving technology. They have developed an unmanned electric sightseeing car, which can reach a maximum tested speed of 52 km/h, making it very suitable for closed park scenarios. This achievement provides practical solutions for the application of autonomous driving technology in specific scenarios.

1.1.3 Introduction to the Competition

The Formula Student Autonomous China [7] is an innovative competition hosted by the China Society of Automotive Engineers (China SAE). It is one of the highest-level events among the series of popular science events initiated by the China SAE and aims to promote the research and development of autonomous driving technology and provide a platform for university students to demonstrate their innovative abilities. The core objective of the competition is to design, manufacture, and drive a fully autonomous electric race car. The participating teams are composed of students from various universities who, through teamwork, use advanced technology and innovative thinking to design and develop autonomous driving race cars capable of autonomously perceiving the environment, making decisions, and driving safely. The competition imposes high technical requirements on the participating vehicles, requiring them to complete multiple tasks in autonomous driving mode, including following a predefined track, avoiding obstacles, performing precise parking, and performing acceleration and braking actions. The safety, stability, and performance of the vehicles are also important evaluation criteria.

The Formula Student Autonomous China competition plays an important role in promoting the development of autonomous driving technology in China and cultivating young talents. It provides students with a stage to exercise and showcase their abilities and also promotes research and innovation in autonomous driving technology in China. Since its establishment in 2017, the Formula Student Autonomous China competition has attracted more and more universities to participate. These universities have demonstrated excellent technical capabilities, innovative thinking, and teamwork spirit in the competition. The competition not only provides students with an opportunity to showcase their talents but also offers a platform for communication, learning, and collaboration with peers. By participating in the Formula Student Autonomous China competition, students can gain in-depth understanding of the cutting-edge development of autonomous driving technology and have opportunities to apply and practice these technologies. At the same time, the competition provides opportunities for industry-academia collaboration, promoting cooperation and communication between the academic and industrial sectors in the field of autonomous driving.

Compared to traditional racing, the autonomous driving competition challenges the research and development capabilities of universities and has obvious breakthroughs and academic exchange. Since its establishment in 2021, the WUTA team from Wuhan University of Technology participated in the competition for the first time in the 2022 season and achieved excellent results, winning the national third prize for the Formula Student Autonomous China competition and the Best Newcomer Award in the NIO Cup. The racing participation photos are shown in Fig. 1.1. The team hopes that the research on the wire-controlled steering system can help our university achieve better results and inject fresh blood into the field of autonomous driving.

Fig. 1.1 WuHan University of technology autonomous driving car (WUTA in 2022)

1.2 Selection of Mechanism

1.2.1 Selection of Steering Motor

As we can see from the Fig. 1.2, the layout of the general power steering motor can be divided into three main types: C-EPS (Column Electric Power Steering), P-EPS (Pinion Electric Power Steering), and R-EPS (Rack Electric Power Steering) [8]. For the wire-controlled steering motor, we have chosen the C-EPS layout due to the requirements of FSAC regulations and the limitations of the vehicle's body structure. This layout allows the steering motor to be positioned on the steering column, ensuring sufficient space in the front compartment. As shown in Fig. 1.3, we are using the first-generation wire-controlled steering system from Tian Shang Yuan (DWSBW2001), which consists of a wired steering control unit, angle sensor, torque sensor, and actuator components. It has a compact size, simple structure, low cost, and is suitable for modification and design in autonomous Formula Student racing. The system integrates angle and torque sensors and uses CAN communication for signal transmission, meeting communication requirements. Additionally, the motor's dynamic parameters are well-matched for the application.

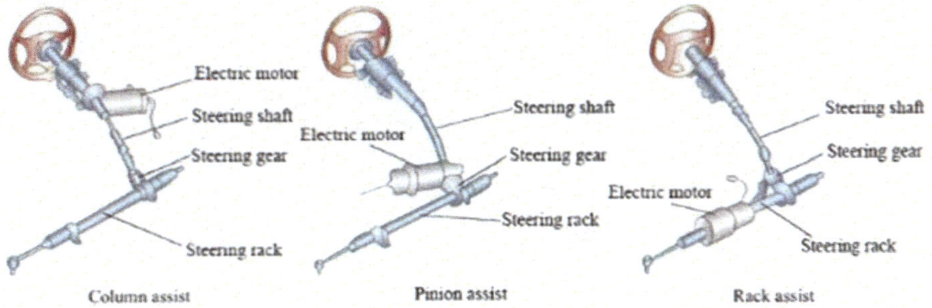

Fig. 1.2 Electric power steering system

Fig. 1.3 Electrically controlled steering motor

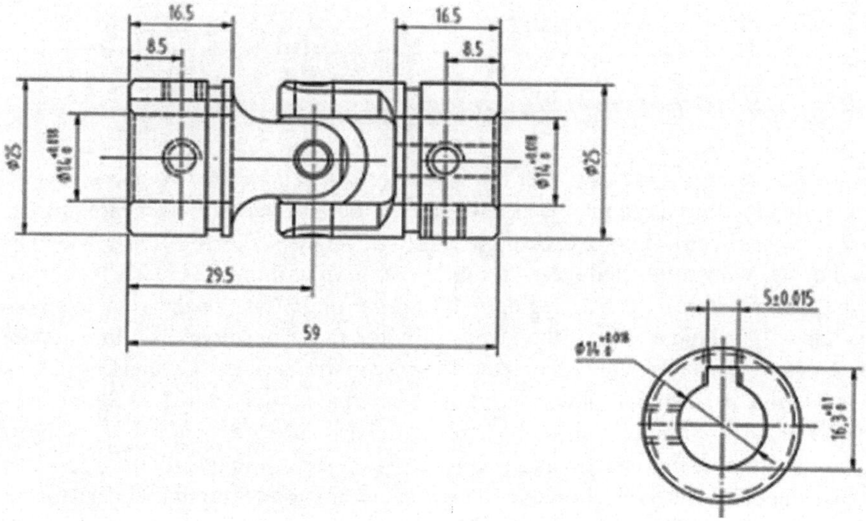

Fig. 1.4 Improved universal joint blueprint

1.2.2 Universal Joint Selection

This season, we will be using an improved double-cross-axis universal joint for power transmission. Compared to the universal joint used in the previous season, this type of universal joint has smaller clearances and more stable transmission performance. It features keyway slots and two threaded tops, making it easier for connection. With these improvements, we can partially avoid excessive free travel and enhance the stability of power transmission. Figure 1.4 shows the improved Universal Joint Blueprint.

1.2.3 Steering Gear Selection

For this season, we have chosen a rack and pinion steering gear, specifically designed for passenger vehicles and small commercial vehicles with independent front suspension. Compared to the straight-tooth rack and pinion steering gear used in the previous season, the rack and pinion gear with helical teeth offers several key advantages. It provides smooth transmission, low noise, high meshing accuracy, strong load-bearing capacity, long service life, and is free from backlash and impact phenomena. Despite the higher manufacturing cost, we have incorporated a clearance adjustment mechanism to ensure precise transmission and proper meshing clearance between the gears and the rack.

The tooth profile, steering mechanism geometry, and gear-rack parameters are shown in Figs. 1.5 and 1.6, and Table 1.1, respectively.

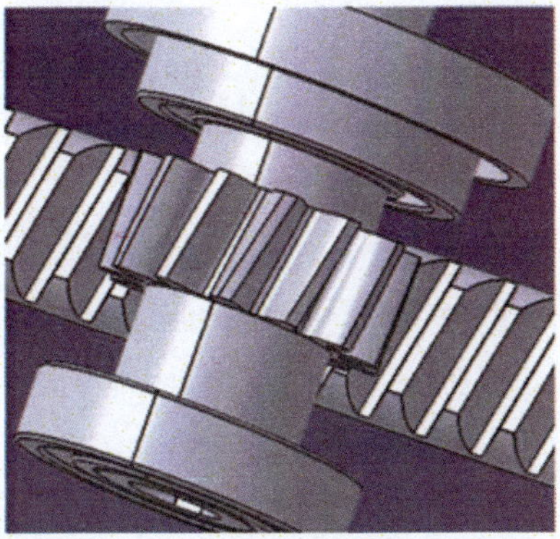

Fig. 1.5 Steering gear pinion and rack

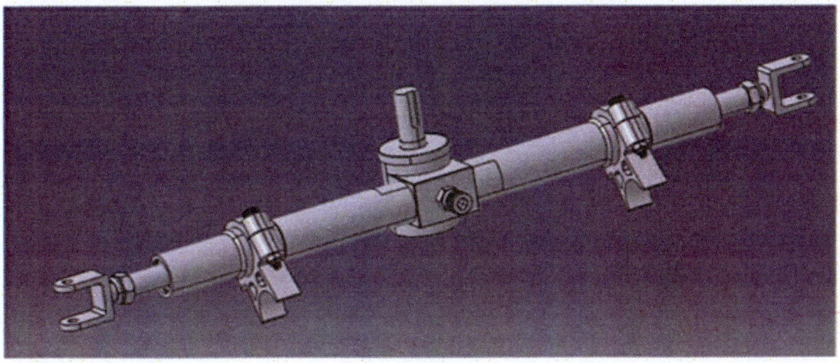

Fig. 1.6 Steering mechanism

Table 1.1 Gear parameters

Name	Code	Gear	Rack
Normal module	m_n	2	2
Number of teeth	z	13	15
Pressure angle	α	20°	20°
Helix angle	β	12°	12°

1.3 Steering System Parameters Calculation

The vehicle parameters are listed in the Table 1.2.

Table 1.2 Vehicle technical specifications

Design parameters	Code	Value
Gross vehicle weight	M	350 kg
Wheelbase	L	1614 mm
Track width	B	1254 mm
Kingpin offset distance	c	50 mm
Front wheel camber angle	θ	2.28°

1.3.1 The Calculation of the Maximum Steering Angle

According to the competition regulations, the minimum radius of curvature for a standardized track is 4 m. Understeering may occur, so the minimum turning radius is set to 3.5 m. The formula for calculating the maximum steering angle of the outer wheel is:

$$R_0 = \frac{L}{\sin\theta_o} + c \tag{1.1}$$

Among them: R_0 represents the turning radius (mm); L is the wheelbase (mm); θ_o denotes the steering angle of the outer wheel (°), and c represents the kingpin offset distance (mm). According to Eq. (1.1), the maximum steering angle $\theta_{omax} = 27.9°$. Therefore, the designed maximum steering angle for the outer wheel is set to 28°.

1.3.2 Calculation of Angular Gear Ratio

The ratio between the change in steering wheel rotation and the corresponding change [9] in the steering angles of the two sides is known as the steering system gear ratio. Since the inner and outer wheel angles differ, the incremental steering angles of both sides are represented by their average value, which is denoted as:

$$i_\omega = \frac{\Delta\varphi_h}{\frac{\Delta\theta_i + \Delta\theta_o}{2}} \tag{1.2}$$

Among them: i_ω represents the steering system gear ratio; $\Delta\varphi_h$ is the change in steering wheel rotation; $\Delta\theta_i$ and $\Delta\theta_o$ are the changes in the steering angles of the inner and outer wheels, respectively.

The steering wheel range for reference in conventional vehicle models is typically ±140°. According to the racing regulations [7], it is not allowed to cross hands while turning the steering wheel. Based on literature [10] and vehicle conditions, the steering gear ratio is designed as 5:1, meaning that when the outer wheel reaches its maximum steering angle of 28°, the steering wheel rotation will be 140°. A certain margin is kept during the design process, so the designed steering wheel rotation angle is greater than 140°.

1.3.3 Steering System Torque Transmission Ratio

The definition of steering system torque transmission ratio is as follows:

$$i_T = \frac{T_{Rt}}{T_h} \tag{1.3}$$

i_T represents the torque transmission ratio of the steering system; T_{Rt} is the torque that the steering joints need to overcome during steering; T_h is the torque applied by the driver on the steering wheel.

In practice, the torque transmission ratio is generally calculated using the following equation:

$$i_T = i_\omega \times \eta_{SG} \times \eta_{SL} \tag{1.4}$$

where i_ω represents the steering system gear ratio, η_{SG} is the steering gear efficiency, and η_{SL} is the steering linkage efficiency.

Assuming a gear ratio of 5:1, a steering gear efficiency of approximately 90% for a rack and pinion steering system, and a steering linkage efficiency of 85%, the torque transmission ratio of the steering system is calculated to be 3.825.

1.3.4 Calculation of Rack Travel and Trapezoidal Arm Length

For conventional passenger cars, the gear module of rack and pinion steering systems typically ranges between 2 and 3 mm. Considering the light overall weight of the FSAC race car and the smaller steering torque, a gear module of 2 mm is chosen. The gear is selected to have 13 teeth, a pressure angle of 20°, and a helix angle of 12°. The single linear travel distance of the rack is

$$L_{R1} = \frac{\alpha_{max}}{360} \times \frac{\pi m_n z}{\cos\beta} \tag{1.5}$$

$L_{R1} = 32.5$ mm, based on this information, the total linear travel distance of the rack is 65.0 mm.

To calculate the length of the trapezoidal arm, considering the rack travel distance and the maximum steering angle:

$$L_s = \frac{L_R}{2sin\theta_{max}} \tag{1.6}$$

$L_s = 69.2$ mm. $\theta_{max} = 28°$, $L_R = 65.0$ mm.

The length of the trapezoidal arm is 70.0 mm. Further optimization will be performed based on the steering trapezoid.

1.3.5 Determination of Steering System Calculation Load

Calculation of Steering System Load.

The calculation formula is as follows:

$$M_R = \frac{f}{3}\sqrt[2]{\frac{{G_1}^3}{p}} \tag{1.7}$$

M_R (Nm) represents the resistance moment for the car to perform stationary steering; f is the sliding friction coefficient between the tires and the road surface, usually taken as 0.7; G_1 (N) is the vertical load on the steering wheel; p (MPa) is the tire inflation pressure.

Taking the vertical load on the front axle as 1332.8 N and the tire inflation pressure as 0.08 MPa, the calculation yields: M_R is 40.14 Nm.

Steering Wheel Effort

$D_{s\omega}$ Represents the steering wheel diameter (mm); i_ω is the steering system angular gear ratio; η_{SG} is the steering gear positive efficiency.

$$F_h = \frac{2M_R}{D_{s\omega} i_\omega \eta_{SG}} \tag{1.8}$$

With a steering wheel diameter of 210 mm, a steering system angular gear ratio of 5, and a steering gear positive efficiency of 85%, the calculation yields: F_h is 89.95 N, which falls within the range of the national standard specification for automotive steering systems [11].

Steering Gear/Load Calculation

F_s Represents the magnitude of the steering gear output force, M_R is the resistance moment for the car to perform stationary steering (Nm); L_s is the length of the steering trapezoid arm (mm); θ is the kingpin inclination angle; η_s is the positive efficiency of the steering trapezoid mechanism, assumed to be 85%.

$$F_s = \frac{M_R}{L_s \times cos^2\theta \times \eta_s} \tag{1.9}$$

Given a resistance moment for stationary steering of 40.14 Nm, a steering wheel moment of 9.44 Nm, and a trapezoid arm length of 70.0 mm, solving the equation yields $F_s = 675.7$ N.

1.3.6 Selection of the Disconnected Point

According to the suspension geometry parameters, the outer point position is determined, and the breaking point is determined based on the Three-Heart Theorem. For a double wishbone independent suspension, a commonly used graphical method (based on the Three-Heart Theorem) is employed to determine the position of the lateral pull rod breaking point. The coordinates of the breaking point are obtained through CATIA R20, a powerful computer-aided design software, as shown in Fig. 1.7. In this figure, the preliminary distance for the breaking point is set to 540 mm.

The coordinates of the breaking point are obtained through CATIA drawing, as shown in the figure, with a preliminary distance of 540 mm for the breaking point.

1.4 Optimized Design of the Steering Trapezoid

1.4.1 Selection of the Steering Trapezoid

The Integral-Type Steering Trapezoid

The integral-type steering trapezoid consists of a steering lateral pull rod, steering trapezoid arm, and front axle, as shown in Fig. 1.8, the advantages of this design are its simplicity, low manufacturing cost, and ease of adjusting the front toe. However, a drawback is that the jounce of the left and right steering wheels can affect each other.

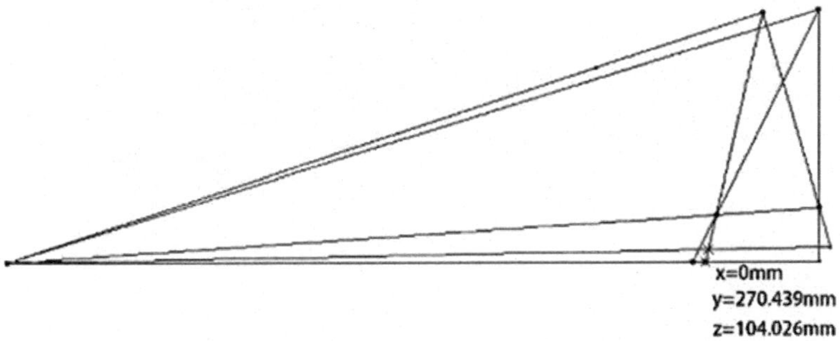

Fig. 1.7 Steering break point coordinates

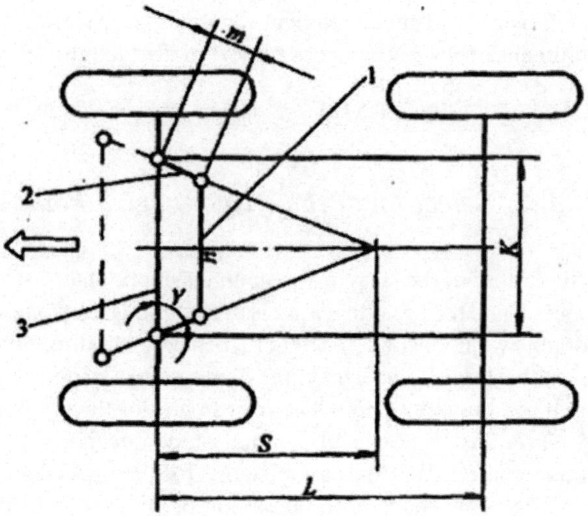

Fig. 1.8 The integral-type steering trapezoid

Disconnected Steering Trapezoid

This race car adopts a disconnected steering trapezoid, suitable for double wishbone independent suspension. As can be seen from the Fig. 1.9, the characteristic of this structure is that the up-and-down movements of the left and right wheels do not affect each other. Although this structure is complex, difficult to manufacture, and has higher costs, and adjusting the front toe is more challenging, it is necessary to carefully arrange the components to avoid interference and maintain sufficient clearance to prevent obstacles on the road from affecting the steering system. In conclusion, this paper has chosen the disconnected steering trapezoid with a front-mounted arrangement.

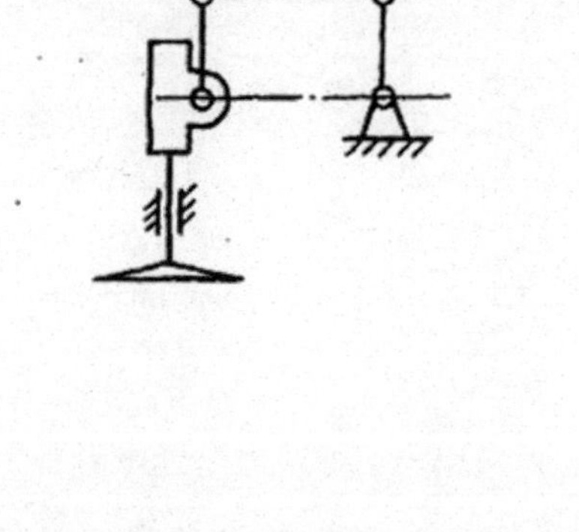

Fig. 1.9 Disconnected steering trapezoid

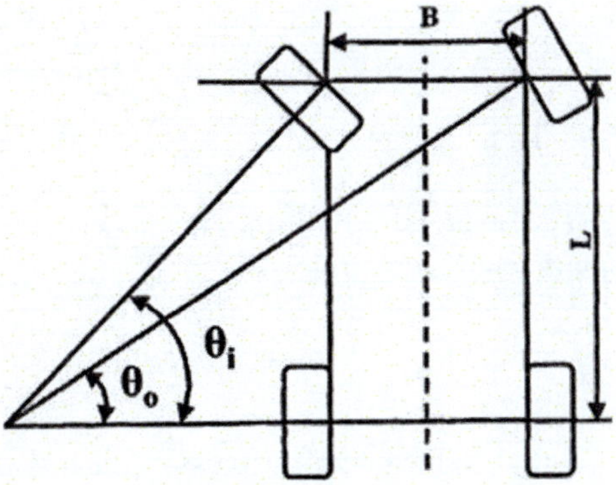

Fig. 1.10 Trapezoidal relationship of standard Ackermann steering

1.4.2 Geometry Model of the Steering Trapezoid

As shown in the Fig. 1.10, when θ_i is the inner steering wheel angle, θ_o is the outer steering wheel angle, L is the wheelbase, and K is the distance between the extended centerlines of the two kingpins and the intersection point on the ground, the steering trapezoid mechanism should ensure the following relationship between the angles of the inner and outer steering wheels:

$$cot\theta_o - cot\theta_i = \frac{K}{L} \tag{1.10}$$

Given θ_o, θ_i can be obtained as follows:

$$\theta_i = \text{arccot}(\text{cot}\theta_o - \text{K/L}) \tag{1.11}$$

1.4.3 Selection of Ackermann Correction Factor

FSAC race car is equipped with many unmanned devices, making it heavier than other race cars, and it often takes corners at high speeds. In order to maintain traction performance, FSAC race car uses tires with a larger aspect ratio, which results in a greater lateral slip angle. Due to weight transfer during cornering, the lateral slip angles of the tires are not equal. Therefore, using the standard Ackermann coefficient [12] can lead to severe tire wear issues.

To address the above problem, the method of using Ackermann correction coefficients is commonly employed. Different Ackermann correction coefficients are used considering the lateral slip conditions of different vehicle models to counteract the adverse effects caused by tire lateral slip. The definition of the Ackermann correction coefficient is denoted as K:

$$K = \frac{\theta_i' - \theta_o}{\theta_{it} - \theta_o} \times 100\% \tag{1.12}$$

θ_i is the actual inner wheel angle, θ_o is the actual outer wheel angle, and θ_{it} is the inner wheel angle corresponding to θ_o under the standard Ackermann condition.

The expression for the inner wheel angle with the Ackermann correction coefficient K is as follows:

$$\theta_{i1} = K arctan \frac{L tan\theta_o}{L - B tan\theta_o} + (1 - K)\theta_o \tag{1.13}$$

Based on the lateral stiffness of the tire, the expression for the actual inner wheel angle θ_{i2}, considering the lateral slip angle and the redistribution of vertical load during turning, can be derived.

$$\theta_{i2} = \mathrm{arccot}\left[\cot(\theta_o + \sigma_f)\frac{cot\sigma_r - c}{cot\sigma_r + c} - 2c\frac{cot\sigma_r}{cot\sigma_r + c}\right] - \sigma_f \tag{1.14}$$

As θ_o varies, θ_{i1} and θ_{i2} should be as close as possible, and the two curves should match as closely as possible. Then, we use MATLAB, a widely-used technical computing and programming environment, provided the necessary tools and capabilities for implementing the algorithm. By utilizing MATLAB 2022b, an optimization algorithm was developed to generate the results depicted in Fig. 1.11.

According to the optimization results, the optimal Ackermann correction coefficient can be taken as 55%.

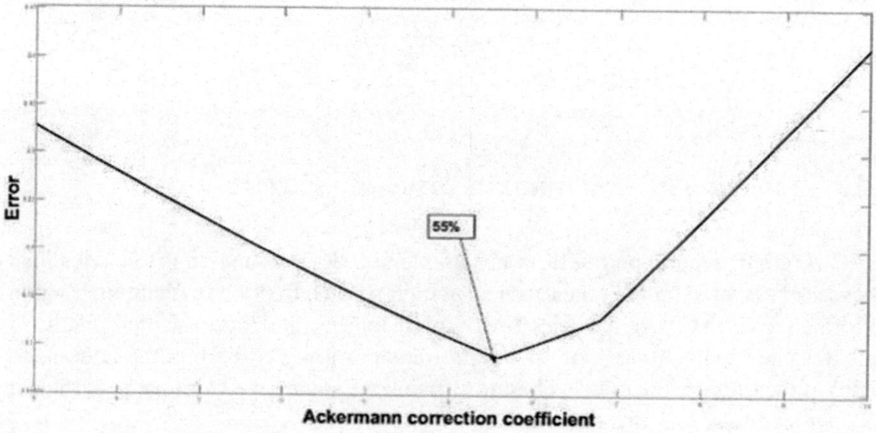

Fig. 1.11 Optimization results for K

1.4.4 Optimization of Steering Trapezoid Parameters

The structure of the front-disconnect steering trapezoid is shown in Fig. 1.12, based on the internal space design of the race car. In the figure, M represents the disconnect point, which is the distance between the inner ends of the two steering tie rods in the actual structure. N represents the distance between the intersection points of the extended centerlines of the two front wheels on the ground. h represents the distance between the front-mounted steering gear and the front axle, which is used to position the steering gear. L_1 represents the length of the trapezoid arm, and L_2 represents the length of the tie rod. λ represents the base angle of the trapezoid.

When the right wheel turns, the outer wheel steering angle is θ_o. According to geometric relationships, it can be determined that:

$$A = \frac{N-M}{2} \tag{1.15}$$

$$l_2 = \sqrt{(A - l_1 cos\lambda)^2 + (l_1 sin\lambda - h)^2} \tag{1.16}$$

$$s = l_1\cos(\lambda - \theta_o) + \sqrt{l_2{}^2 - [l_1 sin(\lambda - \theta_o) - h]^2} - A \tag{1.17}$$

$$\theta_{ir} = arccot\left(cot\theta_o - \frac{K}{L}\right) \tag{1.18}$$

$$\theta_i = -\lambda + arctan\frac{h}{(A-s)} + arccos\frac{l_1{}^2 + h^2 + (A-s)^2 - l_2{}^2}{2l_1\sqrt{h^2 + (A-s)^2}} \tag{1.19}$$

A represents the calculation coefficient; s is the rack travel (mm); l_2 is the tie rod length (mm); θ_i is the actual inner wheel steering angle (°); θ_{ir} is the inner wheel steering angle based on the standard Ackermann relationship; θ_o is the outer wheel steering angle; h is the front-set distance of the steering system. By combining Eqs. (1.15), (1.16), (1.17), and (1.19), the relationship between the actual inner wheel steering angle θ_i and the given θ_o can be obtained.

An optimization algorithm implemented in SIMULINK was used to obtain the optimized parameters for the entire vehicle, as shown in the Table 1.3.

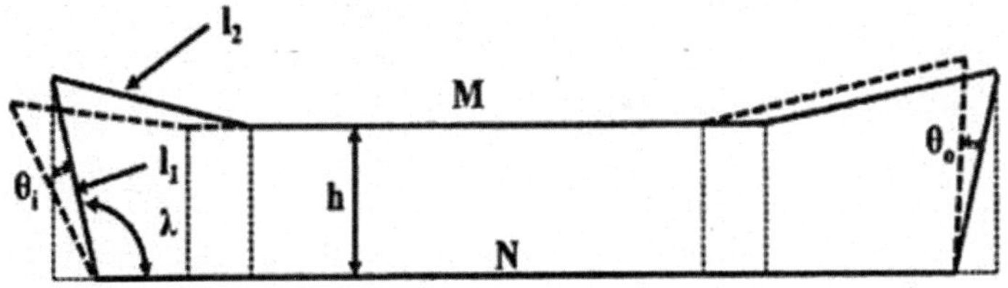

Fig. 1.12 Front-disconnect steering trapezoid

Table 1.3 Vehicle optimization parameters

Project	Value
Wheelbase (mm)	1614
Track width (mm)	1254
Length of trapezoidal arm (mm)	75
Distance to disconnect point (mm)	490
Trapezoidal angle (°)	110
Length of tie rod (mm)	380

By taking the Ackermann correction coefficient K = 55%, the expression for the inner wheel angle with the Ackermann correction factor of 55% for a given outer wheel angle θ_o is obtained as follows:

$$\theta_{i1} = 0.55arctan\frac{Ltan\theta_o}{L - Btan\theta_o} + (0.45)\theta_o \tag{1.20}$$

With the outer wheel angle θ_o varying within the range of 0–28°, the corresponding inner wheel angle θ_i should be as close as possible to θ_{i1}. Therefore, using SIMULINK, an optimization algorithm was implemented, and the result is shown in Fig. 1.13.

From the figure, it can be observed that the actual inner and outer wheel curves closely match the curve with a 55% Ackermann correction coefficient. The optimization results are quite satisfactory.

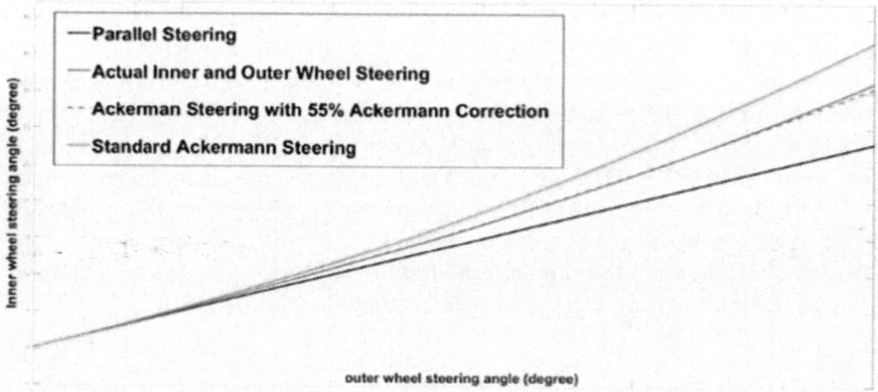

Fig. 1.13 Optimization results

1.5 Simulation and Optimization

1.5.1 Finite Element Analysis

Gear Stress Analysis

The upper end of the gear shaft is connected to the universal joint, which receives the force and torque from the steering wheel transmitted through the universal joint. The gear material is 45Cr with a yield strength of σ_s=835 MPa [13].

Finite element analysis of the gear was performed using ANSYS. Based on the calculated load, a vertical force of 150 N was applied at the pitch circles of two teeth in the model. The stress analysis plot and displacement plot are shown in Figs. 1.14 and 1.15, respectively.

As can be seen, the maximum stress is 11.26 MPa, which is significantly lower than the yield strength of 835 MPa, meeting the requirements.

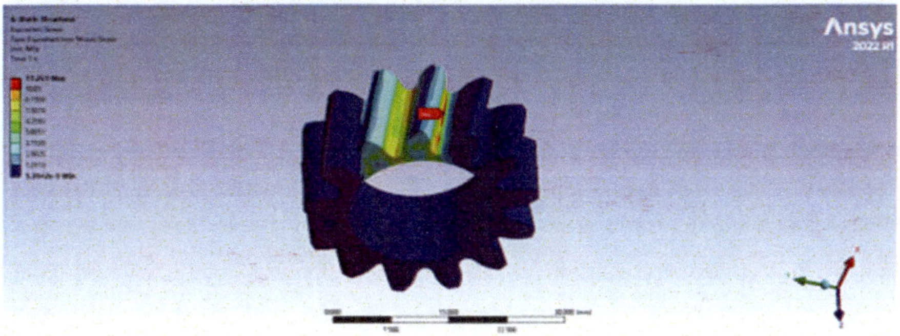

Fig. 1.14 Gear shaft stress profile

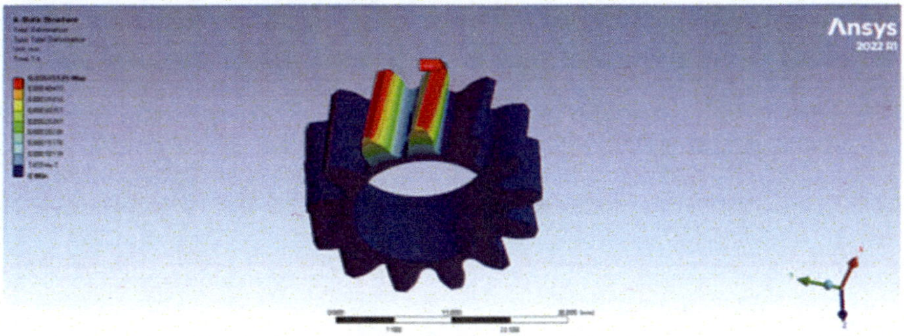

Fig. 1.15 Gear shaft displacement diagram

The maximum displacement of the gear is approximately 0.00046 mm, which meets the design requirements.

Rack Stress Analysis

The rack is made of 45Cr. When a 500 N axial force is applied to two teeth in the same direction on the rack, the displacement and stress distribution of the rack are shown in Figs. 1.16 and 1.17, respectively.

The maximum displacement occurs at the center position of the rack and is approximately 0.02329 mm.

The maximum stress is approximately 63.779 MPa, which is significantly lower than the yield strength of 835 MPa for the 45Cr material. Therefore, it meets the design requirements.

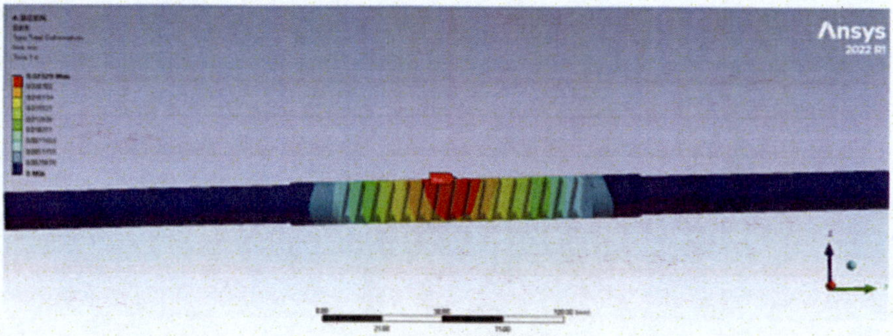

Fig. 1.16 Rack displacement diagram

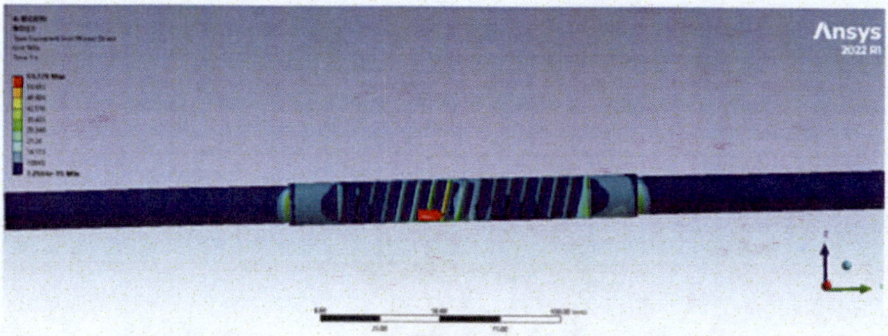

Fig. 1.17 Rack stress profile

1.5.2 ADAMS Car Virtual Prototyping Analysis

Adams Car is an advanced multibody dynamics simulation software that allows for the analysis of front suspension and steering systems. By utilizing Adams Car [14], comprehensive design and analysis of the front suspension and steering system can be conducted to optimize the vehicle's stability and handling performance. This section will provide a comprehensive study of the steering system, covering modeling, simulation, result analysis, and optimization.

Below is the Table 1.4 of key hardpoints for the front suspension and steering system:

Template Creation and Simulation.

First, when creating the front suspension model, it is necessary to have parameters and components that match the actual vehicle, such as the chassis, front suspension, steering system, and tires. The front suspension is a double wishbone suspension with unequal-length arms, including lower control arms, upper control arms, shock absorbers, and springs. By simulating different road conditions and operating conditions, the performance and comfort of the front suspension can be evaluated. The steering system is a rack-and-pinion type, and a gear-rack model is established with the appropriate gear ratio set.

To begin, key hard points are extracted from the front suspension and steering system of the race car. Then, a front suspension assembly model and a steering system model are created in Adams Car, shown in Figs. 1.18 and 1.19.

Table 1.4 Key hardpoints of front suspension and steering system

Hardpoints	Name	Location (x, y, z)
LCA_front	Lower control arm front point	(442, 191, 79)
LCA_rear	Lower control arm rear point	(810, 194, 95)
LCA_outer	Lower control arm outer point	(640, 582, 150)
UCA_front	Upper control arm front point	(459, 250, 288)
UCA_rear	Upper control arm rear point	(838, 255, 293)
UCA_outer	Upper control arm outer point	(661, 567, 385)
TIEROD_inner	Tie rod inner point	(579, 235, 123)
TIEROD_outer	Tie rod outer point	(550, 592, 190)
DAMPER_outboard	External attachment point of the damper	(641, 210, 575)
DAMPER_inboard	Internal attachment point of the damper	(641, 20, 563)
BC_center	Rocker arm center	(641, 216, 522)
PUSHROD_outboard	External attachment point of the push rod	(639, 550, 160)
PUSHROD_inboard	Internal attachment point of the push rod	(641, 283, 560)
WC	Wheel center	(640, 631, 260)

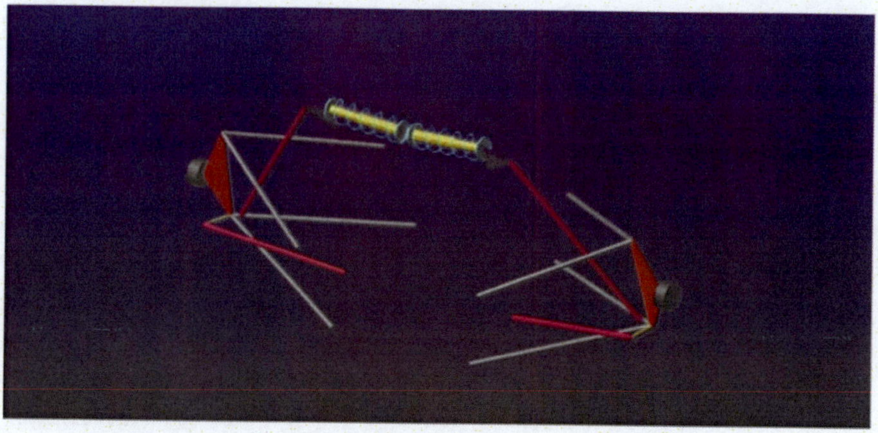

Fig. 1.18 Front suspension assembly model

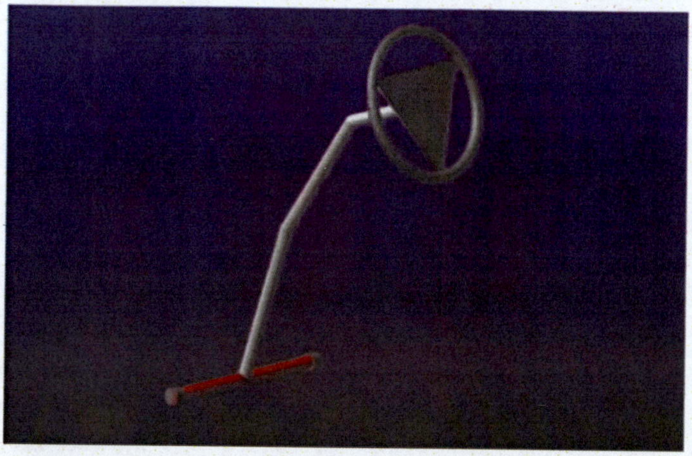

Fig. 1.19 Steering system model

The assembly includes components such as shock absorbers, springs, knuckles, pushrods, upper and lower control arms, uprights, steering tie rods, steering arms, gear-rack, steering column, and steering wheel, as well as motion joints such as ball joints, Hooke's joints, revolute joints, and universal joints.

Next, the front suspension and steering system are assembled into an assembly, as shown in the following Fig. 1.20.

To analyze the performance of the steering system, various parameters of the system need to be set. For the gear-rack system, the gear-rack transmission ratio is set to 0.06, and the trapezoidal pre-alignment distance is set to 75 mm.

To perform the analysis, select the "Steering" module in the "Suspension Analysis" section. Based on the previous design of the steering system, set the maximum

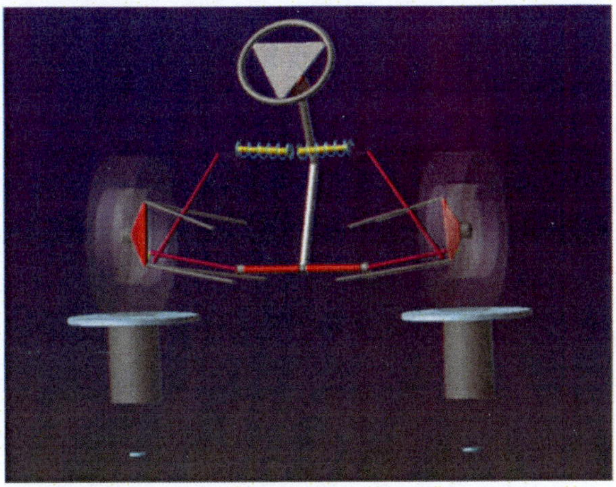

Fig. 1.20 Front suspension assembly model

steering wheel angle to ±140°. The simulation settings are shown in the following Fig. 1.21.

Analysis of Steering System Simulation Results

After conducting the simulation, key output data can be obtained, such as transmission ratio variation, wheel angle vs. steering wheel angle relationship, and Percent Ackermann variation. These data can be used to evaluate the performance and control characteristics of the steering system and make modifications and optimizations to the design.

Relationship between Steering Wheel Angle and Front Wheel Angle

From the Fig. 1.22, it can be observed that as the steering wheel angle rotates from −140° to 140°, the curve representing the relationship between the steering wheel angle and the front wheel angle for the left front wheel is relatively smooth without any fluctuations. The maximum outer wheel angle is close to 25°, while the inner wheel angle is close to 28°, which deviates significantly from the design inner wheel angle of 35°. The smaller wheel angle can result in a larger turning radius. In dynamic racing scenarios, encountering smaller turns with insufficient steering angle can lead to collisions or even going off track. Therefore, optimization measures are needed to address this issue.

Relationship between Percent Ackermann and Steering Wheel Angle

As shown from Fig. 1.23, The Percent Ackermann fluctuates within the range of −100 to 100% as the wheel angle changes. As the wheel angle increases, the Percent Ackermann also increases, reaching a maximum of around 50%. The fluctuation amplitude is almost 200% near 0°. In dynamic racing scenarios, precise steering control is required, and significant fluctuations in the Percent Ackermann can lead to unstable steering response, compromising vehicle stability. Additionally, it can

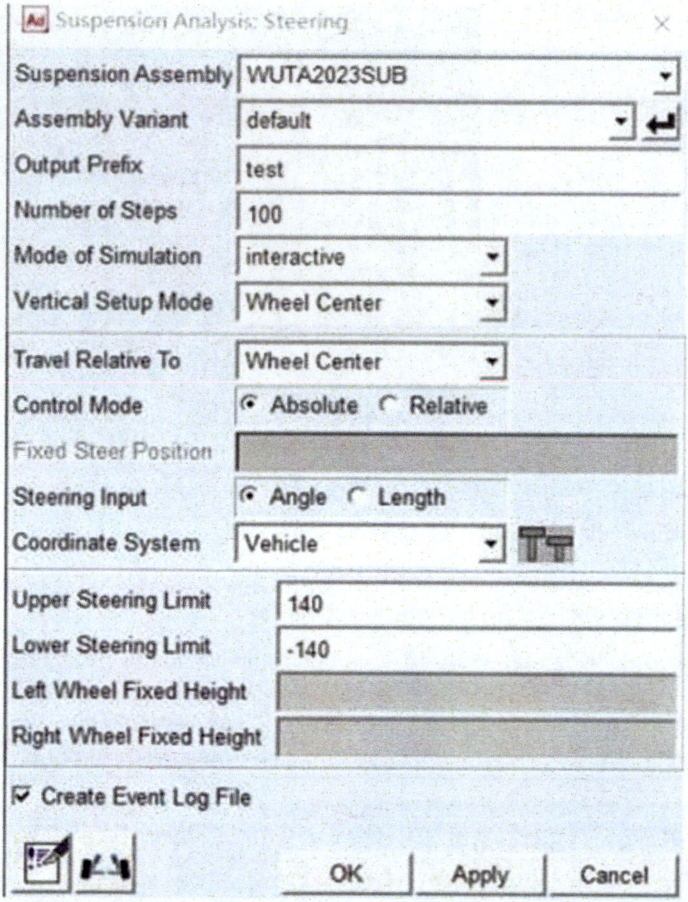

Fig. 1.21 Suspension analysis: steering

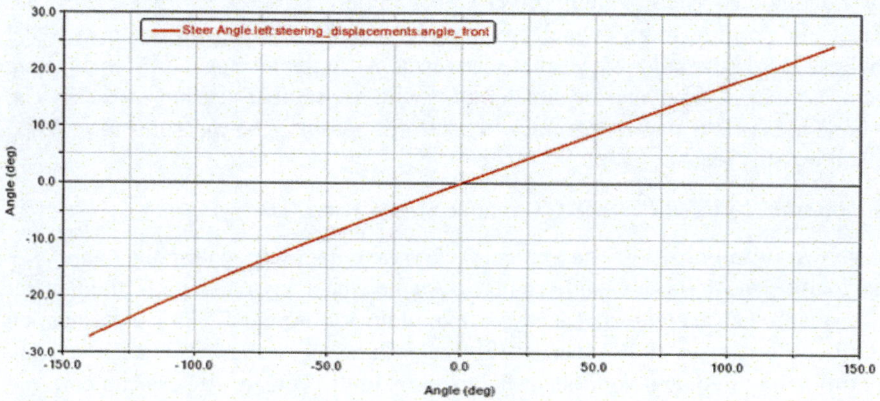

Fig. 1.22 Left road wheel angle versus steering wheel angle

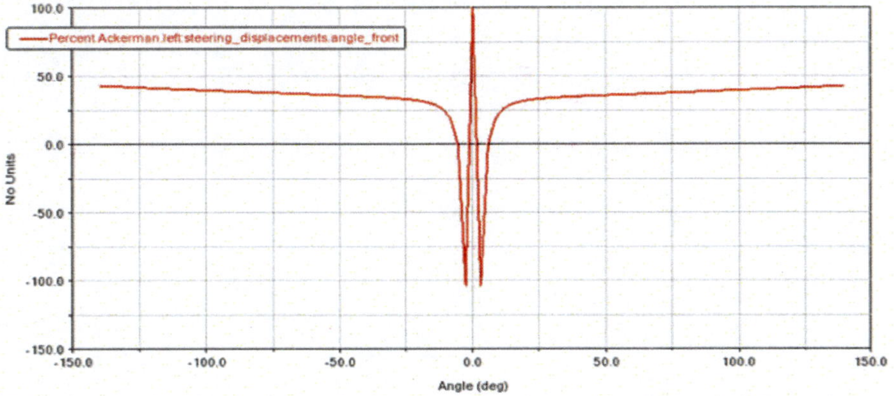

Fig. 1.23 Percent Ackerman versus steering wheel angle

cause uneven tire wear. Therefore, further optimization is needed to address these issues.

Relationship between Steering Ratio and Steering Wheel Angle

The designed steering ratio in previous discussions is 5:1. During the wheel turning process, the steering ratio may vary within a certain range due to design and installation errors. Figure 1.24 shows fluctuations within the range of 4.5–5.75. Although there is room for optimization, the difference is relatively small.

Maximum Turning Diameter

Considering that the minimum radius of the track in dynamic racing is 4.5 mm, the race car needs to have a smaller turning radius of R = 3.5 mm to ensure smooth passage. However, based on the Fig. 1.25, when the steering wheel angle is at its maximum, the minimum turning diameter is 9.8 mm, which does not meet the

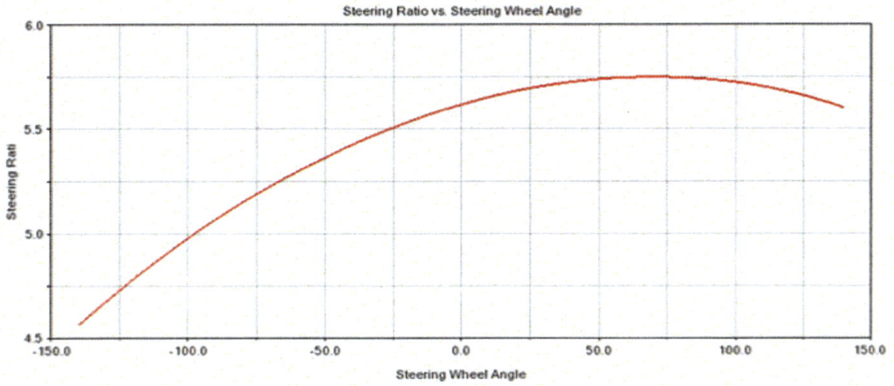

Fig. 1.24 Steering ratio versus steering wheel angle

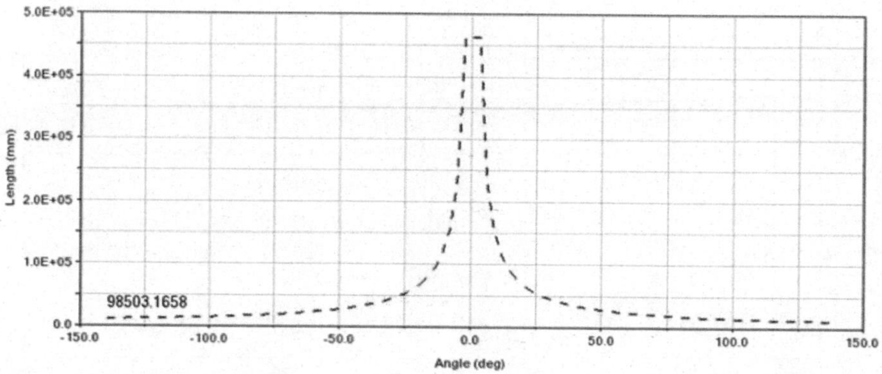

Fig. 1.25 Outside turn diameter

Table 1.5 Optimization results

Optimization parameters	Results
Gear-rack transmission ratio	0.05
TIEROD_inner	(583, −245, 118)
TIEROD_outer	(540, −620, 191)
Trapezoidal front track distance	80 mm

design concept. Therefore, optimization is required to achieve the desired turning performance.

Steering System Optimization

Adams Car provides various methods for optimizing the steering system, including parameter optimization and steering geometry optimization. In this study, a combined approach of these two methods will be used to optimize the gear rack ratio, disconnect point coordinates, steering arm contact point coordinates, and the trapezoidal pre-alignment distance.

The feedback and iterative design method is a design approach based on real-time feedback information and continuous iterative improvements. It is used to optimize the steering system. After multiple evaluations and iterations based on the actual feedback information, the optimized results of the steering system are obtained as shown in Table 1.5, Figs. 1.26, 1.27 and 1.28.

From the optimization results and figures above, it can be observed that the maximum inner wheel steering angle is around 35°, and the fluctuation range of the Ackermann percentage is smaller compared to before optimization. The maximum turning radius is approximately 3.8 m, which is more in line with the design objective.

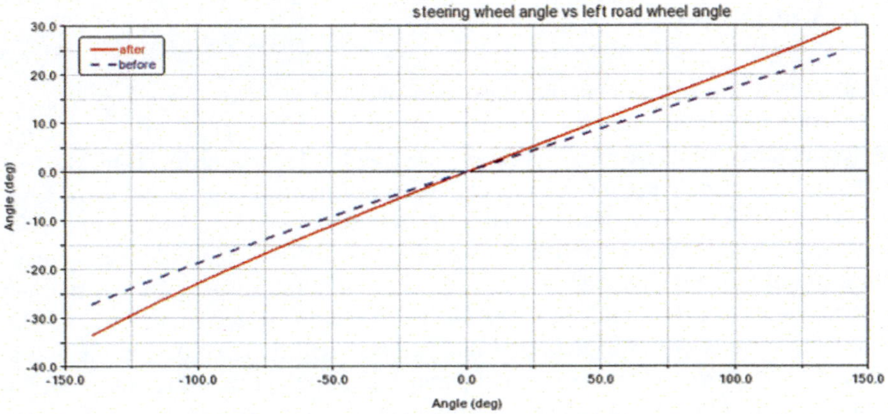

Fig. 1.26 The optimized results of front wheel steering angle variations

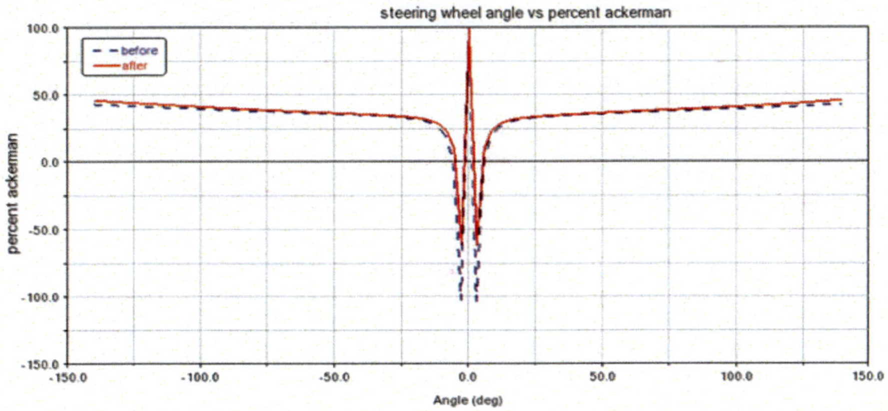

Fig. 1.27 The optimized results of Ackermann percentage variations

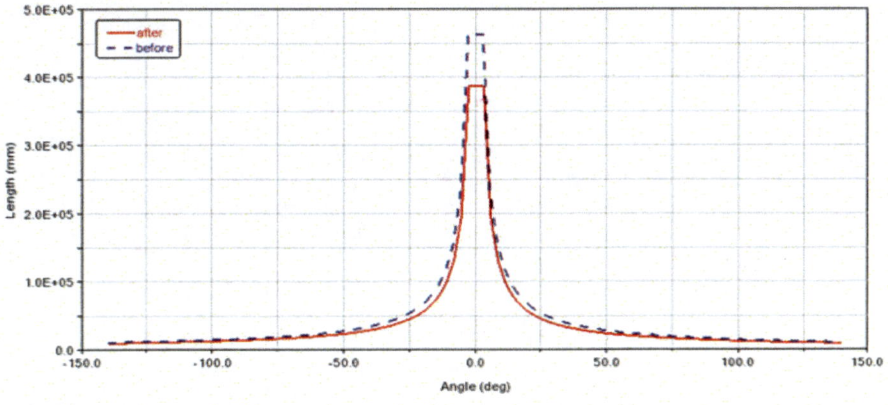

Fig. 1.28 The optimized results of outside turn diameter variations

1.6 Conclusion

This paper begins by summarizing the experience from the previous season's racing events and determines the design process and optimization goals for the steering system in the current season. The process starts with selecting the mechanical structure and then calculating the various parameters of the steering system. MATLAB is used to optimize the steering trapezoid. Ansys is employed to perform finite element analysis on key components to verify their performance under actual operating conditions. Finally, Adams Car is used to build the suspension and steering system model, and the simulation is conducted based on the design parameters. The simulation results are evaluated, and the feedback and iterative design method is utilized to optimize the key points and parameters. The optimized results meet the design requirements and contribute to improving the overall vehicle handling and stability.

One limitation of this study is the absence of a complete vehicle model for dynamic simulation, as only the suspension model is used for simulation, which may result in deviations from the actual performance. In the future, it would be beneficial to develop a complete vehicle model integrated with the racing track to obtain more accurate simulation results.

References

1. Levinson, J., Askeland, J., Becker, J., Dolson, J., Held, D., Kammel, S., Kolter, J.Z., Langer, D., Pink, O., Pratt, V.: Towards fully autonomous driving: systems and algorithms. In: IEEE Intelligent Vehicles Symposium (IV), Baden-Baden, Germany, pp. 163–168 (2011)
2. Seetharaman, G., Lakhotia, A., Blasch, E.P.: Unmanned vehicles come of age: the DARPA grand challenge. Computer **39**(12), 26–29 (2006)
3. Baranova, T., Khalyapina, L., Yakhyaeva, C.: Google products as a source of students' autonomy in content and language integrated learning. In: 2019 12th International Conference on Developments in eSystems Engineering (DeSE), Kazan, Russia, pp. 383–387 (2019)
4. Liu, B.: Verification of rule correctness in autonomous vehicle decision-making systems. Master's thesis, National University of Defense Technology (2015)
5. An, J.: Development of high-precision localization and environment mapping algorithms for autonomous driving in industrial parks. Master's thesis, Jilin University 2020
6. Kang, J.: Research on key technologies for autonomous localization of unmanned vehicles in urban environments. Ph.D. thesis, Chang'an University (2016)
7. Rules of the 2022 China University student formula autonomous driving competition
8. Wei, H., Wang, J., Jian, M., Mei, S., Huang, M.: Steer-by-wire control system based on Carsim and Simulink. In: 2021 IEEE International IOT, Electronics and Mechatronics Conference (IEMTRONICS), Toronto, ON, Canada, pp. 1–5 (2021)
9. Yang, Y., Wang, Y.: Analysis of transmission ratio in automotive mechanical steering system. Internal Combust. Eng. Parts **12**, 59–60 (2021)
10. Zhao, Y., Lin, J., Huang, L., Zhou, J., Zhang, B.: Steering system design for FSC university formula race car. Sci. Technol. Innov. **19**, 49–52 (2022)

11. ISO: ISO 11446:2022, Road Vehicles–Dimensions of Motor Vehicles and Towed Vehicles–Terms and Definitions. International Organization for Standardization (2022)
12. Boyce, M.P.: Automotive Engineering Fundamentals. SAE International (2015)
13. ASM International: Metals Handbook, vol. 8: Mechanical Testing and Evaluation (2000)
14. Smith, J.D.: Adams Car Tutorial: A Comprehensive Guide to Vehicle Dynamics Simulation. ABC Publishing, Los Angeles, CA (2021)

Chapter 2
New Approach for Soil Moisture Prediction Based on Multiple Influencing Factors

Zhujia Zhang, Wenping Jiang, and Ningyuan Xu

Abstract This study explores the intrinsic connection between meteorological characteristics and soil moisture. Based on the measured soil moisture data and meteorological data from 2012 to 2022, this paper screens out 6 key influencing factors among 24 features, and use the Seasonal Auto Regressive Integrated Moving Average (SARIMA) model to predict the regression of the screened key impact factors. The Multi-Layer Perceptron (MLP) model, the Long Short Term Memory (LSTM) model and the Long Short Term Memory optimized by the Sparrow Search Algorithm (SSA-LSTM) model are established to predict the soil moisture. By combining with SARIMA model, the relationship between the impact factors and soil moisture can be obtained. The results show that the LSTM model optimized by Sparrow Search Algorithm has good predictive ability on soil moisture in different soil layers, and R2 values are all above 0.92. Verified by the case, the relative error between the predicted value and the real value is all within 10%, indicating that the method can obtain the predicted value of soil moisture with high accuracy and stability.

2.1 Introduction

Soil moisture prediction is an important research topic in the field of agricultural and environmental sciences. The accurate prediction of soil moisture remains an important task in agriculture and environmental science, as well as effective use of water resources.

Despite the remarkable progress have seen significant advancements in the research related to soil moisture prediction, there are still many challenges to be overcome. Traditional machine learning models are usually more limited in terms of quality and quantity of samples, and require more efforts in data pre-processing and feature extraction. How to handle the large amount of complex soil data and how

Z. Zhang · W. Jiang (✉) · N. Xu
Shanghai Institute of Technology, Shanghai 200235, China
e-mail: jiangwenping@sit.edu.cn

R. Kountchev et al. (eds.), *Multidimensional Signals, Augmented Reality and Information Technologies*, Smart Innovation, Systems and Technologies 374,
https://doi.org/10.1007/978-981-99-7011-7_2

to select appropriate soil moisture prediction models, must require further in-depth research.

Data-driven methods based on the relationship between meteorological elements and soil moisture, which are simple in form and operation and straightforward to obtain parameters, are gradually applied in soil moisture determination and forecasting studies. Acharya et al. [1] explored different machine learning methods for soil moisture prediction in the Northern Red River Valley (RRVN). Their goal was to evaluate the performance of these techniques and to determine the significance of the predictive variables used. Khan et. al developed an interpretable and reusable workflow for predicting oil palm yield by combining multiple data sources with traditional machine learning techniques and automated model selection processes, demonstrating the great potential of machine learning methods for predicting oil palm yield using weather and soil moisture data [2].

In response to the problems that existing studies seldom consider the impact of multiple meteorological factors on soil moisture [3], and that the quantitative relationship between soil moisture and meteorological factors is not well represented by a single model in terms of spatial variability, the paper integrates time series models and neural network models to investigate variations in soil moisture content in grasslands located in the Xilingol region, in an attempt to develop a sound model and predict soil moisture. The paper aims to establish an appropriate model and predict the soil moisture to provide technical support for the regulation of grassland plant-soil water relationship and vegetation construction.

2.2 Data Source and Processing

2.2.1 Study Area Description

The geographical coordinates of Xilin Gol Grassland in Inner Mongolia exist between longitude 110°50′E and 119°58′E, and latitude 41°30′N and 46°45′N, with an elevation ranging from 800 to 1200 m. The grassland belongs to a temperate continental climate with semi-arid to arid characteristics, characterized by cold, windy, and dry conditions. The annual average temperature ranges from 1 to 2 °C, with a frost-free period of 90–120 days. The terrain of Xilin Gol Grassland is relatively flat, with limited precipitation and intensity, and the water cycle is dominated by vertical water exchange. The inflow and outflow of the entire grassland region can be considered equal, although local runoff may occur during heavy rainfall.

2.2.2 *Data Processing*

Basic data on soil and climate of the Xilin Gol grassland are monitored and provided by specialized institutions. The data set contains soil evaporation and soil moisture content for each month from 2012 to 2022, as well as meteorological factors including temperature, rainfall, and different weather conditions such as sea level pressure.

Data Pre-processing

Since the distribution of a portion of numerical characteristics does not conform to a normal distribution, the paper chooses to use box line plots, which are not required for data distribution, to detect outliers for numerical characteristics. The average of the two values before and after the outlier is used to correct the outlier.

After data preprocessing, the values of all features as well as the dependent variable soil moisture are explored to find their intrinsic connections, and scatter plots of the correlation between characteristics and soil moisture at different depths are shown in Fig. 2.1.

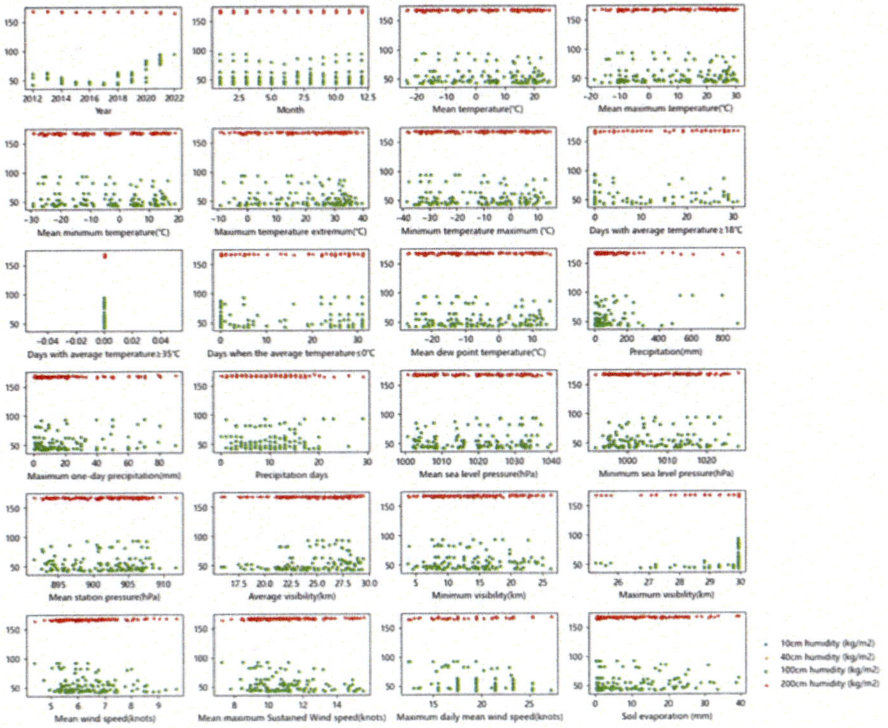

Fig. 2.1 Scatter plot of the relationship

Feature screening

According to the scatter plot in Fig. 2.1, it can be seen that the association between multiple characteristics and soil moisture is not purely linear. Pearson correlation coefficient is a commonly used statistical measure in natural sciences to quantify the strength and direction of the correlation between two variables with values between −1 and 1, calculated as follows.

$$\rho_{X,Y} = \frac{\operatorname{cov}(X, Y)}{\sigma_X \sigma_Y} \tag{2.1}$$

X and Y denote two vectors, cov donates covariance, σ donates standard deviation.

The frequency histograms of Pearson correlation coefficients between features and soil moisture are shown in Fig. 2.2. This indicates that the magnitude value of correlation coefficients between some features and soil moisture is lower than 0.5, while the correlation coefficients between soil moisture and features at the 40 cm layer are all less than 0.5, indicating that there is a high degree of nonlinearity between these 24 features and the dependent variable.

After the above analysis, Max-Relevance and Min-Redundancy [4] combined with random forest [5] is used for feature selection. The mRMR principle is based on identifying the subset of features from the original set that are highly relevant to the final output, while also having minimal redundancy or correlation among themselves. The maximal advantage is that the redundancy between the features is guaranteed to be minimal. The algorithm is used to calculate the correlation between each feature and soil moisture and score it, and then the random forest feature importance is used to evaluate all the features and calculate the weight value of each feature. Combining the two methods it is observed that the year, maximum visibility, and days with the average temperature is greater or less than a certain value have an almost negligible effect on soil moisture, which is removed. The weight bar graph is shown in Fig. 2.3.

Then, the paper verifies the reasonableness of the obtained features from two aspects, heat map and information gain. Taking soil moisture at the 10 cm layer as an example, after feature screening, the heat map of features and soil moisture values are drawn as shown in Fig. 2.4. The goal is to minimize redundancy between the features and maximize the correlation between the features and soil moisture levels.

The analysis proceeds from the perspective of information entropy, which is common to utilize it as a quantitative measure of the information content present in a system, and thus the aforementioned criterion can be utilized as a basis for optimizing system equations or selecting parameters. Information entropy is defined as follows.

$$Ent(X) = -\sum_{x \in X} p(x) \log p(x) \tag{2.2}$$

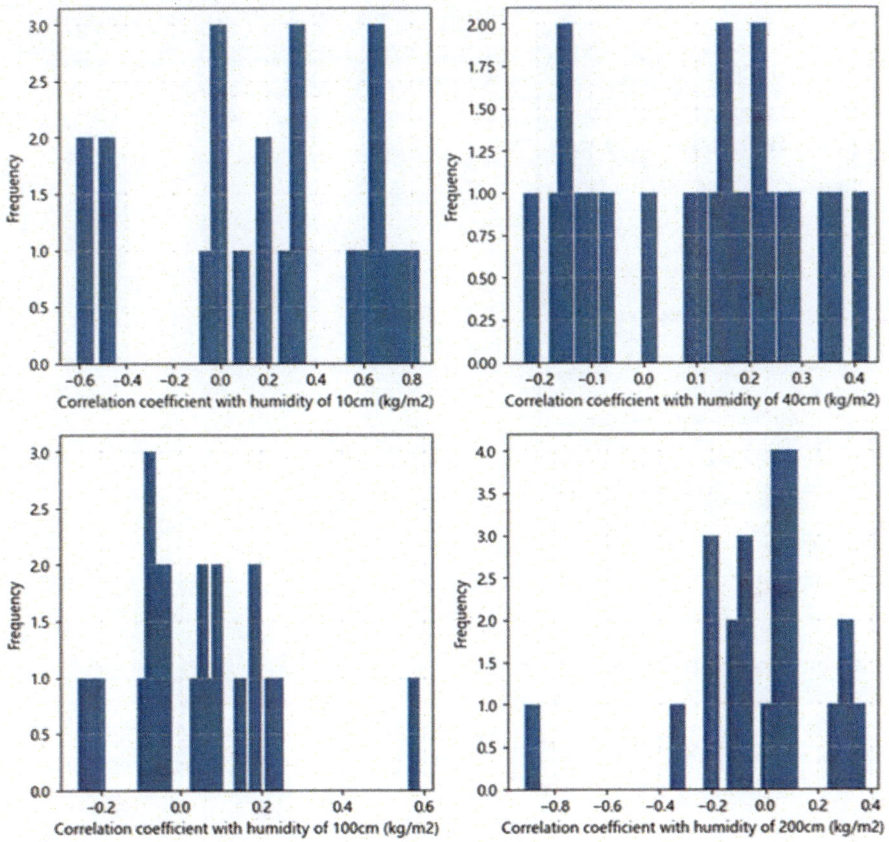

Fig. 2.2 Frequency histogram of correlation coefficient

X denotes the random event, $p(x)$ denotes the probability that the random event X is x, $-\log p(x)$ denotes the amount of information. $Ent(x)$ denotes information entropy.

The criterion of attribute division is to make each branch pure. Preferably, the data of each branch belongs to the same category as much as possible. The information and entropy are equal in size and opposite in meaning, so the elimination of how much entropy is equivalent to the increase of how much information, which is the origin of information gain. Formula 2.3 below shows the specific equation.

$$Gain(X, A) = Ent(X) - Ent(X|A) \tag{2.3}$$

$Ent(X|A)$ denotes the entropy after division A, $Gain(X, A)$ denotes information gain. Formula 2.4 below provides the equation.

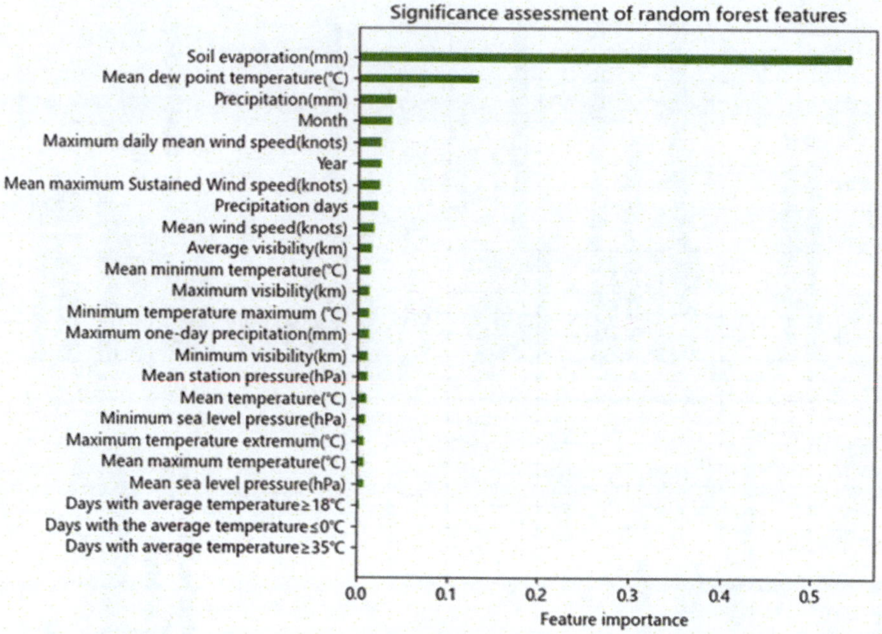

Fig. 2.3 The weight bar graph

$$Ent(X|A) = \sum_{A \in a} (a)^* \left[-\sum_{x \in X} (x|A) \log p(x|A) \right] \tag{2.4}$$

The higher the information gain, the more information the feature carries. After the above series of analysis, the paper finally gets 6 key feature dimensions as the key impact factors.

2.2.3 *Methods*

Variations in soil moisture are strongly linked to alterations in the surrounding environment. Using the numerical characteristics analyzed above, the data of soil moisture at different layers are used as the predicted output values, and the impact factors are used as the input value to train the neural network prediction model to find the nonlinear correlation between the impact factors and soil moisture. Since the future meteorological data and soil evaporation data are not known, it is also necessary to train the time series model to get the predicted values of future impact factors [6]. To forecast future soil moisture levels, the projected values are inputted into the neural network model for prediction.

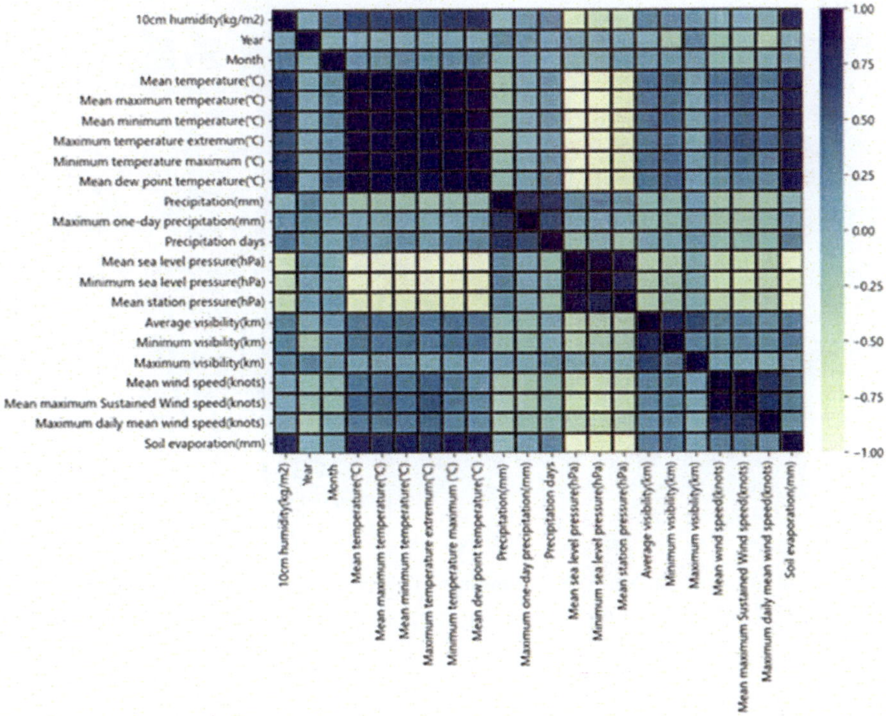

Fig. 2.4 Heat map

Temporal Prediction

The Autoregressive Integrated Moving Average Model is a widely used time series analysis method that splits time series data into an autoregressive term (AR), a difference term (I), and a moving average term (MA), and then predicts future values based on the combination of these terms [7].

To incorporate the seasonal component of the series, univariate data containing both trend and seasonality are analyzed using SARIMA (Seasonal Autoregressive Integrated Moving Average). SARIMA is based on identifying the periodicity of the data through external observation, which is then used to determine the length of the season (s), the order of the seasonal autoregressive term (P, determined by PACF), the order of the seasonal moving average term (Q, determined by ACF), and the order of the seasonal differencing term (D, typically 0 or 1). Combining this model about seasonality with the ARIMA model, the SARIMA model is established.

Multi-layer Perception Machine

MLP (Multi-layer Perceptron), is a forward structured artificial neural network [8], which maps input vectors to output vectors through a series of interconnected layers.

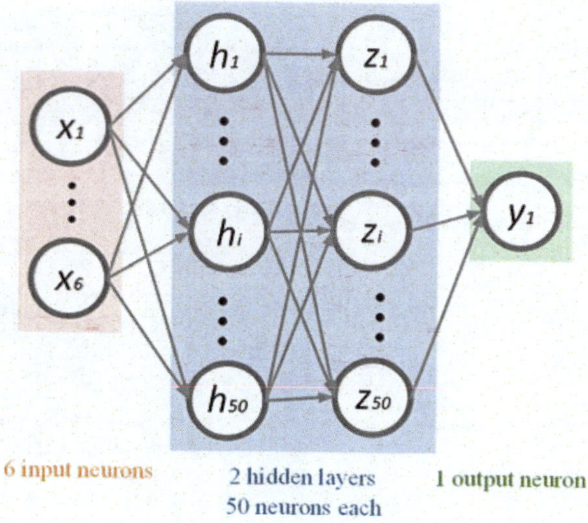

Fig. 2.5 MLP network structure diagram

In addition to the input and output layers, MLP can have multiple hidden layers in the middle. The model structure adopted in this paper is shown in Fig. 2.5.

Long Short Term Memory

The LSTM (Long Short Term Memory) consists of multiple LSTM neurons [9], each of which contains three gates: the forget gate, the input gate, and the output gate. Each LSTM neuron receives the input value from the previous layer and outputs it to the next layer. The input layer feeds the raw data into the neural network, and the LSTM layer updates the current state based on the input and the output state of the previous time step, and outputs it to the next layer. The output layer calculates the prediction based on the output of the LSTM layer. Throughout the process, the LSTM layer memorizes and selects the input data to better capture timing information.

In Fig. 2.6, each LSTM neuron is represented as a rectangle and contains three gates: forget gate, input gate, and output gate. Forget gate controls how much previous state information should be retained in the current state. The input gate manages the extent to which new information should be incorporated into the existing state, while the output gate regulates the degree to which information about the new output state should be produced. Throughout the process, the LSTM neuron updates the current state based on the current input and the output state of the previous time step and passes it to the next layer. Finally, the output layer calculates the prediction based on the output of the LSTM layer.

Sparrow Search Algorithm

SSA (Sparrow Search Algorithm), which is a heuristic optimization algorithm, can be used to solve nonlinear optimization problems. The algorithm divides the sparrows into three roles: producers, followers and danger-aware sparrows, and updates their positions separately to find the optimal solution. Among them, the producer can

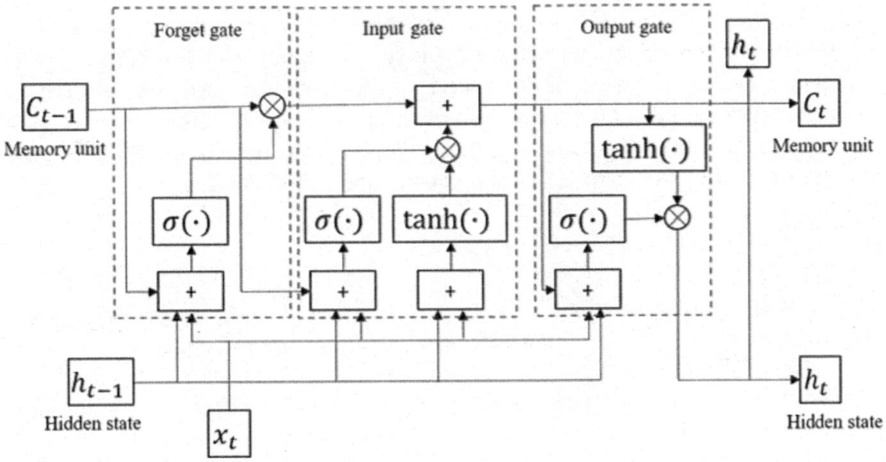

Fig. 2.6 LSTM cell structure diagram

generate new individuals according to its fitness value, the follower looks for food around the best producer, and the sparrow aware of danger changes its position according to its own state. Several parameters are included in the whole algorithm, including learning rate, the number of iterations, and the number of nodes in the hidden layer, etc. The algorithm updates the position of each sparrow in expectation of eventually finding the global optimal solution. Also, to ensure the convergence of the algorithm, the algorithm involves the handling of the boundaries and the calculation of the fitness function.

$$x_{i,j}(t+1) = \begin{cases} x_{bestj}(t) + \beta\left|x_{i,j}(t) - x_{bestj}(t)\right|, & f_i > f_g \\ x_{i,j}(t) + K\left(\frac{\left|x_{i,j}(t) - x_{worstj}(t)\right|}{(f_i - f_w) + e}\right), & f_i = f_g \end{cases} \tag{2.5}$$

$X_{bestj}(t)$ denotes the current global best position, β denotes a step control parameter that follows a normal distribution with a mean of 0 and variance of 1, $K \in [-1,\ 1]$ denotes the direction of sparrow movement, which is also a step control parameter, f_i denotes the current sparrow fitness value, f_i and f_w denote the current global best and worst values, and e denotes a constant to avoid a denominator of 0.

2.3 Model Preparation

2.3.1 Impact Factor Prediction Model

Taking soil evapotranspiration as an example, a smoothness test needs to be used before using time series forecasting. The statistical properties of non-stationary time series data can change over time, making data analysis and prediction more difficult.

The Augmented Dickey-Fuller smoothness test suggests that the time series data has a p-value of 0.56, which is greater than the typical significance level of 0.05. This result indicates that the time series data has a unit root and is non-smooth. To address this issue, differencing can be performed by calculating the difference between the data at the current time point and the data at the previous time point. For example, the first-order difference can be calculated using the following formula 2.6.

$$y(t) = x(t) - x(t-1) \tag{2.6}$$

$y(t)$ denotes the sequence after the difference, $x(t)$ denotes the value of the original sequence at moment t, and $t-1$ denotes the value of the original sequence at moment $t-1$.

If the sequence after first-order differencing is still non-smooth, we can perform second-order or higher-order differencing until we get a smooth sequence. The data in the paper have been differenced by second order and the p-value is much less than 0.05, so the data is already a smooth time series and the analysis and prediction of the data can be performed.

2.3.2 SSA-LSTM Model

The hyperparameters of the neural network model defined by LSTM are optimized by SSA for fitness calculation. The mean squared deviation of the validation set is adopted as the fitness function. The goal is to find a set of hyperparameters that minimize the mean squared deviation of the model on the validation set. By searching the hyperparameter space through the SSA, the optimal combination of hyperparameters can be obtained, which makes the performance of the model optimal. The part of the optimization curve is shown in Fig. 2.7.

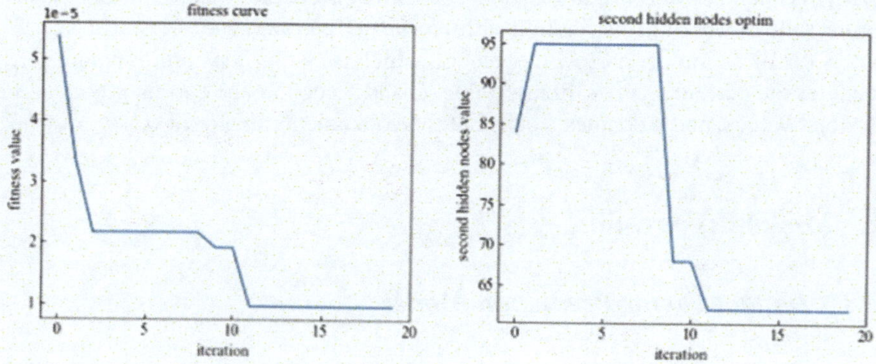

Fig. 2.7 Partial optimization process diagram

It can be seen from the fitness curve that the curve converges to a smaller value, representing a better optimization effect. From the change diagram of the second hidden layer, the optimization algorithm tries different combinations of the number of nodes in the iterative process to find the optimal number of nodes. The curve tends to be stable, so it can be considered that the algorithm has found the optimal number of nodes.

2.3.3 Model Evaluation Indicators

In order to be able to reflect the effect of the model predictions more clearly [10], the following indicators are used as evaluation indicators in the comparison of model prediction results in the paper.

$$RMSE = \sqrt{\frac{1}{N}\sum_{i=1}^{N}\left(y_{P,i} - y_{R,i}\right)^2} \tag{2.7}$$

$$MAE = \frac{1}{N}\sum_{i=1}^{N}\left|y_{P,i} - y_{R,i}\right| \tag{2.8}$$

$$R^2 = 1 - \frac{\sum_{i=1}^{N}\left(y_{P,i} - y_{R,i}\right)^2}{\sum_{i=1}^{N}\left(y_{R,i} - y_{R,N}\right)} \tag{2.9}$$

$y_{R,i}$ denotes the true value of i, $y_{P,i}$ denotes the predicted value of i, N denotes the total number of samples. Generally, the smaller calculated errors (MAE and RMSE) present the greater forecast accuracy and better performance of the calculation method. The larger the value of R^2, the stronger the prediction ability of the model.

2.4 Results and Analysis

2.4.1 Impact Factor Prediction Model

In the paper, the data before December 2019 are used as the training set and the data afterwards are used as the test set. Importing data into the model for the prediction, the part of the impact factors obtained are fitted to the Fig. 2.8.

SARIMA curve fit is much better than ARIMA. AIC is an indicator adopted to assess the quality of a statistical model, with a smaller value indicating a better model. The AIC of the SARIMA model is 580.574, which is lower than the ARIMA model.

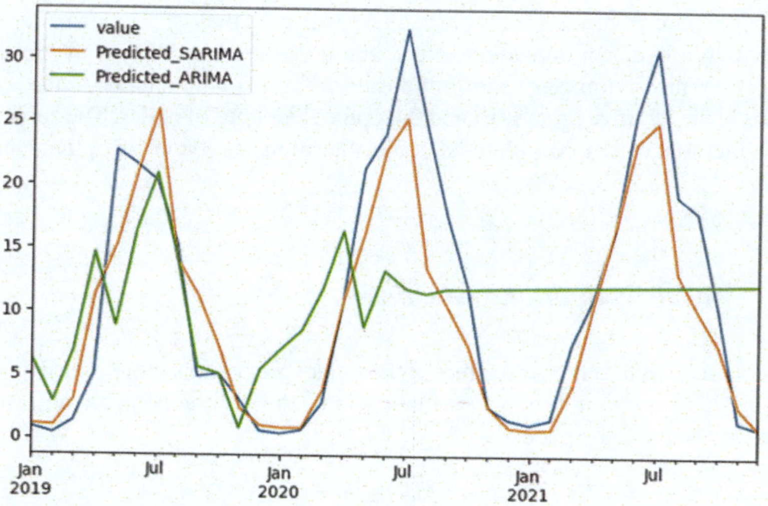

Fig. 2.8 Forecast result graph

The seasonal term in the SARIMA model is significantly effective, and the results indicate that SARIMA can better predict the seasonality in the time series.

2.4.2 *Soil Moisture Prediction Model*

In the soil moisture prediction model, impact factors such as soil evaporation and precipitation are used as inputs to the prediction model. The independent variables are predicted by multiple dependent variables, and future data changes are predicted by combining some factors that affect the variables to be predicted. The fitting effects of diverse models for different soil layers are shown in Figs. 2.9, 2.10, 2.11 and 2.12.

From the results of the figures, it can be seen that in predicting soil moisture, the predicted values of SSA-LSTM model can fit well with the real values in different soil layers, while the predicted value curve of MLP model has a large deviation from the real value curve when the soil layer is deeper. The prediction accuracy of LSTM gradually decreases with the increase of soil depth. However, the main reason for the prediction effect at the 200 cm layer increases instead of decreasing which may be caused by the difficulty of soil moisture collection at this layer and the small amount of data.

In order to further explain the prediction ability of each model in different soil layers, Table 2.1 is drawn. It can be seen from the table that both LSTM and SSA-LSTM have good prediction accuracy for soil water content, but SSA-LSTM is better than LSTM. MLP was the worst predictor, especially for deep soil moisture. The performance of all models in the 10 cm soil layer is almost consistent. In the deep soil moisture (100 cm) prediction, the SSA-LSTM model has 11% higher R^2

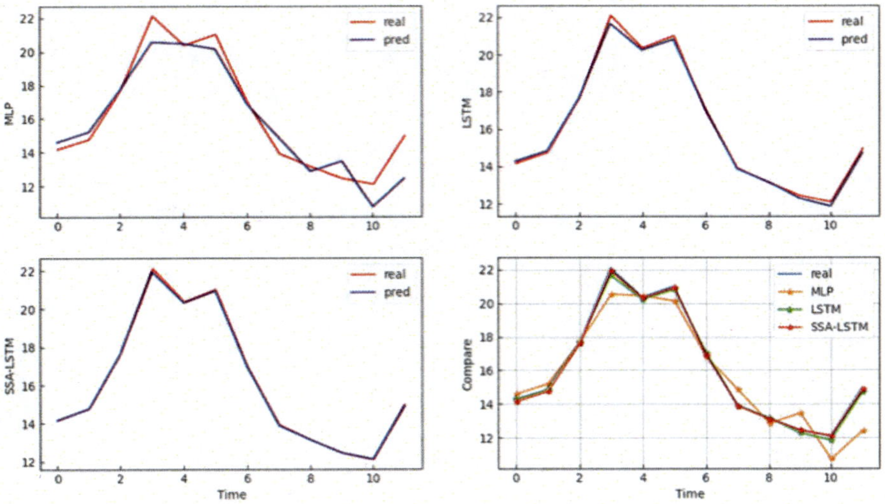

Fig. 2.9 Comparison results of 10 cm soil layer

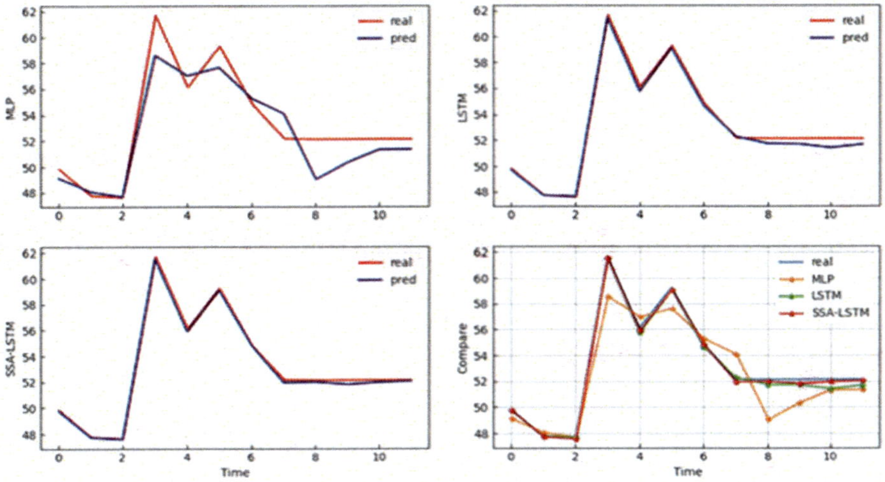

Fig. 2.10 Comparison results of 40 cm soil layer

value compared to the LSTM model. It is observed that the SSA-LSTM model has better optimization results in both RMSE and MAE performance metrics. R^2 is also improved compared to the most commonly used base model.

This study can conclude that the SSA-LSTM prediction model has greater advantages over the single LSTM model as the depth of the soil layer increases. Overall, the performance of SSA algorithm is excellent, both in terms of convergence speed and convergence accuracy. In short, the overall vertical prediction based on SSA-LSTM performs better in terms of spatial variation of soil moisture.

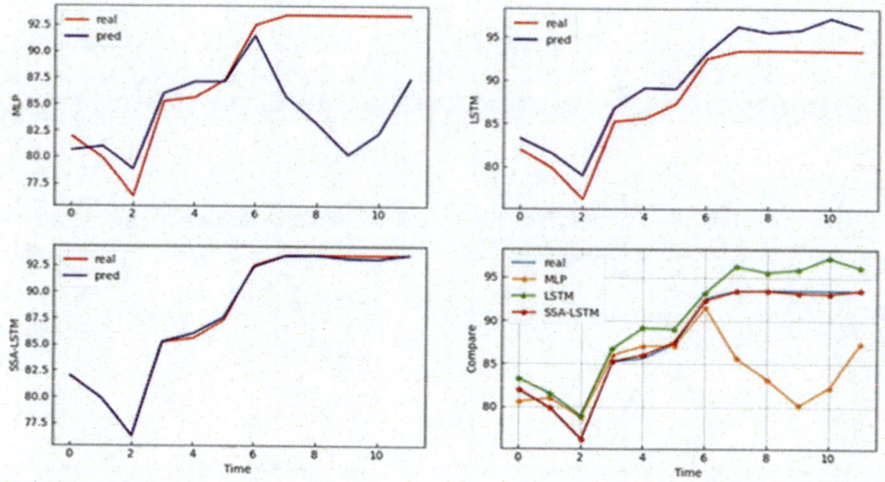

Fig. 2.11 Comparison results of 100 cm soil layer

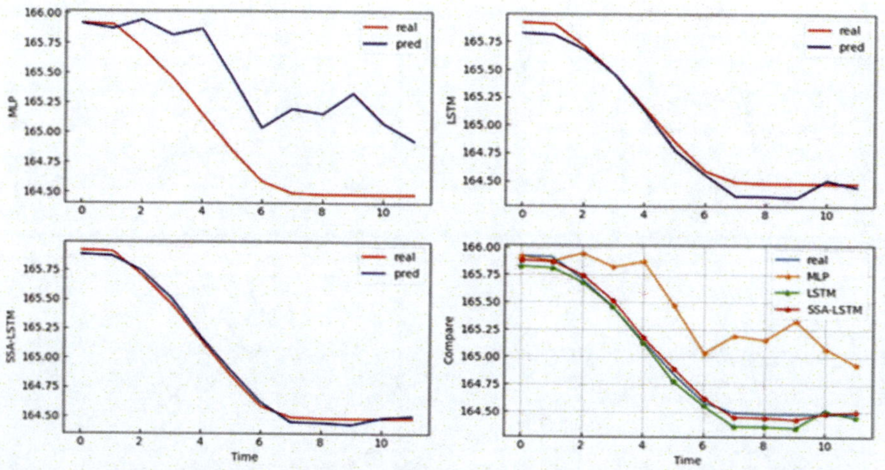

Fig. 2.12 Comparison results of 200 cm soil layer

2.4.3 Model Verification

The predicted values of soil moisture obtained by substituting the predicted results of the impact factors into the model are shown in the following Table 2.2.

RV denotes the real value, PV denotes the predicted value, RE denotes the relative error.

According to Table 2.2, the relative error between the predicted and actual values is obtained by combining the SARIMA and SSA-LSTM models over the next 4 months

Table 2.1 Precision comparison of different models in different soil layers

Indicators	10 cm soil moisture			40 cm soil moisture			100 cm soil moisture			200 cm soil moisture		
	RMSE	MAE	R^2	RMSE	MAE	R^2	RMSE	MAE	R^2	RMSE	MAE	R^2
MLP	0.796	0.513	0.902	2.196	1.591	0.707	1.857	6.910	−0.972	8.332	1.813	−9.497
LSTM	0.389	0.339	0.971	0.638	0.527	0.963	0.078	1.886	0.856	2.251	0.064	0.913
SSA-LSTM	0.056	0.045	0.986	0.062	0.056	0.976	0.041	0.125	0.964	1.581	0.030	0.926

Table 2.2 Verification results of soil moisture series model in each layer %

Indicators	10 cm soil moisture			40 cm soil moisture			100 cm soil moisture			200 cm soil moisture		
	RV	PV	RE	RV	PV	RE	RV	PV	RE	RV	PV	RE
2021.12	13.14	14.23	8.29	52.14	53.69	2.97	93.43	96.84	3.65	164.85	167.86	1.82
2022.01	12.45	11.39	−8.51	52.14	49.31	−5.43	93.45	98.10	4.87	164.59	166.73	1.36
2022.02	12.10	11.13	−8.01	52.14	51.28	−1.65	93.45	95.64	2.24	164.49	163.96	−0.32
2022.03	14.96	15.81	5.68	52.13	55.71	6.87	93.44	96.52	3.30	164.48	160.87	−2.19

The relative margin of their error is small, ranging from −8.51 to 8.29% for the 10 cm soil layer, −5.43 to 6.87% for the 40 cm layer, 2.24 to 4.87% for the 100 cm layer, −2.19 to 1.82% for the 200 cm layer, all of which is less than 10%. Therefore, the method introduced in the paper is considered to be effective in predicting soil moisture. It is also effective in predicting soil moisture in different soil layers, which makes it possible to understand soil moisture using meteorological data.

2.5 Conclusion

Three different models are adopted to predict soil moisture in the Xilin Gol grassland region. The results show that SSA-LSTM has better prediction accuracy at different depths of soil layers. Based on environmental factors, soil moisture prediction can be achieved by means of SSA-LSTM, which finally provides a reference scheme for soil moisture prediction in similar application situations.

According to the method proposed in the paper, effective prediction of soil moisture can be achieved by using environmental factors, which not only enables non-intrusive detection of soil layers, but also can be extended to agricultural prediction, effectively reducing the cost of agricultural monitoring systems and providing macroscopic prediction support for small-area land plots. It can also provide advance measures for fertilization, irrigation, drought resistance, flood prevention and other aspects.

References

1. Acharya, U., Daigh, A.L.M., Oduor, P.G.: Machine learning for predicting field soil moisture using soil, crop, and nearby weather station data in the Red River Valley of the North. Soil Syst. **5**(4), 57 (2021)
2. Khan, N., Kamaruddin, M.A., Ullah Sheikh, U., et al.: Prediction of oil palm yield using machine learning in the perspective of fluctuating weather and soil moisture conditions: evaluation of a generic workflow. Plants **11**(13), 1697 (2022)
3. Matei, O., Rusu, T., Petrovan, A., et al.: A data mining system for real time soil moisture prediction. Proc. Eng. **181**, 837–844 (2017)
4. Toğaçar, M., Ergen, B., Cömert, Z., et al.: A deep feature learning model for pneumonia detection applying a combination of mRMR feature selection and machine learning models. Irbm **41**(4), 212–222 (2020)
5. Belgiu, M., Drăguţ, L.: Random forest in remote sensing: a review of applications and future directions. ISPRS J. Photogramm. Remote. Sens. **114**, 24–31 (2016)
6. Nocita, M., Stevens, A., Noon, C., et al.: Prediction of soil organic carbon for different levels of soil moisture using Vis-NIR spectroscopy. Geoderma **199**, 37–42 (2013)
7. Rivero, A., Alayón, C.A.M., Ferro, R., Iglesia, D., Secades, V.A.: Network traffic modeling in a wi-fi system with intelligent soil moisture sensors (WSN) using IoT applications for potato crops and ARIMA and SARIMA time series. Appl. Sci. **10**(21), 7702 (2020)
8. Nguyen, T.T., Ngo, H.H., Guo, W., et al.: A low-cost approach for soil moisture prediction using multi-sensor data and machine learning algorithm. Sci. Total Environ. **833**, 155066 (2022)

9. Filipović, N., Brdar, S., Mimić, G., et al.: Regional soil moisture prediction system based on long short-term memory network. Biosys. Eng. **213**, 30–38 (2022)
10. Jin, H.N., Zhang, X.L., Liu, H.J., et al.: Soil moisture predicting model based on spectral absorption characteristics of the soil. Acta Pedol. Sin. **53**(3):627–635 (2016)

Chapter 3
Fuzzy PID Control of Brushless DC Motor Based on the Improved Bat Algorithm

Jie Li, Wentian Zhao, Jiahang Bai, Liu Yang, Tianxu Zhang, Yuan Gao, Man Hu, Ziyuan Wang, Yuhao Liu, and Jingzhi Lv

Abstract This paper focuses on researching the control system of brushless direct current motor (BLDCM), and proposes a control strategy of fuzzy PID, which is based on the improved bat algorithm (IBA). The BA has some disadvantages in the later stage of iteration, including slow convergence, low convergence precision and easy to converge to local minimum. In this paper, the convergence speed is increased by adaptive strategy, the Gaussian mutation is used to improve the population diversity, and the exploration ability is improved by dynamic assignment of frequency. Results of the simulation show that the IBA has favorable performance in the BLDCM control system.

3.1 Introduction

BLDCM has the characteristics of high controllability, fast dynamic response and simple construction, which is widely used in automation instruments, medical equipment, aerospace technology and other fields. BLDCM control system is a complex nonlinear time-varying system with multiple variables and strong coupling [1]. The primitive PID control is unable to adaptively adjust parameters, therefore, it is ineffective in the nonlinear system. The fuzzy control makes up for disadvantages of the primitive PID control, but the dynamic response ability and control accuracy are insufficient [2].

At present, the application of the intelligent algorithm in fuzzy control has been widely studied [3–5]. Reference [6] introduced the improved ant colony algorithm

J. Li · W. Zhao (✉) · J. Bai · L. Yang · T. Zhang · Y. Gao · M. Hu · J. Lv
Transformer Maintenance Department, State Grid Tianjin High Voltage Company, Tianjin, China
e-mail: 1509156150@qq.com

Z. Wang · Y. Liu
Operation and Maintenance Department, State Grid Tianjin High Voltage Company, Tianjin, China

R. Kountchev et al. (eds.), *Multidimensional Signals, Augmented Reality and Information Technologies*, Smart Innovation, Systems and Technologies 374,
https://doi.org/10.1007/978-981-99-7011-7_3

and LM algorithm into the training of fuzzy neural network, and proposed a control strategy for fuzzy PID based on the RBF neural network. In reference [7], a new speed sensorless vector control scheme for induction motor is obtained by combining the improved feedback matrix algorithm with the improved speed adaptive law algorithm. Reference [8] proposed a fuzzy controller design based on the MFO algorithm, which improved the control accuracy and reduced the control system overshoot.

The improved algorithm in the above literatures lack the mutation mechanism to improve the global search ability, which makes it easy to fall into the local minimum in the iteration.

In this paper, the bat algorithm is improved by optimizing the speed, frequency and position, which enhance accuracy and speed of BLDCM control system.

3.2 BLDCM Control System

3.2.1 Mathematical Model

To facilitate the analysis of the operating characteristics of BLDCM, the following assumptions are made [9–11]:

(1) Three-phase voltage is balanced and distribution of stator current and rotor field is symmetrical.
(2) The permeability of the Nd-Fe-B permanent magnet is identical to that of air. The equivalent air gap length of BLDCM is a constant.
(3) Ignoring hysteresis loss, eddy current loss, and the influence of slot effect, magnetic saturation, armature reaction, etc.
(4) Waveform of air-gap field is rectangular with flat width of 120°.
(5) On the inner surface of stator, armature winding is uniformly and continuously distributed.

Under the above assumptions, the balance equation of BLDCM three-phase winding voltage can be expressed as follows:

$$\begin{bmatrix} u_a \\ u_b \\ u_c \end{bmatrix} = \begin{bmatrix} R_a & 0 & 0 \\ 0 & R_b & 0 \\ 0 & 0 & R_c \end{bmatrix} \begin{bmatrix} i_a \\ i_b \\ i_c \end{bmatrix} + \frac{d}{dt} \begin{bmatrix} L & M & M \\ M & L & M \\ M & M & L \end{bmatrix} \begin{bmatrix} i_a \\ i_b \\ i_c \end{bmatrix} + \begin{bmatrix} e_a \\ e_b \\ e_c \end{bmatrix} \tag{3.1}$$

where

u_a, u_b, u_c——stator voltage, V;
R_a, R_b, R_c——stator winding resistance, Ω;
i_a, i_b, i_c——stator current, A;
L——stator winding self-inductance, H;
M——stator winding mutual inductance, H;
e_a, e_b, e_c——stator winding back EMF, V.

According to the energy conservation law, the equation of BLDCM electromagnetic torque is as follow:

$$T_e = \frac{e_a i_a + e_b i_b + e_c i_c}{\omega} \tag{3.2}$$

where
T_e——electromagnetic torque, N·M;
ω——motor angle velocity, rad/s.
The motion equation of BLDCM is as follows:

$$T_e = T_L + B\omega + J\frac{d\omega}{dt} \tag{3.3}$$

where
T_L——load torque, N·M;
B——damping coefficient, N·M·rad/s;
J——motor inertia, kg/m^2.

3.2.2 *Simulation System*

BLDCM control system is established and simulated in MATLAB/Simulink, according to its operation principle. Figure 3.1 shows the block diagram of the BLDCM control system.

The control system of BLDCM is designed as a double closed-loop. The outer loop can control the output current through detecting and comparing the output speed with the given speed, which ensures the dynamic and static tracking ability. The inner loop follows the change of output speed and controls the maximum current in the process of dynamic speed regulation to ensure the system stability.

Running the simulation system. Figure 3.2 shows the three-phase stator current waveform. Figure 3.3 shows the three-phase back EMF waveform.

Both waveforms are trapezoidal, which verifies the correctness of the simulation system.

3.3 Optimization of IBA

3.3.1 *Fuzzy PID Control*

ID control system consists of three parameters: proportion (P), integral (I) and differential (D), which have the advantages of simple principle and strong robustness, but the three parameters cannot be changed during system operation.

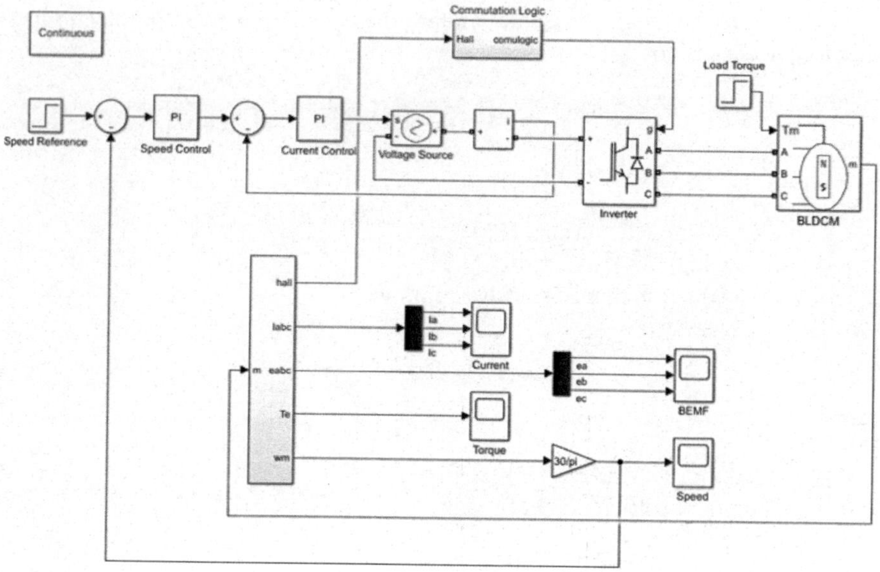

Fig. 3.1 BLDCM control system

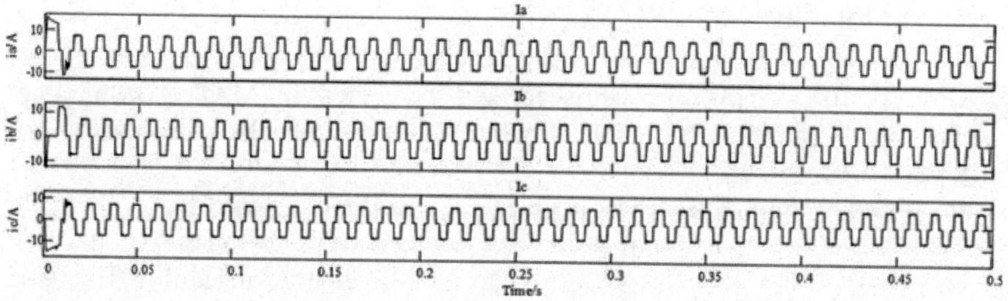

Fig. 3.2 Three-phase stator current waveform of BLDCM

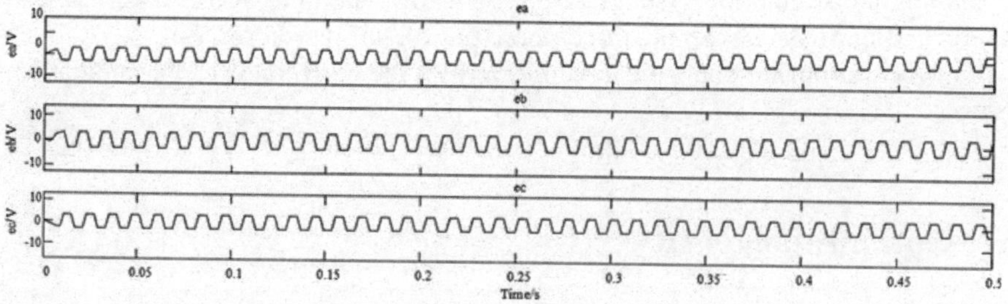

Fig. 3.3 Three-phase back EMF waveform of BLDCM

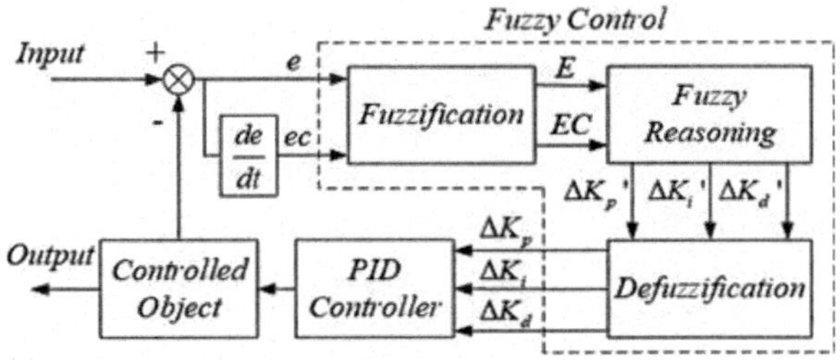

Fig. 3.4 Control principle of fuzzy PID

Table 3.1 Fuzzy control parameters

	Input		Output		
	e	*ec*	$\triangle K_p$	$\triangle K_i$	$\triangle K_d$
Basic domain	[−3, 3]	[−3, 3]	[−3,3]	[−1,1]	[−1,1]
Fuzzy subset	{NB, NM, NS, ZE, PS, PM, PB}				
Fuzzy domain	[−6, 6]				

Fuzzy PID control combines fuzzy mathematics theory with automatic control, which is based on fuzzy set theory, linguistic variable and logical reasoning [11]. It makes up the deficiency of the primitive PID control. Figure 3.4 shows the control principle of fuzzy PID.

3.3.2 *Fuzzy Control Design*

Parameters of fuzzy control are set according to actual requirements. The fuzzy subset contains seven elements, which are NB (negative big), NM (negative middle), NS (negative small), ZE (zero), PS (positive small), PM (positive middle), PB (positive big). Table 3.1 shows the parameters.

The triangle membership function (TRIMF) has the advantages of simple, stable and fast calculation, so it is used for processing variable of the input and output. The weighted average method is selected for defuzzification, which can fully utilize all membership information of the fuzzy subset. Table 3.2 shows the fuzzy control rules.

3.3.3 *Improvement of Bat Algorithm*

Professor Yang proposed the bat algorithm (BA) based on swarm intelligence, which is a bionic heuristic search optimization method [12]. Interaction and influence among individuals determine the optimization capability of BA. But bat individuals lack

Table 3.2 Fuzzy control rules

E	EC = NB	EC = NM	EC = NS	EC = ZE	EC = PS	EC = PM	EC = PB
ΔK_p							
NB	PB	PB	PB	PM/	PS	PS	ZE
NM	PB	PB	PB	PM	PS	ZE	ZE
NS	PM	PB	PM	ZE	NS	NM	NM
ZE	PM	PS	PS	NM	NB	NB	NB
PS	PS	PS	ZE	NS	NM	NM	NM
PM	ZE	ZE	NS	NM	NM	NB	NB
PB	ZE	NS	NS	NM	NB	NB	NB
ΔK_i							
NB	NB	NB	NB	NM	NM	ZE	ZE
NM	NM	NM	NM	NM	NS	ZE	ZE
NS	NM	NM	NM	ZE	ZE/NM	ZE	ZE
ZE	NS	NS	NS	ZE	PS	PS	ZE
PS	NS	ZE	ZE	ZE	PS	PM	PM
PM	ZE	ZE	PS	PM	PM	PM	PM
PB	ZE	ZE	PS	PM	PB	PB	PB
ΔK_d							
NB	PS	PS	ZE	ZE	ZE	PM	PS
NM	PS	NM	NS	NS	ZE	PM	PS
NS	NB	NM	NM	NS	NM	NM	PM
ZE	NM	NS	NS	ZE	NS	PS	PM
PS	NB	NS	NS	NS	PS	PM	PS
PM	NM	NS	PS	PS	PS	PM	PS
PB	PS	ZE	ZE	ZE	ZE	PB	PB

mutation mechanism, therefore, BA is easy to converge to local minimum in the later stage of iteration [13–15]. This paper improves the shortcomings of BA.

3.3.3.1 Speed

The bat speed is updated according to the adaptive strategy. The bat speed is updated as follows:

$$v_i^{t+1} = \left[1 - \frac{1}{2}\left(\frac{t}{tmax}\right)^2\right] v_i^t + \left(x_i^t - x^*\right) f \tag{3.4}$$

where

t——current iteration times;
$tmax$——maximum iteration times;
x^*——current best location.

The bat speed is faster in the early stage of iteration, which improves the exploration ability and convergence speed. The bat speed decreases in the later stage of iteration, which subtilizes the exploration.

3.3.3.2 Location

The bat location is updated based on Gaussian variation. The probability of Gaussian mutation of the bat individual is defined as:

$$P_t = \sqrt{\frac{t}{tmax - 1}}, t = 1, 2, 3...tmax \tag{3.5}$$

In each iteration, a random number N, which is uniformly distributed in [0, 1], is generated and compared with Pt. When N < Pt, the bat position is Gaussian mutated as follows:

$$x_i^{t+1} = (x_{max} - x_{min}) \times N \tag{3.6}$$

where
x_{max}——bat maximum location;
x_{min}——bat minimum location.

At the beginning stage of iteration, Pt << 1, the probability of Gaussian mutation in bats is very low, and the bats have a large space to seek for the optimal results. In the later stage of iteration, Pt is close to 1, and the probability of Gaussian mutation of bats increases, which improves the diversity of population.

3.3.3.3 Frequency

An updating method of the bat frequency based on dynamic assignment is proposed. The bat frequency is updated as follows:

$$f_i = f_{min} + \frac{\sqrt{(D_i - D_{imin})^2}}{d} \times (f_{max} - f_{min}) \tag{3.7}$$

$$D_i = \sqrt{(x_i - x^*)^2} \tag{3.8}$$

$$d = D_{imax} - D_{imin} \tag{3.9}$$

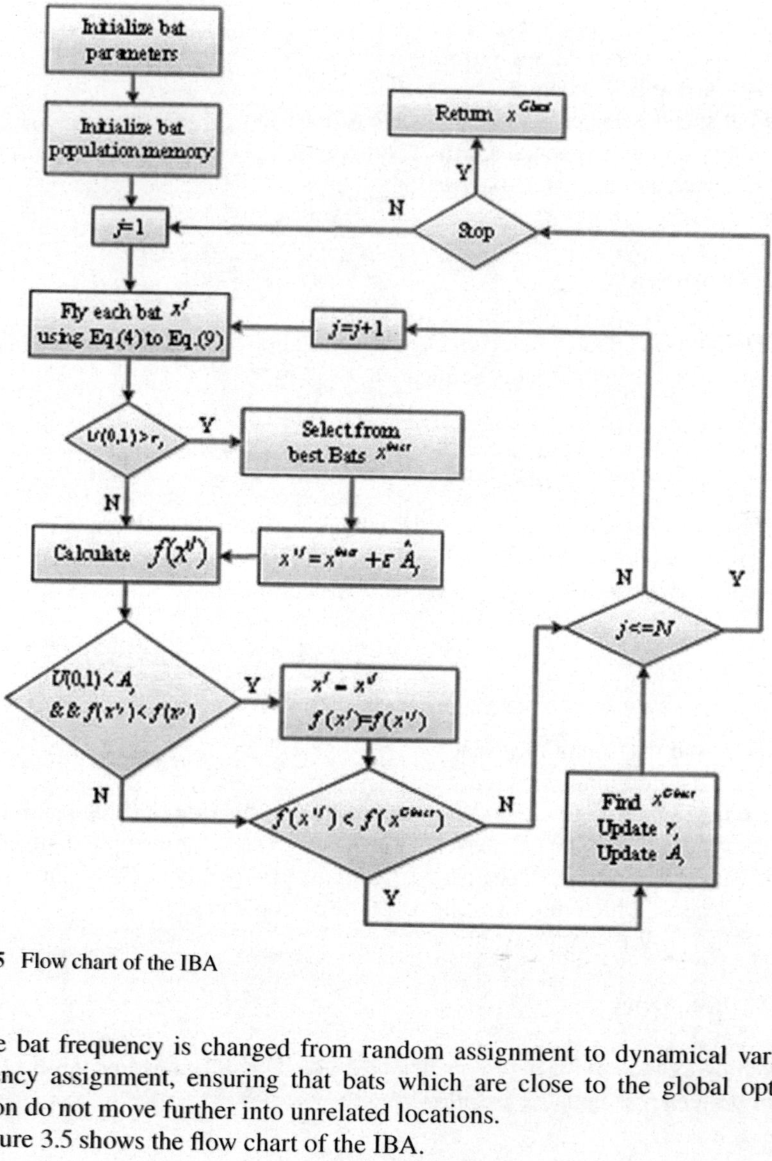

Fig. 3.5 Flow chart of the IBA

The bat frequency is changed from random assignment to dynamical variable frequency assignment, ensuring that bats which are close to the global optimal solution do not move further into unrelated locations.

Figure 3.5 shows the flow chart of the IBA.

3.4 Results

The technical data of IBA: maximum iteration times tmax = 200, population size N = 50, initial volume A = 0.25, pulse emissivity r = 0.5, frequency f = [0, 2], enhancement factor $\gamma = 0.05$, attenuation factor $\alpha = 0.95$. The ITAE is selected as

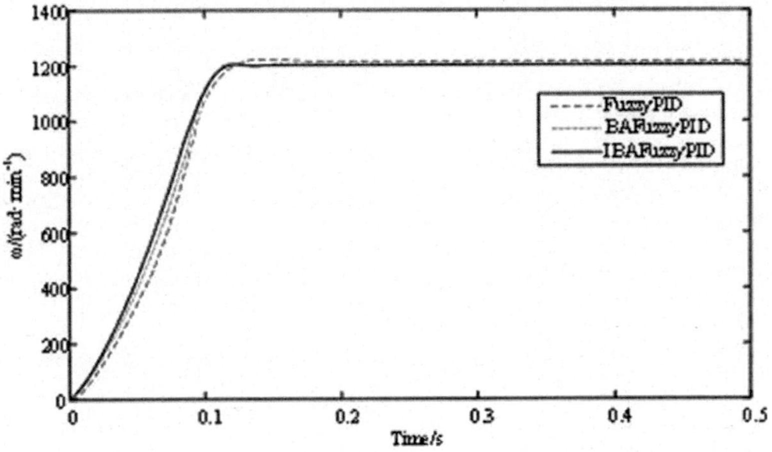

Fig. 3.6 Output speed curves

Table 3.3 Fuzzy control parameter

Type	Time to reach steady-state	Stable value	ITAE
Fuzzy PID	0.196	1216.231	4.312
BA fuzzy PID	0.162	1209.684	3.849
IBA fuzzy PID	0.143	1201.463	2.741

the fitness function of IBA and the initial speed of BLDCM is 1200 revolutions per minute. The operation results of fuzzy PID, the BA fuzzy PID and the IBA fuzzy PID control system are compared. The output speed curves are shown in Fig. 3.6. The selected operating indicators are shown in Table 3.3.

Results show that the IBA Fuzzy PID control system has a shorter time to reach steady-state, a smaller steady-state residual error, and a better control performance, which meets the expectations.

3.5 Conclusion

In this article, the simulation pattern of BLDCM is established, and a fuzzy PID control strategy based on the IBA is proposed. The performance of fuzzy PID, the fuzzy PID and the IBA fuzzy PID in BLDCM is compared in simulation software, and the experiment results display that the IBA fuzzy PID has better control effect in convergence speed and convergence precision.

Speed control has a great influence on the performance of BLDCM. Future research will continue to optimize the anti-interference ability of IBA at a low speed of BLDCM based on the existing conclusions.

References

1. Fang, W., Zhang, H., Liu, X.D.: Design of the brushless DC motor control system of double-loop. J. Power Supply **2**, 35–42 (2014)
2. Yang, X.H., Liu, C.W.: Fuzzy PID control design of brushless DC motor based on MATLAB. Instrum. Tech. Sens. **11**, 105–108 (2019)
3. Geng, W.B., Zhou, Z.A.: Research on improved particle swarm optimization for BLDCM speed control system. Control Eng. China **26**(9), 1636–1641 (2019)
4. Wen, M.F.: Study of Optimizing Parameters of Improved Fuzzy Controllers by Genetic Algorithm. Changchun University of Science and Technology, Changchun (2016)
5. Mirjalili, S.: Moth-flame optimization algorithm: a novel nature-inspired heuristic paradigm. Knowl.-Based Syst. **89**, 228–249 (2015)
6. Gu, W.B., Yang, S.S., Wang, X.L., Yuan, M.H.: PID control of brushless DC motor based on fuzzy RBF neural network. Comp. Technol. Dev. **32**(8), 15–19 (2022)
7. Jin, A.J., Shao, F.X., Yan, Z.G.: Vector control of induction motor based on adaptive fuzzy neural network. Electron. Sci. Tech. **35**(9), 65–73 (2022)
8. Liu, Y.H., Liao, P.: Fuzzy control design of brushless DC motor based on MFO algorithm. Instrum. Tech. Sens. **4**, 107–111 (2021)
9. Wang, C.X., Yan, D.C., Yin, H.J.: Motor speed control based on fuzzy PID. Comp. Simul. **32**(10), 430–434 (2015)
10. Wang, W., Zhang, Y.K., Liu, P.P., Lin, G.S.: Study and simulation of fuzzy-PID control system of BLDCM. Comp. Simul. **29**(4), 196–199 (2014)
11. Hu, B.G., Ying, H.: Review of research and development of fuzzy PID control technology and some important problems. Acta Autom. Sin. **27**(4), 567–584 (2001)
12. Yang, J.Q., Zhang, D.M.: K-means clustering algorithm based on improved BA algorithm. Appl. Res. Comp. **35**(5), 1454–1457 (2018)
13. Wang, Y.B.: Research and Implementation of KFCM Algorithm Based on Bat Algorithm. Nanchang University, Nan Chang (2021)
14. Li, Y.G., Lin, H.: Multi-objective mobile energy storage scheduling based on improved bat algorithm. J. North China Electr. Power University 1–9 (2022). http://epub2.hknsspj.cn/kcms/detail/13.1212.TM.20220616.1432.002.html
15. Yue, X.X., Zheng, Y.S., Lin, J.T.: Bat algorithm with adaptive mutation. Comp. Meas. Control **23**(2), 516–519 (2015)

Chapter 4
Monocular Stereo Learning Based on Hybrid Attention Transformer

Jie Yang, Shicong Chen, Vantoi Vu, and Pengfei Zheng

Abstract Monocular stereo matching estimation network is an important research direction in the field of artificial intelligence. In this paper, a monocular stereo learning network based on Hybrid Attention Transformer for joint estimation of optical flow and depth for monocular sequences is proposed. Firstly, spatial window and channel group attention are used to enhance the ability of the network to extract local and global features, optimizing the learning framework. Secondly, cross attention is used to realize the cross-window connection, which makes the details of image estimation more rich. Finally, we discuss the effect of the two different connection modes of attention in series and parallel on the network estimation tasks. Our experiment results on KITTI datasets show that, compared with the previous methods, our multi-task learning framework has more significant improvements in optical flow and depth.

4.1 Introduction

Stereo vision networks are important methods to obtain the motion information and structure of 3D objects by reconstructing images obtained from different perspectives. According to the different ways of obtaining images, they can be mainly divided into monocular and multi-view stereo vision networks. Multi-view stereo vision networks calculate the matching cost between pixels by using the normalized cross-correlation and absolution difference, and further reflect the correlation between adjacent pixels through cost aggregation, so as to determine the disparity

J. Yang (✉) · S. Chen · V. Vu
College of Information and Communication Engineering, Harbin Engineering University, Harbin, China
e-mail: hrbeu_yj@126.com

P. Zheng
College of Computer and Communication Engineering, University of Science and Technology Beijing, Beijing, China

R. Kountchev et al. (eds.), *Multidimensional Signals, Augmented Reality and Information Technologies*, Smart Innovation, Systems and Technologies 374,
https://doi.org/10.1007/978-981-99-7011-7_4

value of each pixel. However, multi-view stereo vision networks will result in abundant artifacts in the weak texture area and repeated texture area of scene, which makes it difficult to recover 3D information.

The monocular stereo vision networks solve the problem that multi-view stereo networks must predict on multi-view pixels by estimating the motion information of monocular image sequences obtained from monocular cameras calibrated with fixed parameters. DeMoN [1] is the first model that used deep learning network to estimate jointly optical flow and depth. Its coding network used FlowNetS [2] to estimate the optical flow of the input images, and the decoding part further decomposed it into depth and camera poses. CodeSLAM [3] used automatic encoding and decoding to optimize the overall network, and simultaneously estimated optical and depth information by geometric errors and minimizing photometric errors. Although monocular stereo vision networks have been gradually improved, the following limitations still exist. First, fixed monocular camera parameters make the input wide-angle images need to be cut, resulting in the loss of edge features information extraction. Secondly, layered and multi-tasking networks lack the ability to distinguish the importance of features. Finally, the monocular stereo vision networks need to be repeatedly iterated to update and optimize the optical flow and depth.

In this article, we propose a monocular stereo vision network based on Hybrid Attention Transformer (HAT) [4]. With FMDN [5] as the baseline, the network can extract more feature information by using HAT, and fuse images with multiple target information from the front-end federated network to refine the depth estimation. In our proposal, the Hybrid Attention Block (HAB) based on Spatial Window Attention (SWA) and Channel Group Attention (CGA) is introduced into the feature extraction of joint estimation network to selectively extract and enhance the features of input monocular image sequences. In addition, we introduce the Cross Attention (CAT), which is a slide window mode based on Swin Transformer [6]. It can realize cross-window attention connection and efficiently capture long-range and short-range information. Our network is trained on the unlabeled datasets of KITTI raw [7], and the optical flow estimation and depth estimation are tested respectively. Our experiments show that, compared with the baseline used, our proposed framework has a better performance in optical flow and depth estimation network, which optimizes the baseline network. The proposed monocular stereo vision network has obvious competitive advantages.

4.2 Related Work

4.2.1 Monocular Stereo Vision Networks

The design methods of monocular stereo vision networks are divided into two kinds, based on traditional knowledge and data driven. In the traditional monocular stereo vision networks, sparse point matching method was used to match images. The basic matrix and coordinates of image pixels are used to find the corresponding polar lines of each point in the image and then correct them. However, the traditional knowledge-driven monocular stereo has an extremely complex process of acquiring the motion information of objects, which requires a large amount of calculation for each pixel. Therefore, many methods of designing monocular stereo vision networks using deep learning have been proposed.

MVDepthNet [8] proposed to extract and encode the multi-view information into the cost volume, and used data geometric augmentation in network training, which solved the problem of fixed camera parameters and restrictions on the number of input images. To reduce the optical flow search space, FMDN [5] proposed a combined method of optical flow and came poses estimation network, eliminating the need for iterative refinement of the network, and the network is further expanded by fusing multiple image information with target information, which improved the accuracy and speed of the network. DROID-SLAM [9] was based on RAFT [10] and used geometric constraints to alternately optimize camera poses and depth information, which improved the robustness of the network. CMDF [11] proposed a compressed structure for cross-task collaborative learning, which enhanced the ability of the network to estimate motion information of edge objects. LOFD [12] performed accurate deep triangulation on results generated by the optical flow estimation network, which uses the geometric consistency loss function and cross-task time loss function to improve multi-task learning. UFSD [13] integrated optical flow and depth estimation into a framework for pixel-intensive prediction to make the model lightweight, and used cross attention to improve the ability of feature extraction and cross-task transfer. However, these networks unify joint tasks into a pyramid extraction network for pixel feature extraction, which leads to insufficient feature information extraction in the image and neglects important information for each part of the estimation task. Therefore, in this paper, we choose the FMDN [5] as the baseline and achieve joint estimation of optical flow and depth by two independent encoding and decoding networks.

4.2.2 Transformer

Transformer [14] was first used in natural language processing as an attention mechanism model, and improved network performance by using it instead of RNN (Recurrent Neural Network) to build the whole framework. Transformer is a module of encoding and decoding architecture, which can segment the input images, input these images to the attention activation layer for calculation, and finally conduct normalization processing. The encoder generates the input sequence, and the decoder takes all the encoding and uses combined context information to generate output sequence. With the gradual development of Transformer, the combination of Transformer and image processing has solved many downstream tasks, such as remote sensing image, medical image segmentation, human pose estimation, etc.

In optical flow estimation, GMA [15] was the first network that combined Vision Transformer [16] with optical flow. It calculated the features of similar pixels between adjacent frames by introducing a global aggregation module, and decoded the residual flow by GPU, which solved the bias caused by occlusion. GMFlow [17] proposed to use matrix multiplication to calculate the global match between adjacent frames, making optical flow estimation no longer limited to local correlation. FlowFormer [18] embedded the Vision Transformer encoder into the cost volume so that the cost volume information of each layer was aggregated into cost tokens. In addition, a circular cost volume decoder was used to dynamically and repeatedly decode the cost features to iteratively refine the optical flow.

In terms of depth estimation, Adabins [19] proposed to use the mini Vision Transformer (ViT) to divide the prediction image into cells and perform global information processing at high resolution, so that the final depth prediction result was a linear combination of depth predictions in the cells. STTR(STereo TranformeR) [20] replaced the cost volume structure of dense pixel matching with self and cross-attention, while it added uniqueness constraints in the matching process to reduce the error of depth estimation networks. In this paper, we propose a novel approach to combine optical flow and camera pose estimation networks with HAT, which can improve the performance of the estimation for each part of the network.

4.3 Network Architecture

4.3.1 Overall Network

The proposed monocular stereo vision network is shown in Fig. 4.1, and the optical flow estimation and depth estimation network constitute the overall network architecture. In the optical estimation network, the monocular image sequences are input into the network coding of the optical flow and camera poses. The input characteristics of each layer are refined by the optical flow results upsampled from the previous layer, and the network regularizes each pixel along the corresponding polar line using the fixed calibration parameter of the camera at the decoding end of the last three layers of the network, reducing the search space for calculating the cost volume.

In the six-layer pyramid optical flow network, the upsampled results of each layer are input into the HAT to capture the local and global feature information through SWA and CGA respectively. Firstly, HAT partitions in a non-overlapping manner input features into feature tokens to compute self-attention mechanism within local windows. Secondly, we transpose each token and use the channel attention mechanism to calculate global feature correlation. Finally, Relying on CAT, the network can effectively aggregate feature information to increase the ability of feature selection and extraction.

For the depth estimation network with code and decoder structure, it can predict the depth information of the input original image by inputting the predicted optical

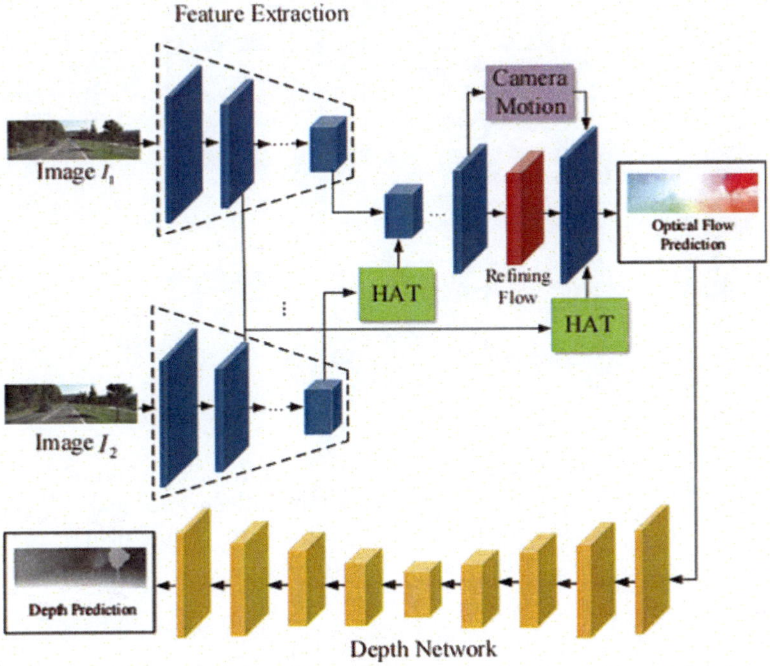

Fig. 4.1 The architecture of the overall network

flow results into the triangulation layer. The coding part performs eight-channel triangulation on each predicted pixel, while extending the processing of multiple images with target objects for fusion. The fusion network method adopts multi-layer pooling with different resolutions, which makes the network acquire more sufficient edge information. In addition, since the local and global detail features of the estimated image are preserved in the estimation on optical flow network, triangulation in each predicted pixel can obtain accurate depth values, optimizing the depth estimation network.

4.3.2 Mechanism of Attention

The structures of HAB based on spatial window attention and channel group attention and CAT are respectively shown in Fig. 4.2. The SWA divides the feature into several local windows through cross-spatial position, each window is labeled with multi-heads to calculate internal self-attention, and local feature information is refined by performing fine-grained interactions in each local window. For spatial dimension segmentation, we define the spatial dimension to represent the range of the divided sequences, and the channel dimension to represent the feature dimension of the

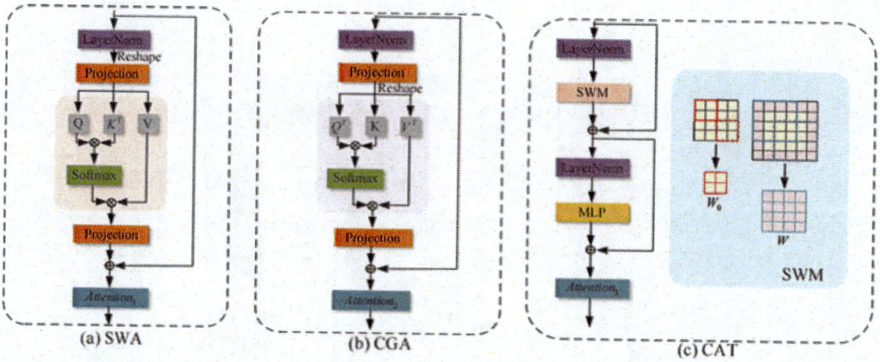

Fig. 4.2 The three methods of the attention module calculation

divided sequences. Given an input feature map $F_1 \in R^{N_1 \times C_1}$, where $N_1 = H_1 W_1$ represents the number of tokens split across spatial dimensions, and H_1, W_1, and C_1 denote the height, width, and channel size of the input feature map respectively. Self-attention is calculated in each window as follows (4.1)

$$\text{Attention}_1(Q, K, V) = \sum_{i=0}^{N_1} \text{soft}\max\left(\frac{Q_i K_i^T}{\sqrt{d}}\right) V_i \tag{4.1}$$

where $Q_i = F_{1_i} W_i^Q$, $K_i = F_{1_i} W_i^K$, and $V_i = F_{1_i} W_i^V$ represent the Query, Key, and Value for calculating the sequence of multi-input feature graphs. W_i^Q, W_i^K, and W_i^V respectively represent the trainable weigh, and d is the dimension of the Q, K, and V. The SWA is only used to calculate the local attention within each divided window, lacking the ability to interact with the attention mechanism for global information.

The CGA achieves global information by segmenting the channel dimension of the input feature. For channel dimension segmentation, we define channel dimension to represent the range of the partitioned tokens, and the space dimension to represent the feature dimension of the partitioned tokens. CGA achieves high dimensional embedding of each pixel marker by projecting the segmented feature tokens to high dimensional space, and then inputs them into the attention calculation layer for further global information extraction, adding a global residual connection to fuse local attention features. Given an input feature map $F_2 \in R^{N_2 \times C_2}$, where $N_2 = H_2 W_2$ represents the feature size of tokens, and C_2 is the number of groups representing cross-channel dimension segmentation. The calculation of channel group is as follows (4.2)

$$\text{Attention}_2(Q, K, V) = \sum_{i=0}^{C_2} \text{soft}\max\left(\frac{Q_i^T K_i}{\sqrt{d}}\right) V_i^T \tag{4.2}$$

where Q_i, K_i, and V_i can all be obtained by the linear projection of input features. HAB captures the long-range and short-range visual dependencies of feature images by combining local and global attention. CGA fuses the representation information of the image and passes it to the next layer to realize the dynamic feature fusion between levels, in contrast, SWA performs fine-grained interaction within each window in an efficient computational manner, while promoting the modeling of channel attention.

CAT provides a cross-window connection of attention, consisting of a shift window module (SWM) and an MLP layer. Before the SWM and MLP, there is a LayerNorm, which is connected by the residual. In the SWM, the window size is set to 8, and is a constant to control the overlapping size. The moving window size W is as follows (4.3)

$$W = (1 + 2\partial)W_0 \tag{4.3}$$

by sliding the window to realize the activation and normalization calculation of the input long-range feature sequence, the attention of the l layer is obtained through the calculation of the $l - 1$ layer, as follows (4.4)

$$\text{Attention} = MLP(LN(SWM(LN(F_{l-1})) + F_{l-1})) \tag{4.4}$$

where LN is layernorm, MLP represents multi-layer perceptron, and F_{l-1} is the feature of layer $l - 1$, SWM is a shift window module. We combined HAB and CAT in different ways: serial connection (SC) and parallel connection (PC), as shown in Fig. 4.3. Comparing with the impact of the two methods on network performance, in part IV, we conclude that the accuracy of the result in the optical flow estimation is improved significantly by SC, while the effect of depth estimation is more obvious by PC.

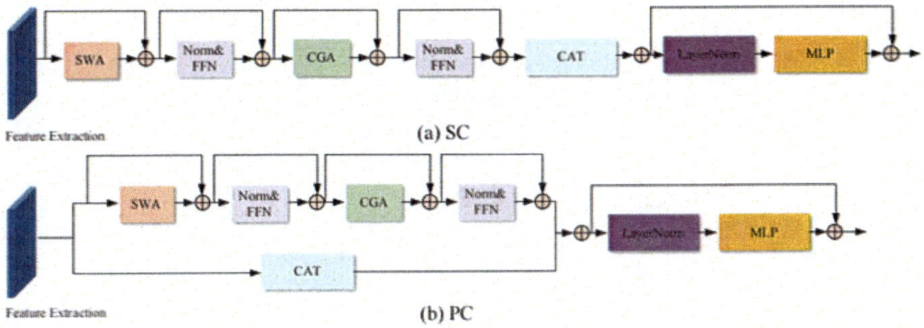

Fig. 4.3 Two connection methods of series (SC) and parallel (PC)

4.3.3 Function of Loss

Our network can learn the motion of the object from the input monocular images in self supervised way, and enable each sub-network to jointly complete downstream tasks by defining the multi-task loss function. The total loss function includes optical flow loss L_{oe}, camera pose loss L_{ce}, and depth loss L_{de}, λ_{oe} and λ_{ce} respectively represent the loss proportion in joint optical flow and camera pose estimation. The total loss function is expressed as follows (4.5)

$$L_{total} = \lambda_{oe} L_{oe} + \lambda_{ce} L_{ce} + L_{de} \tag{4.5}$$

In optical flow estimation, the photometric loss function can reconstruct the difference between the predicted value estimated by the network and the ground truth in the input images. Given two adjacent I_1 and I_2, we can obtain the reconstructed image $\hat{I}_1$ by warping the optical flow I_2 predicted by the previous layer. Equation (4.6) shows that the photometric loss function can be obtained by summing pixel-by-pixel p before and after the warped image.

$$L_{oe} = \sum_{p} \left\| I_1(p) - \hat{I}_1(p) \right\|_2 \tag{4.6}$$

In the camera pose estimation, camera parameters are primarily used for triangulation to estimate pose from the last third layer to the last first layer of the joint networks. In the following (4.7), the weight parameter is represented λ, and the true values of the camera rotation and translation are respectively represented $\tilde{\gamma}$ and $\tilde{t}$.

$$L_{ce} = \lambda \sum_{l=1}^{3} \|\gamma_l - \tilde{\gamma}\|_2 + \sum_{l=1}^{3} \left\| t_l - \tilde{t} \right\|_2 \tag{4.7}$$

In the depth estimation network, the smoothness loss function can optimize the continuity of disparity map. $D_t(p)$ denotes the disparity map estimated at time t, ∇ denotes gradient operator, and β is smooth driver factor, as shown in the following (4.8)

$$L_{de} = \sum_{p} |\nabla D_t(p)| e^{-\beta |I_1(p)|} \tag{4.8}$$

4.4 Experiments

4.4.1 Implementation Details

We used the datasets of KITTI raw for training and evaluation, including 42,949 pairs of training images and 1000 verification images to learn the optical flow and depth tasks. KITTI Raw provides two different test segmentation methods, KITTI Split [21] and KITTI Eigen [22]. For optical flow estimation, we chose KITTI Split to test the joint optical flow network, because it provided the ground truth of 200 optical flow estimation results. In order to prevent the network overfitting, we selected 697 images generated by KITTI Eigen covering 28 scenes for depth estimation.

Our network was trained by the Pytorch library, and optimized iteratively using Adam [23] with $\alpha_1 = 0.9$, $\alpha_2 = 0.999$, and a learning rate of 2×10^{-4}. During the experiment, the batch size was set to 5, and the number of iterations was 500 k. In each iteration, the weights λ_{oe}, λ_{ce} in (4.5) were dynamically determined, which made the optical flow and camera pose network dynamic balance. Our experiment first studied the network of each task independently, then optimized the network performance by implementing HAT, and finally, the comparison and ablation experiments showed that the proposed network has some advantages.

4.4.2 Optical Flow

We used the end-point-error (EPE) to measure the performance of the network in optical flow prediction, which reflects the Euclidean distance between the predicted value and the ground truth on all pixels. In addition, we also used the ratio of optical flow outliers (F1-all) to evaluate the error of network prediction, which can better reflect the optical flow error prediction ratio. Our visual results of the experiment are shown in Fig. 4.4.

It can be seen from Fig. 4.4 that our optical flow estimation network can clearly predict the motion information and position of some objects. In the optical flow map estimated by FMDN, the optical flow estimation of the rear of the occluded vehicle is

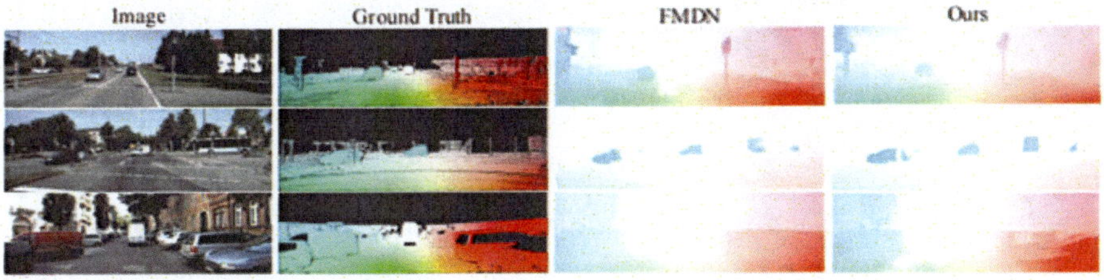

Fig. 4.4 Visual comparison of optical flow estimation results

seriously missing, and the motion boundary is blurred, which includes a lot of noise, and the target object and the background image are partially connected. Compared with the results of baseline network prediction, our proposed network can accurately predict the outline information of the object, and enrich the detailed pixel information of the object, which shows that we have improved the precision of the baseline in optical flow estimation. For the occluded objects, we can predict the fine occluded area and the direction of motion, and keep a sharp motion boundary at the rear of the occluded vehicle. Compared with the ground truth, although our network lacks accurate prediction for objects in the distant region, the predictions for objects in the close region show competitive accuracy, and it can distinguish the edge information of multiple objects well. According to the above comparative experiments, it can be concluded that our optical flow network can preserve more motion details and motion boundary information.

In Table 4.1, compared with PWC [24], the network error of our experiment is reduced, the main reason is that the combing optical flow and camera poses tasks can be mutually supervised to optimize optical flow results. Compared with the other networks, the reason our optical flow network estimation performs better is that we apply HAT to it, providing pixel-level modeling ability to clearly estimate object boundaries. Our network performs better than CMDF [11] in optical flow estimation because it requires to calibrate the internal and external parameters of the camera in advance for camera reconstruction, and only makes a simple cost average estimation in the feature, while our proposed network has global and local attention effects during training. Using the fixed parameters to construct the loss function for optimizing the network. For UFSD [13], the effect of our network on optical flow is not significant because UFSD uses the GRU iterative refinement network architecture of RAFT [10], adopting 4D correlation volume to update the operator, but the network reasoning speed is lower and the model is larger. However, our network can improve the network reasoning speed even if we add the HAT, and the time consumed by a forward calculation to obtain the optical flow is 287ms. Through the comparison of experiments, it is found that the SC mode has better optical flow estimation results than PC mode at the feature extraction.

Table 4.1 Comparisons with other models for optical flow estimation

Method	Error metric↓	
	EPE	F1-all%
PWC[24]	10.35	33.70
CodeSLAM [3]	8.80	28.94
LOFD [12]	7.04	24.21
FMDN [5]	7.01	20.92
CMDF[11]	6.84	19.88
UFSD [13]	**6.67**	17.37
Ours (SC)	6.73	**17.35**
Ours (PC)	6.81	18.23

4.4.3 Depth Prediction

We used the average absolute error (Abs Rel), mean square error (Sq Rel), root mean square error (RMSE), and function log to evaluate the performance of the depth estimation network, at the same time, the accuracy of the forecast compared to the ground true was measure by using the precision evaluation term δ. We provided 697 image sequences in the KITTI Eigen datasets for depth testing. Our experimental results are compared as shown in Fig. 4.5.

As can be seen from Fig. 4.5, our proposed depth estimation network can accurately predict the outline information of target objects at close and long distances, while separating the street markers from the background. The FMDN lacks the ability to distinguish objects with multiple overlapping areas. Especially in the overlapping areas of the jungle, the depth information estimated by FMDN has significant errors. In addition, when distant objects appear in the scene, such as roadside railing, the FMDN will lose the depth prediction information of these objects. However, Compared with the results of baseline network prediction, when there are more similar objects in the scene, our proposed method can successfully distinguish the detail of depth, and can more completely predict the depth of all objects in the scene, because the HAT can combine local and global attention to model at the pixel level. Compared with the truth of the predicted map, our innovative method in optical flow estimation network successfully estimates the distant objects in the scene by the cross-attention connection, which shows the innovation and advantage of our network in depth estimation. According to the above comparative experiments, we can conclude that our proposed network can effectively extract the disparity detail information of the target position in the scene and retain clearer and sharper target boundary information.

As shown in Table 4.2, our experimental results are lower in error than some networks, and the accuracy is significantly improved. DPSNet [25] aimed to reconstruct the image from the motion of the camera to obtain the depth information, which used a plane sweep algorithm to construct the cost volume from the depth feature, and then regulated the cost volume through context cost volume into aggregation. In this paper, our proposed network can effectively encode the cost volume into cost tokens for compact context aggregation, which makes the performance of our network estimation task better than other networks. However, compared with the

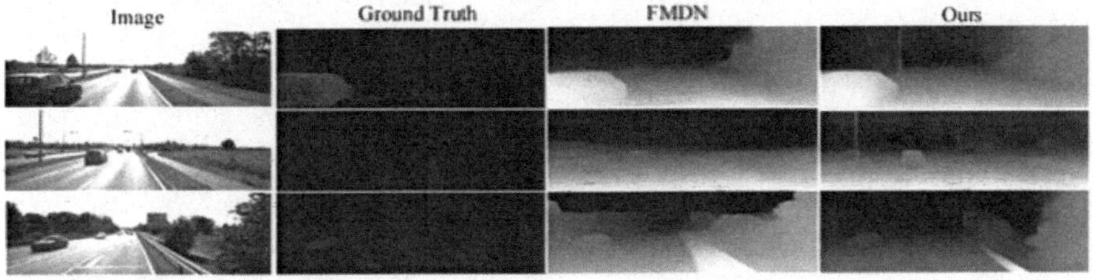

Fig. 4.5 Visual comparison of depth estimation

Table 4.2 Comparisons with other models for depth estimation

Method	Error metric↓				Accuracy metric↑		
	Abs Rel	Sq Rel	RMSE	RMSE log	$\delta < 1.25$	$\delta < 1.25^2$	$\delta < 1.25^3$
DEMoN [1]	0.191	1.120	1.059	0.240	0.733	0.898	0.951
BA-Net [26]	0.161	0.092	0.346	0.214	–	–	–
DPSNet [25]	0.147	0.107	0.427	0.191	0.893	0.950	0.976
DeepV2D [27]	0.062	0.091	0.403	0.174	0.885	0.990	**0.996**
FMDN [5]	0.075	0.082	0.396	0.150	0.867	0.992	0.991
IIB [28]	0.099	–	0.393	–	0.902	–	–
UFSD [13]	**0.050**	0.069	0.336	0.146	**0.911**	–	–
Ours (SC)	0.059	0.073	0.341	0.149	0.904	0.991	0.990
Ours (PC)	0.061	**0.065**	**0.334**	**0.142**	0.910	**0.994**	0.992

UFSD, our experiment has a decrease in error, but the improvement in accuracy is limited. The main reason is that UFSD performs deep estimation from 3D convolution display matching, while our depth estimation network is based on 2D convolution matching, which requires us to first train the optical flow network on monocular image sequences, and then extend the training to multi-view fusion network. From the experimental comparison, it is found that the PC mode has better results in depth estimation than SC mode, which is contrary to the optical flow results. In our opinion, the cycle of inference is relatively short, and the strong correlation between the pixels of partial feature of optical flow can be more efficiently completed by parallel attention, allowing the back-end network to accurately estimate the depth information.

4.4.4 Ablation Study

In this section, we evaluate the availability of our proposed method through ablation experiments, as shown in Table 4.3. CAT and HAB are applied to the feature extraction of the front-end network respectively. Compared with the baseline, both CAT and HAB can improve the evaluation indicator of the network, but HAB makes the performance of the overall network better. We analyze the reasons as follows. Firstly, this is because HAB can capture global interactions and representations by considering all spatial locations through channel attention when calculating attention scores, while CAT puts more emphasis on building connections across windows. Secondly, the two attention mechanisms in HAB are complementary to each other and have high computational efficiency, while adding CAT alone to the estimation network will lack dynamic feature fusion for the overall feature map.

In addition, we compare the CAT and HAB with the baseline by SC and PC, and it can be found that, compared to adding the attention mechanism alone, the connection

Table 4.3 Ablation study on kitti datasets

Method	Depth optical flow				
	Abs Rel	Sq Rel	RMSE	RMSE log	EPE
Baseline	0.075	0.082	0.396	0.150	7.01
Baseline + CAT	0.070	0.081	0.388	0.150	6.96
Baseline + HAB	0.063	0.079	0.375	0.149	6.92
Baseline + HAT (SC)	0.059	0.073	0.341	0.149	**6.73**
Baseline + HAT (PC)	**0.061**	**0.065**	**0.334**	**0.142**	6.81

methods of two different attention join produce more accurate results for our overall network. However, in the optical flow estimation, the SC has a better improvement on the network index, while in the depth estimation, the PC improves the network performance better. We analyze this interesting phenomenon as follows. Firstly, the attention is connected serially in the feature pyramid of the hierarchical structure. This network connection method makes the optical flow estimation to be a unified framework, which is beneficial to the refinement of optical flow in the previous layer. Secondly, although the method by parallel connection of attention is not obvious to reduce the error result of the optical flow estimation, it improves the computational efficiency of the network while enhancing the ability of the network to focus on the selection of the feature area, so the depth estimation results of the network is more obvious. Our experiments show that the additional attention added to the optical flow estimation can improve and optimize the accuracy of the depth estimation network, making the estimated results of the overall network more accurate and more robust.

4.5 Conclusion

In this paper, we propose a monocular stereo matching network based on Hybrid Attention Transformer, and introduce two different internal attention composition methods of series and parallel. The key of this paper is that the joint task of optical flow estimation and depth estimation can be completed efficiently by introducing HAT, which overcomes the limitations of existing multi-task methods. HAB combines the SWA and CGA to realize the local and global attention, and CAT can realize cross-window connection between pixels to enhance pyramid feature extraction. In this article, we introduce the calculation methods of different attention mechanisms, and evaluate the optical flow and depth network by two different connection methods of SC and PC. The experimental results show that both different connection methods play a vital role in the performance improvement of the network estimation, indicating the novelty of the proposed method. The proposed network finally shows extremely competitive performance on KITTI datasets, and is more simple and efficient in model design. Considering the limited size of the network, we lack innovation in the back-end of the network, and how to balance the inference time on the basis of increasing the size of the model will be our further research direction. In the future, we hope that the Transformer attention mechanism can be more used in machine vision to handle more network tasks.

References

1. Ummenhofer, B., Zhou, H., Uhrig, J., Mayer, N., Ilg, E., Dosovitskiy, A., Brox, T.: DeMoN: depth and motion network for learning monocular stereo. In: The IEEE Conference on Computer Vision and Pattern Recognition (CVPR) (2017)
2. Ilg, E., Mayer, N., Saikia, T., Keuper, M., Dosovitskiy, A., Brox, T.: Flownet 2.0: evolution of optical flow estimation with deep networks. In: IEEE Conference on Computer Vision and Pattern Recognition (CVPR), pp. 1647–1655 (2017)
3. Bloesch, M., Czarnowski, J., Clark, R., Leutenegger, S., Davison, A.J.: CodeSLAM learning a compact, optimisable representation for dense visual SLAM. In: The IEEE Conference on Computer Vision and Pattern Recognition (CVPR) (2018)
4. Chen, X., Wang, X., Zhou, J., et al.: Activating more pixels in image super-resolution transformer. arXiv e-prints https://doi.org/10.48550/arXiv.2205.04437 (2020)
5. Wang, K., Shen, S.: Flow-motion and depth network for monocular stereo and beyond. IEEE Robot. Autom. Lett. **5**(2), 3307–3314 (2021)
6. Liu, Z., Lin, Y., Cao, Y., et al.: Swin transformer: Hierarchical vision transformer using shifted windows. In: Proceedings of the IEEE/CVF International Conference on Computer Vision, pp: 10012–10022 (2021)
7. Geiger, A., Lenz, P., Stiller, C., Urtasun, R.: Vision meets robotics: the KITTI dataset. Int. J. Robot. Res. **32**(11), 1231–1237 (2013)
8. Wang, K., Shen, S.: MVDepthNet: real-time multiview depth estimation neural network. In International Conference on 3D Vision (3DV) (2018)
9. Teed, Z., Deng, J.: Droid-slam: DROID-SLAM: deep visual SLAM for monocular, stereo, and RGB-D cameras. Adv. Neural. Inf. Process. Syst. **34**, 16558–16569 (2021)
10. Teed, Z., Deng, J.: Raft: recurrent allpairs field transforms for optical flow. arXiv preprint arXiv: 2003.12039 (2020)
11. Chen, Y., Cao, X., Lin, X., et al.: A compacted structure for cross-domain learning on monocular depth and flow estimation. arXiv preprint arXiv:2208.11993 (2022)
12. Guizilini, V., Lee, K.H., Ambruş, R., et al.: Learning optical flow, depth, and scene flow without real-world labels. IEEE Robot. Autom. Lett. **7**(2), 3491–3498 (2022)
13. Xu, H., Zhang, J., Cai, J., et al.: Unifying flow, stereo and depth estimation. arXiv preprint arXiv:2211.05783 (2022)
14. Dosovitskiy, A., Beyer, L., Kolesnikov, A., Weissenborn, D., Zhai, X.H., Houlsby, N., et al.: An image is worth 16x16 words: transformers for image recognition at scale. arXiv preprint arXiv:2010.11929v2 (2020)
15. Jiang, S., Campbell, D., Lu, Y., et al.: Learning to estimate hidden motions with global motion aggregation. In: Proceedings of the IEEE/CVF international conference on computer vision, pp 9772–9781 (2021)
16. Liu, Y., Wu, Y.H., Sun, G., et al.: Vision transformers with hierarchical attention. arXiv 2106 03180 (2021)
17. Xu, H., Zhang, J., Cai, J., et al.: GMFlow: learning optical flow via global matching. In: Proceedings of the IEEE/CVF Conference on Computer Vision and Pattern Recognition, pp. 8121–8130 (2022)
18. Huang, Z., Shi, X., Zhang, C., et al.: FlowFormer: a transformer architecture for optical flow. arXiv preprint arXiv:2203.16194 (2022)

19. Bhat, S.F., Alhashim, I., Wonka, P.: Adabins: depth estimation using adaptive bins. In: Proceedings of the IEEE/CVF Conference on Computer Vision and Pattern Recognition, pp. 4009–4018 (2021)
20. Li, Z., Liu, X., Drenkow, N., et al.: Revisiting stereo depth estimation from a sequence-to-sequence perspective with transformers. In: Proceedings of the IEEE/CVF International Conference on Computer Vision, pp. 6197–6206 (2021)
21. Eigen, D., Puhrsch, C., Fergus, R.: Depth map prediction from a single image using a multi-scale deep network. In: Proceedings of Neural Information Processing Systems (NIPS), pp. 2366–2374 (2014)
22. Geiger, A., Lenzp, U.R.: Are we ready for autonomous driving? The KITTI vision benchmark suite. In CVPR (2012)
23. Kingma, D., Ba, J.: Adam: a method for stochastic optimization. Comp. Sci. (2014)
24. Sun, D., Yang, X., Liu, M.Y., et al.: Pwc-net: Cnns for optical flow using pyramid, warping, and cost volume. In: Proceedings of the IEEE conference on computer vision and pattern recognition, pp. 8934–8943 (2019)
25. Im, S., Jeon, H.G., Lin, S., Kweon, I.S.: DPSNet: end-to-end deep plane sweep stereo. In: The International Conference on Learning Representations (ICLR) (2020)
26. Tang, C., Tan, P.: Ba-net: Dense bundle adjustment network. arXiv preprint arXiv:1806.04807 (2018)
27. Teed, Z., Deng, J.: Deepv2d: video to depth with differentiable structure from motion. In The International Conference on Learning Representations (ICLR) (2020)
28. Yifan, W., Doersch, C., Arandjelovi, R., Carreira, J., Zisserman, A., et al.: Input-level Inductive Biases for 3D Reconstruction. In: Proceedings of the IEEE/CVF Conference on Computer Vision and Pattern Recognition, pp. 6176–6186 (2022)

Chapter 5
Prediction of Tool Remaining Useful Life of NC Machine Tool Based on DTW Algorithm and LSTM Neural Network

Kechang Zhang and Jicheng Duan

Abstract Based on dynamic time warping (DTW) algorithm and long short-term memory (LSTM) neural network, this paper carried out research on the prediction of tool remaining useful life (RUL) of NC machine tools, evaluated the accuracy of the health index prediction model of each tool by using root mean square error, and introduced a scoring mechanism to score the predicted results of tool residual service life based on the specified prediction starting point. The method can realize the prediction of the tool remaining useful life.

5.1 Introduction

Under the wave of global industrial reform with intelligent manufacturing as the main body set off by Industry 4.0, China, as a major manufacturer, formulated the "Made in China 2025" plan in 2015 to promote China to become a "manufacturing power". NC machine tools in the process of high-speed cutting will appear in a variety of abnormal conditions leading to poor quality of machining parts, precision is not up to standard and even become waste parts, and serious machine damage occurs safety accidents. As the key parts of NC machine tool, the tool is also the most prone to wear and failure. As the tool directly contacts the workpiece being processed, its wear condition directly affects the machining quality, accuracy, and equipment operation efficiency of the workpiece [1]. It is very easy to produce defective products and waste products when the tool works in abnormal state, so it is very important to carry out the prediction of the tool remaining useful life of NC machine tools.

K. Zhang · J. Duan (✉)
Hunan Railway Professional Technology College, ZhuZhou 412001, China
e-mail: 305748030@qq.com

R. Kountchev et al. (eds.), *Multidimensional Signals, Augmented Reality and Information Technologies*, Smart Innovation, Systems and Technologies 374,
https://doi.org/10.1007/978-981-99-7011-7_5

5.2 The Significance of Tool Remaining Useful Life Prediction

To carry out the prediction of tool remaining useful life of NC machine tools, that is, to realize the online monitoring of tool wear, the key is to automatically find the appropriate tool change time according to the actual situation and requirements of on-site processing, real-time monitoring of tool state, and its importance is self-evident. An aviation enterprise due to the lack of tool monitoring management system, machining precision hard parts, in the case of tool breakage and unknown continued processing parts damage, resulting in millions of losses; According to statistics, 22.4% of the total downtime of machine tools is caused by tool failure [2]. According to the research statistics of HK Tonshoff in Germany, the production efficiency of NC machine tools with online monitoring of tool wear is increased by 10–60%, the downtime of fault of 75% of machine tools is reduced, and the utilization rate is increased by more than 50% [3]. A survey by Kennametal, an American company, found [4] that the application of online tool wear monitoring can save up to 30% of the machining cost of NC machine tools.

Due to the complex and changeable machine tool processing environment and too many factors causing tool wear, the tool change work is completed by the site workers, they rely on years of experience and the number of workpiece machining site tools and processing noise, choose the approximate time to complete the tool change. The accuracy of this highly subjective way is not high, and it is inevitable that the following two situations will occur: (1) The replaced tool does not reach the preset service life, causing the waste of tool resources and increasing the production cost. (2) Beyond the preset service life, the replaced tool wears seriously, resulting in poor machining quality of the workpiece, does not meet the processing accuracy of the problem, and even leads to machine failure and shutdown, resulting in a huge loss of production benefits [5]. Therefore, the realization of online monitoring of tool wear to ensure processing quality and accuracy is of great significance for reducing machine downtime, improving production efficiency, and maximizing the rational use of tools to save production costs and reduce economic losses [6]. On the other hand, many real-time data collected during tool processing are not well used, resulting in idle waste of data resources. Based on such real-time processing data, tool wear monitoring realizes online automatic identification of wear state and intelligent real-time prediction of wear amount [7]. Therefore, effective tool wear monitoring technology has become the core of promoting the development of automation and intelligence in the manufacturing industry. Therefore, it is of vital significance to carry out the prediction of tool residual service life of NC machine tools.

5.3 Research Status of Tool Residual Service Life Prediction

At present, research on tool wear and life prediction models are in full swing among domestic and foreign researchers. Based on various models in numerous literatures [8, 9], model research can be divided into three main categories: mechanism modeling, numerical simulation, and data-driven.

(1) Mechanism modeling method: It refers to the method that uses prior knowledge and tool wear mechanism of machine tool to establish a mathematical model to describe the relationship between wear value and various parameters in cutting (such as workpiece material, cutting depth, speed). The commonly used mechanism models are Taylor formula model and wear rate model. The mechanism-based modeling method must consider many machining parameters, which can only be determined by massive field experiments. Moreover, a model only targets at one kind of parameter in the process of processing. Once any of the parameters are changed, it is necessary to conduct tests again to obtain and establish a new model. The whole process consumes huge human resources, material resources, and financial resources, the model is specific but not universal, and the scope of application is narrow. In addition, due to the complexity of actual processing and various influencing factors, the established models often fail to consider all the factors and are idealized to a certain extent, which is too restrictive.
(2) The method based on numerical simulation: it refers to the method of finite element analysis, under the condition of known machining parameters and wear model, intuitively and visually simulate the tool wear process, change different parameters to analyze their effects on tool wear, and establish the numerical simulation model of tool wear through simulation analysis. Nayak and Bartarya et al. established a two-dimensional finite element simulation model to simulate the cutting temperature distribution, determine the critical temperature at which the transition results in the formation of the white layer, and finally achieve the simulation of the temperature distribution on the tool surface and the prediction of the white layer [8]. The research results based on numerical simulation are more extensive, but the accuracy of this method depends on the performance of the wear model to a large extent. The simplified model used in the practical application of this method cannot accurately and completely reflect the complex environment of the real field, so the error between the simulated value and the real value is large [9, 10].
(3) Data-driven method: it refers to the use of many monitoring signal data collected in the indirect method when cutting tools, the use of big data mining technology to obtain effective information, build a variety of models to complete the status discrimination and life prediction. In recent years, experts and scholars at home and abroad have invested a lot of effort in data-driven methods and produced a lot of results. The data-driven approach can be summarized into the following

three types of models: probability graph model, machine learning model, and deep learning model [11–13].

① Probability graph model: It mainly includes Hidden Markoc Model (HMM), Kalman Filter (KF), Particle Filter (PF), Bayesian Network (BN), etc.
② Machine learning model: There are many machine learning models, among which support vector machine (SVM), Random Forest (RF), Back Propagation Neural Network (BPNN) and so on are commonly used in tool wear.
③ Deep learning model: build multi-layer machine learning network structure, adaptive feature extraction, and the established nonlinear model. Deep learning models commonly used in tool wear monitoring include deep confidence network (DBN), convolutional network (CNN), gated cycle network (GRU), etc. Xu and Wang et al. designed CNN to extract features from vibration data for tap tool monitoring, then developed expansion convolution residuals to complete wear prediction with a full connection layer, and finally verified the robustness and accuracy of the model in the tapping process of the engine cylinder head production line [12].

5.4 Tool Health Curve Constructed Based on DTW Algorithm

Based on the building steps of tool health index of NC machine tools in literature [14], DTW algorithm model is built by using MATLAB program language, and the health index of each tool is calculated respectively. The final output results are shown in Fig. 5.1. As can be seen from the figure, with the continuous wear degradation of the tool, the overall HI curve presents a gradual upward trend, indicating that the tool health curve constructed based on DTW algorithm can better characterize the wear degradation process of the tool.

Health indicators (HI) quantify the ongoing degradation of tool wear by extracting information from acquired signal signature data, and the quality of the HI curve greatly influences the validity of the health assessment and the predictive accuracy of subsequent remaining useful life. However, the above health curve shows a lot of different degrees of local fluctuation phenomenon, No. 1 tool and No. 3 tool HI curve in the middle of the small amplitude fluctuation, No. 2 tool HI curve in the last section of the larger fluctuation, this situation will cause a certain degree of impact on the remaining service life prediction of the tool. This situation is related to the noise contained in the extracted signal features. However, if the signal features are filtered and denoised, information loss related to tool wear will be caused. Therefore, to reduce the impact of all kinds of noise on health indicators, you can directly use five points cubic smoothing method to smooth the HI curve.

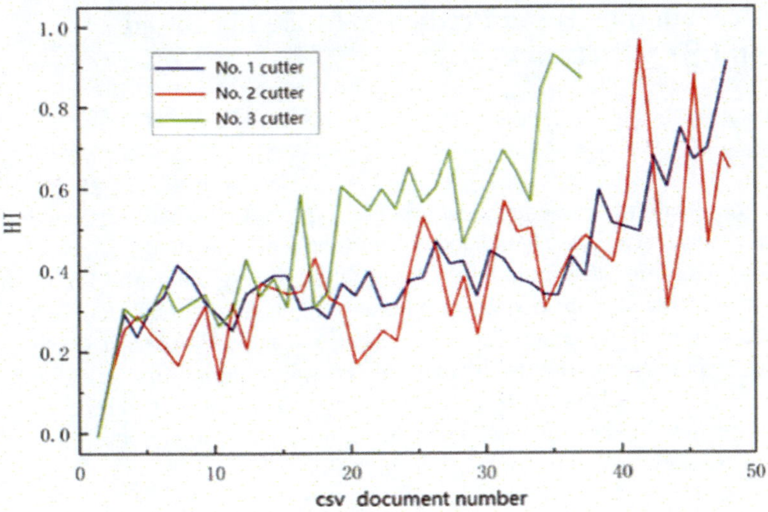

Fig. 5.1 Health index curve of cutting tool

The five-point cubic smoothing method is also used to smooth the health indicators of No. 1, No. 2, and No. 3 cutting tools. The monotonicity and trend of the HI curve before and after the treatment are shown in Table 5.1.

As can be seen from Table 5.1, before and after smoothing, the monotonicity and trend of HI curve of No. 2 tool and No. 3 tool show the same changes as that of No. 1 tool, indicating that the five-point cubic smoothing algorithm is applicable to the smoothing of tool health index, providing a guarantee for the accuracy of the subsequent remaining service life prediction of the tool.

Table 5.1 Comparison of monotonicity and trend of HI curve before and after smoothing treatment

	Monotonicity index value		Trend index value	
	Before smoothing	After smoothing	Before smoothing	After smoothing
No.1 tool	0.02128	0.36171	0.79974	0.81895
No.2 tool	0.06384	0.27660	0.73663	0.93139
No.3 tool	0.16667	0.55556	0.18375	0.20708

5.5 Degradation Trend Prediction Modeling Based on LSTM Network

5.5.1 Definition of RUL

Known tool health index sequence HI $= \{x_1, x_2, \ldots, x_n\}$, where n is the length of the time series, and x_n is the health index value corresponding to time n, which is also defined as the failure threshold here. The problem of residual useful life prediction is essentially to predict the degradation trend of the health indicator curve.

If you know the sequence $\{x_1, x_2, \ldots, x_n\}$, t < n, x_t is forecast starting point of RUL, Input the model to predict the changing trend of health indicators $\{x_{t+1}, x_{t+2}, \ldots, x_{t+k}\}$, while $x_{t+k} \geq x_n$ and $x_{t+k-1} < x_n$, k is the time difference between the predicted starting point and the failure point, which is also the predicted remaining service life of the tool, as shown in Fig. 5.2.

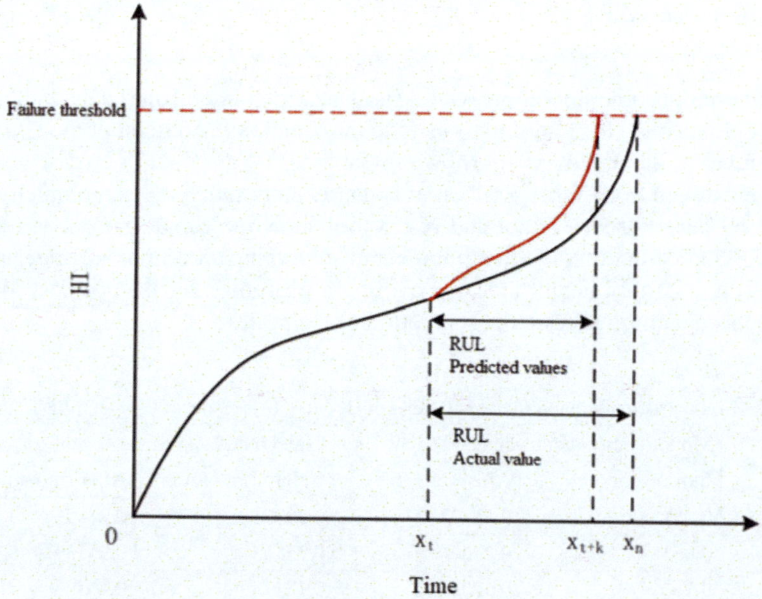

Fig. 5.2 Schematic diagram of RUL prediction

5.5.2 Modeling Steps of Degradation Trend Prediction Based on LSTM Network

The modeling steps of degradation trend prediction based on LSTM network are as follows:

Step 1: Sample construction: The input of the LSTM network model based on health indicators is the sequence of tool health indicators, and the output is the changing trend of health indicators in the future. Therefore, it is necessary to reconstruct the existing health indicator data by sliding window. As shown in Fig. 5.3, according to a certain window length, slide the window with step size of 1, the next HI value of the window is the corresponding label of the window, and finally, get the reconstructed data matrix and corresponding label;
Step 2: Divide the training data set and test data set, input the training data set into the established LSTM network model, and judge whether the model training is over according to whether the training loss function is small enough.
Step 3: Set the starting point of prediction, input the test data set into the trained LSTM network model, and predict the health index value at the next moment.

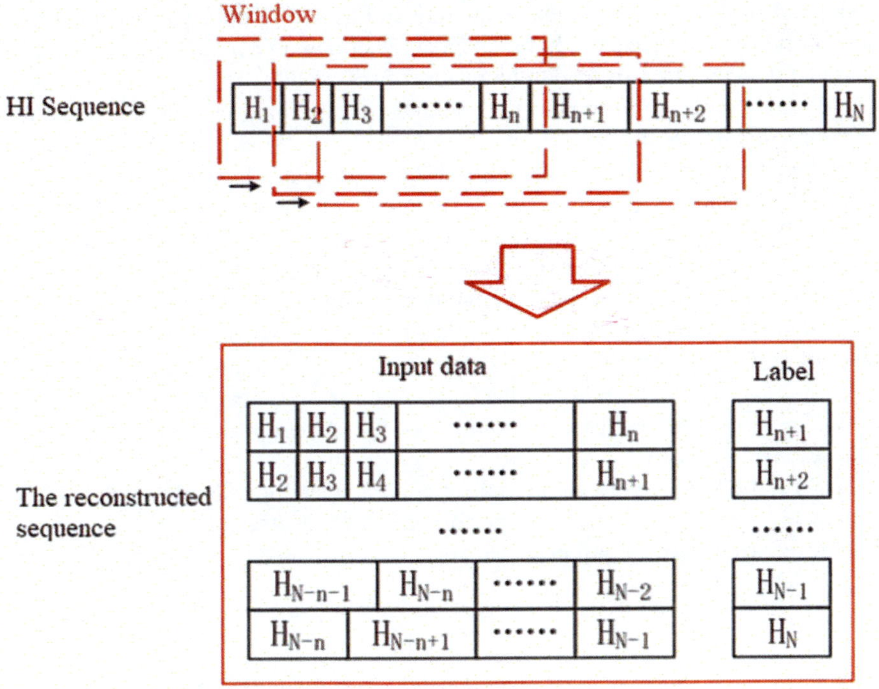

Fig. 5.3 Diagram of sample reconstruction of health indicators

Step 4: Set the tool wear failure threshold, judge whether the current health index value reaches the threshold, and calculate the time difference between the prediction starting point and the failure point as the tool RUL prediction result.

5.5.3 *Analysis of Prediction Results*

The prediction result of the tool remaining useful life is of practical significance. If the prediction result is too early, it will cause unnecessary waste of resources; if the prediction result is too late, it will cause equipment damage and reduce the machining accuracy of the workpiece. Therefore, this chapter introduces the scoring mechanism and root mean square error as two evaluation criteria to evaluate the prediction accuracy of the model.

The comparison of the two evaluation criteria is shown in Fig. 5.4. As can be seen from the figure, RMSE evaluation index is axisymmetric, which indicates that for RMSE evaluation criteria, premature prediction and late prediction have the same degree of impact on model research objects, which is obviously not the most reasonable in practical application. In the Score function curve, the slope of the score curve changes slowly and gently when the score is predicted too early, while the slope of the score curve changes quickly and the curve changes sharply when the score is predicted too late. In practical applications, to avoid greater economic losses and to avoid the late prediction results of the model, the scoring mechanism of Score function is in line with this goal and can optimize the prediction model towards premature prediction.

Combined with root mean square error and scoring mechanism, the prediction results of each tool health index were evaluated, and the evaluation results were shown in Table 5.2. The RMSE values of tool No. 2 and tool No. 3 are small. However, because tool No. 2 is predicted too late, which will seriously harm the workpiece and machine tool, its RUL prediction score is low; while tool No. 3 is predicted too early, and its harm is lower than that of tool No. 3. Therefore, its RUL prediction score is relatively high. No. 1 tool and No. 2 tool are predicted too late. However, due to a large deviation in the late prediction of its health index, the failure threshold point is predicted too late, so the RMSE and Score values of the predicted results are relatively low.

The DTW-LSTM network model can better predict the future change trend of tool health index, and has good predictive performance, which is suitable for the research of tool residual service life prediction.

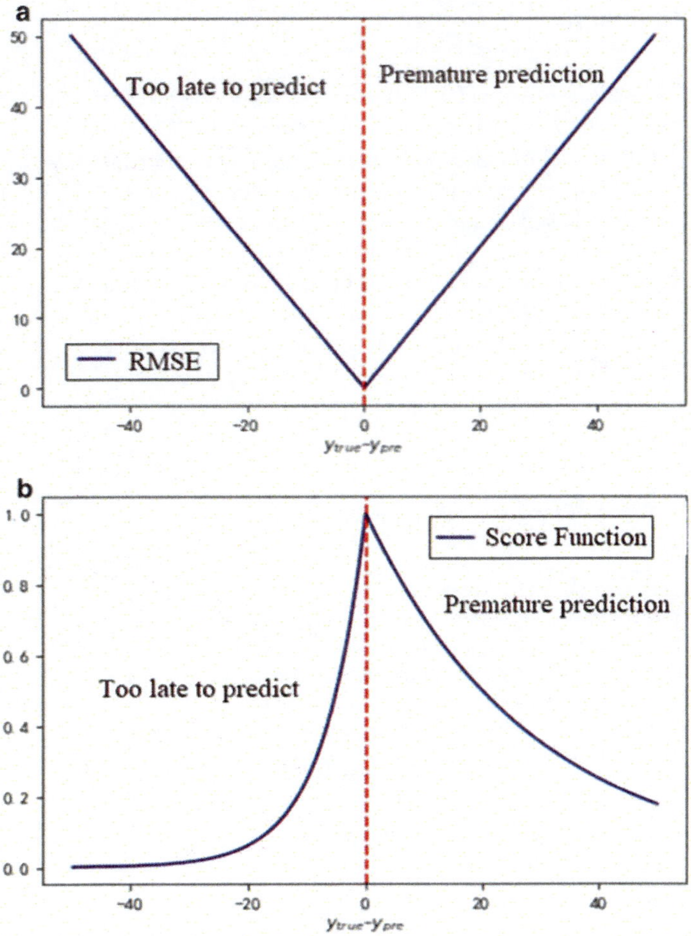

Fig. 5.4 Comparison of root mean square error and scoring mechanism **a** Root mean square error graph **b** Scoring mechanism chart

Table 5.2 Evaluation of prediction results of health indicators of each tool

	RMSE	Score
No.1 tool	0.04789	0.63
No.2 tool	0.02245	0.77
No.3 tool	0.03868	0.88

5.6 Conclusion

Based on DTW algorithm and LSTM neural network, the prediction research of tool residual service life of NC machine tools was carried out. The accuracy of the health index prediction model of each tool was evaluated by using root mean square error. The scoring mechanism was introduced to score the residual service life prediction results of each tool based on the specified prediction starting point.

Acknowledgements This work was supported by Scientific Research Fund of Hunan Provincial Education Department (No. 22C1097).

References

1. Qin, W., Chen, X., Zhang, Y.S.: Strong machine tools lead to strong manufacturing: strategy of strengthening China from "Made in China 2015." Equip. Manuf. **2016**(07), 29–31 (2016)
2. Rehorn, A.G., Jiang, J., Orban, P.E.: State-of-the-art methods and results in tool condition monitoring: a review. Int. J. Adv. Manuf. Technol. **26**, 693–710 (2005)
3. Tonshoff, H.K., Wulfsberg, J.P., Kals, H.J.J., van Luttervelt, C.A.: Developments and trends in monitoring and control of machining processes. CIRP Ann. **37**(2), 611–622 (1988)
4. Tlusty, J., Andrews, G.C.: A critical review of sensors for unmanned machining. CIRP Ann. **32**(2), 563–572 (1983)
5. Lin, Y.: Research on Tool Wear Condition Monitoring Technology Based on Deep Learning. University of Science and Technology of China (2017)
6. Peng, R.T., Pang, H.L., Jiang, H.J., Hu, Y.B.: Study of tool wear monitoring using machine vision. Autom. Control. Comput. Sci. **54**(3), 259–270 (2020)
7. Lins, R.G., de Araujo, P.R.M., Corazzim, M.: In-process machine vision monitoring of tool wear for Cyber-Physical Production Systems. Robot. Comput.-Integr. Manuf. **61**(February 2020), 101859 (2020)
8. Nayak, R.K., Bartarya, G., Sahoo, M.R.: Numerical analysis of the effect of tool wear oil surface integrity during hard turning. J. Mech. Sci. Technol. **35**, 1215–1222 (2021)
9. Simon, G.D., Deivanathan, R.: Early detection of drilling tool wear by vibration data acquisition and classification. Manuf. Lett. **21**(August 2019), 60–65 (2019)
10. Gomes, M.C., Brito, L.C., da Silva, M.B., Duarte, M.A.V.: Tool wear monitoring in micro milling using Support Vector Machine with vibration and sound sensors. Precis. Eng. **67**(January 2021), 137–151 (2021)
11. Dou, J.M., Xu, C.W., Jiao, S.J., Li, B.D., Zhang, J.L., Xu, X.X.: An unsupervised online monitoring method for tool wear using a sparse auto-encoder. Int. J. Adv. Manuf. Technol. **106**(5), 2493–2507 (2020)
12. Xu, X.W., Wang, J.W., Ming, W.W., Chen, M., An, Q.L.: In-process tap tool wear monitoring and prediction using a novel model based on deep learning. Int. J. Adv. Manuf. Technol. **112**(1), 453–466 (2021)
13. Huang, Z.W., Zhu, J.M., Lei, J.T., Li, X.R., Tian, F.Q.: Tool wear predicting based on multi-domain feature fusion by deep convolutional neural network in milling operations. J. Intell. Manuf. **31**, 953–966 (2020)
14. Zhang, K.C., Yang, M.Q., Shi, J.Y.: Construction of the tool health index of NC machine tool based on DTW algorithm. In: The 5th International Conference on Information Technologies and Electrical Engineering (ICITEE2022), pp. 534–538. Association for Computing Machinery, New York (2022)

Chapter 6
Research on Online Detection Device and Method of Corn Grain Crushing Rate

Liang Yang, Xiaoping Bai, Zhuo Wang, Yongjia Zhao, and Sijia Wang

Abstract For the direct grain harvesting combined harvester, the grain crushing rate is one of its important parameters, but in the actual harvest, the corn grain crushing rate is difficult to detect online. Aiming at the research difficulty of online detection of corn grain crushing rate, this paper proposes a complete solution and designs and develops a prototype of the principle of corn grain online detection device. Machine vision technology combined with deep learning method is used to detect and analyze the corn grain crushing rate online. The online detection device for rice grain crushing rate is installed at the elevator of the grain direct harvesting combined harvester to carry out field experiments, and the field experiment results show that the corn grain crushing rate online detection device can be effectively realized, sampling and resampling on the grain direct harvesting combined harvester, and achieving a better single-layer tiling effect. Using deep learning methods, it can effectively identify broken corn kernels.

6.1 Introduction

Corn, as one of the three major food crops in the world, has a long history of cultivation and a wide range of cultivation. China is a major country in corn cultivation, with planting areas almost covering all parts of the country. Corn is a raw material in many fields such as industry and feed industry in China. However, due to the manufacturing

L. Yang (✉) · X. Bai · Z. Wang · Y. Zhao
Shenyang Institute of Automation Chinese Academy of Sciences, Shenyang 110016, China
e-mail: yangliangsia@sina.cn

Institutes of Robotics and Intelligent Manufacturing, Chinese Academy of Sciences, Shenyang 100019, China

Key Laboratory of Agricultural Equipment Intelligent Technology, Liaoning Province, Shenyang 110016, China

S. Wang
College of Engineering, Shenyang Agricultural University, Shenyang 110866, China

R. Kountchev et al. (eds.), *Multidimensional Signals, Augmented Reality and Information Technologies*, Smart Innovation, Systems and Technologies 374,
https://doi.org/10.1007/978-981-99-7011-7_6

process and raw material level of China's corn combine harvesters, there is still a certain gap compared to developed countries. During the harvesting process of corn machinery, it is easy to cause corn grains to break. Broken corn will lead to a decrease in the comprehensive utilization value of corn and cannot meet production needs. Therefore, the online detection device and method for corn kernel crushing rate remains a hot and difficult research topic in the agricultural field of China.

With the development of artificial intelligence, there have been new breakthroughs and developments in the research of detection devices and methods based on machine vision. Machine vision technology refers to the technology of constructing a machine vision system to achieve non-contact observation and measurement of the target being tested [1, 2]. The application of machine vision in agricultural harvesting is currently mainly aimed at crops with larger individual crops and sparse fruit density, such as apples, peaches, etc. Machine vision is equivalent to capturing the eyes of a robot. After identifying the target to be picked through machine vision, the control system controls the picking execution mechanism to reach the target area and complete fruit picking [3–6]. Li Xinping [7–9] studied the impact of the external mechanical shock on the internal stress cracks of the corn grain when the corn grain is detached, and analyzed it in combination with the ANSYS finite element technology. The research results show that there is a close relationship between the corn grain crushing and the moisture content, the grain feeding form, and the speed of the threshing drum. On this basis, a speed threshing method was designed, which significantly reduced the generation of corn grain cracks. Yang Liang [10], using machine vision technology, used the color difference that the broken part of broken corn grain was milky white while the whole corn grain was yellow, combined with K-Means color image segmentation algorithm, to distinguish the complete corn grain and broken corn grain; Chen Jin [11–13], used the two-step method of "chromatography training verification" to identify the complete and broken rice grain, The segmentation method of grain, branch and stem in the image of harvested rice grain based on U-Net depth learning model is proposed. The improved U-Net network is used to increase the network depth and add a Batch Normalization layer to obtain more abundant Semantic information on small data sets, solving the problem of image training data shortage and training overfitting. Ding [14] used the curvature index, radius index, and continuous symmetry index as comparative parameters between intact and crushed corn kernels. This method can effectively distinguish the contour of crushed corn kernels from that of intact corn kernels. Reid [15] used the linear features of the mechanical crack area in corn kernels as the discrimination basis, combined with the Hough algorithm, to detect corn kernels with a detection accuracy of 91.8%. Valiente Gonzá lez [16], based on the methods of outlier detection and principal component analysis, combined with machine vision technology, built a corn grain separation system to distinguish complete corn grains from those with dents on the surface. The experimental results show that under the HSV color space, the detection accuracy can reach 92%.

6.2 Study on Online Detection Device for Corn Grain Crushing Rate

The online detection device for corn kernel breakage rate aims to achieve real-time sampling of corn kernels during the harvesting process, and achieve single-layer tiling of kernels through institutional processing to reduce occlusion between kernels. The real-time sampling of grain images is achieved through the establishment of an image acquisition platform, and an online detection method for corn kernel breakage rate is developed to achieve online detection of corn kernel breakage rate. In order to ensure the normal operation of the detection device in harsh field conditions, with good working performance, and to meet the normal working requirements, the material of the detection device is stainless steel with better properties. The entire detection device shell is designed through sheet metal processing of stainless steel plates. The device is mainly composed of a device box, blanking trough, buffer trough, master/slave Timing belt wheel, Timing belt, light source, camera bracket and other mechanisms. The feeding trough is connected by bolts and installed on the box cover of the device box. After being drained by the feeding trough, the external corn seeds enter the detection device, and the lower end of the feeding trough is a buffer trough. The function of buffer groove is to reduce the impact of grains on the Timing belt, prevent the Timing belt from fluctuating, and ensure the image quality. The front end of buffer groove is equipped with a brush. When corn grains flow through the brush, due to the effect of slight resistance, it is not easy for corn grains to stack, and effectively realizes the corn grains to be a single layer. At the same time, the buffer tank is in a trapezoidal shape, with the front end closed, which can effectively gather corn seeds and make them fall into the center area of the belt as much as possible, making it easier for CCD industrial cameras to collect as many experimental samples as possible and obtain more experimental data. The area below the buffer groove is the core of the detection device, namely the Timing belt transmission system. The driving wheel is driven by the stepping motor, and the driven wheel is driven by the Timing belt to rotate, so as to realize real-time dynamic collection of corn grain samples. The CCD industrial camera is installed in the middle area of the belt movement path to collect images of corn grain samples flowing through it. The CCD industrial camera has light sources on both sides, which are symmetrically distributed on both sides of the CCD industrial camera, reducing the impact of shadows caused by a single light source and facilitating subsequent data analysis of collected images. The lower side of the Timing belt is the tensioning wheel. Through the up and down adjustment of the tensioning wheel, the Timing belt is in the best working state to prevent slip and other failures and ensure the working efficiency of the detection device. The box and cover of the Timing belt type corn grain crushing rate online detection system are designed by sheet metal, which makes the detection device easy to disassemble and carry. The sheet metal at the bottom is designed to be inclined, and an opening is left at the end to ensure that the tested corn grains are released, prevent corn grains from piling up inside the detection device, and ensure the smooth and normal operation of the detection device. The online monitoring device for grain crushing rate is installed on the grain elevator, as shown in Fig. 6.1. The online device for corn grain crushing rate is installed at the elevator of the direct harvesting combine harvester, as shown in Fig. 6.2.

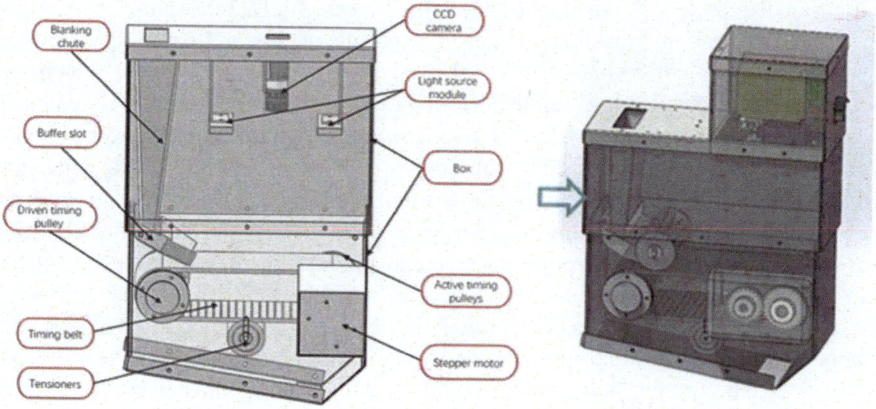

Fig. 6.1 Design of online detection device for corn grain crushing rate

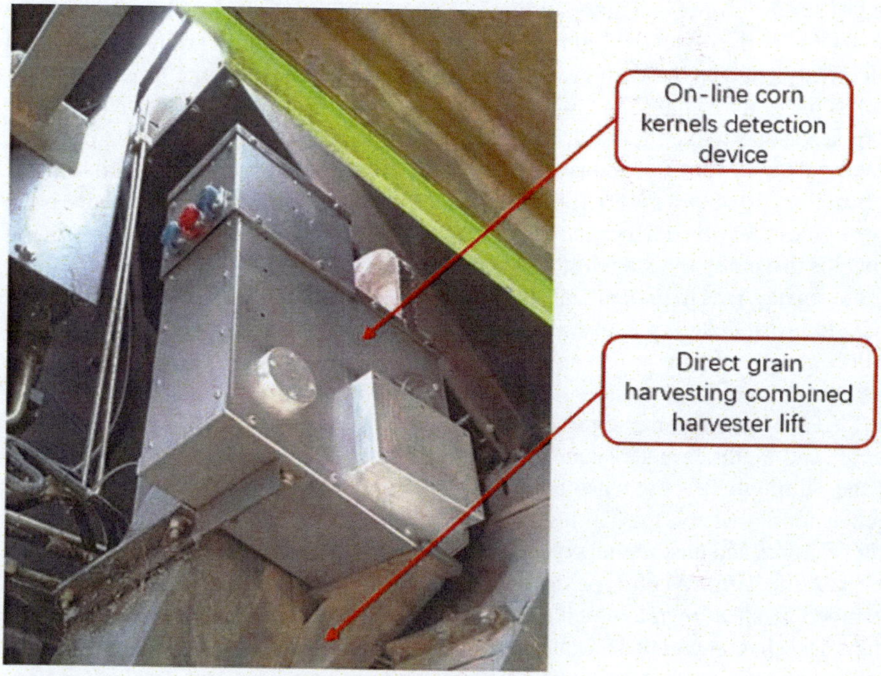

Fig. 6.2 Actual installation of online detection device for corn grain crushing rate

6.3 Online Detection Method for Corn Grain Crushing Rate

Aiming at the problem that there are great differences in the type, size and morphology of corn kernels, machine vision combined with deep learning method is used to realize the self-learning of corn completion. Firstly, the collected sample image is binarized, and the connected region is extracted from the binary image. When extracting the connected region, each region is assigned a unique number representing the region, and the pixel value of all pixels in the area is assigned to the number to realize the marking of the image. The traditional image processing method is used to label the single seed, and then the convolutional neural network model is used to train and recognize the grain, and the overall research scheme is shown in Fig. 6.3.

6.3.1 *Two-Dimensional Image Connectivity Area Marker*

A two-dimensional image connected region is an image region composed of pixels with the same pixel value and adjacent locations, that is, in a connected set, there is a path between any two pixels that are completely composed of elements in this set.

If two target pixels P and Q are connected, if and only if there is one path P_1, P_2,P_3,......P_n, such that $P_1 = P$, $P_n = Q$, for any$1 \leq i \leq n\text{-}1$,P_i is adjacent to P_{i+1}, where P_1, P_2, P_3,......P_n are target pixels.

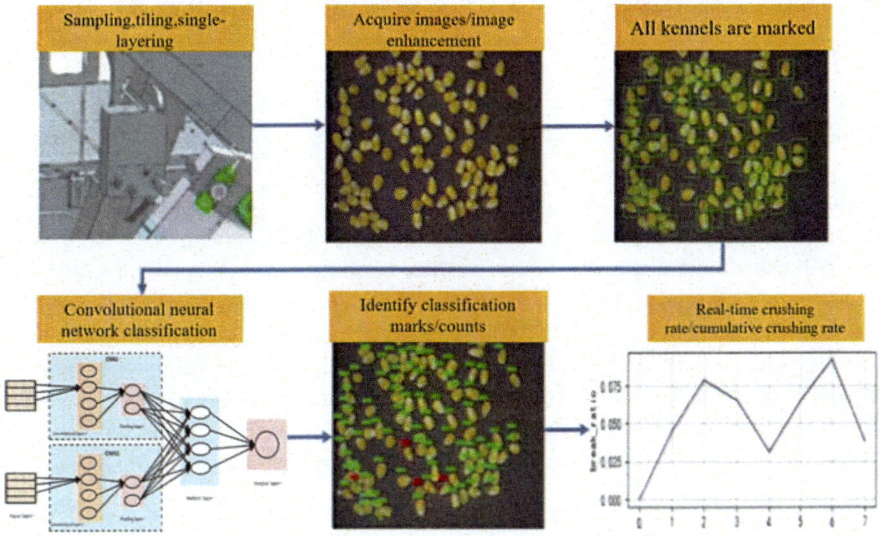

Fig. 6.3 Research scheme for online detection of corn kernel crushing

There are mainly 4 connected and 8 connected domains, and this topic uses 8 connected regions as a mark.

If the two pixels $P_1(x_1,y_1)$ and $P_2(x_2,y_2)$ are satisfied

$$1 \le (x_1 - x_2)^2 + (y_1 - y_2)^2 \le 2 \tag{6.1}$$

In the known image, the 8 connected area notation method is used, and each pixel in the image is scanned in turn in the order from left to right and top to bottom, and the black point with a gray value of 0 is found and marked to determine whether the right point, the lower right point, and the lower edge point are black points with a gray value of 0, and if it is a black point, it is uniformly numbered, otherwise, the number is increased by 1 and the scan continues until the end of the scan.

The principle is to extract the collection of pixels with a gray value of 0 in the 8 pixels adjacent to each other, and mark them with the same number, so that the labeled image can be obtained.

6.3.2 *Dataset Construction*

In this paper, crushed kernels and complete kernels were randomly placed in the background of black cardboard under natural lighting, and a total of 17,611 samples were constructed from the training set and test set, and the constructed partial data set is shown in Fig. 6.4.

6.3.3 *Convolutional Neural Network Models*

The burst data is sent in the form of a competitive time slot (CAP), and the competitive time slot convolutional neural network is composed of three structures: convolution, activation, and pooling, in identifying grain crushing, we define it as a binary task, i.e. broken and complete. Therefore, we combine convolutional neural network with fully connected neural network, and use the backward propagation algorithm to complete the training task of weights. The overall neural network structure model is shown in Fig. 6.5.

In this paper, transfer learning is used to train the grain crushing rate model. Transfer learning is the simple adjustment of a trained model on a problem to apply it to a new problem. Corn recognition training basically consists of five parts, you need to manually select corn kernels on the existing corn sample pictures, and then encapsulate the sample data, use the existing model for training, and view the recognition effect after training. After 5000 iterations of training, the accuracy curve converges and the performance of the algorithm reaches stability. The results of experiments using kernels are shown in Fig. 6.6.

(a) Full grain dataset

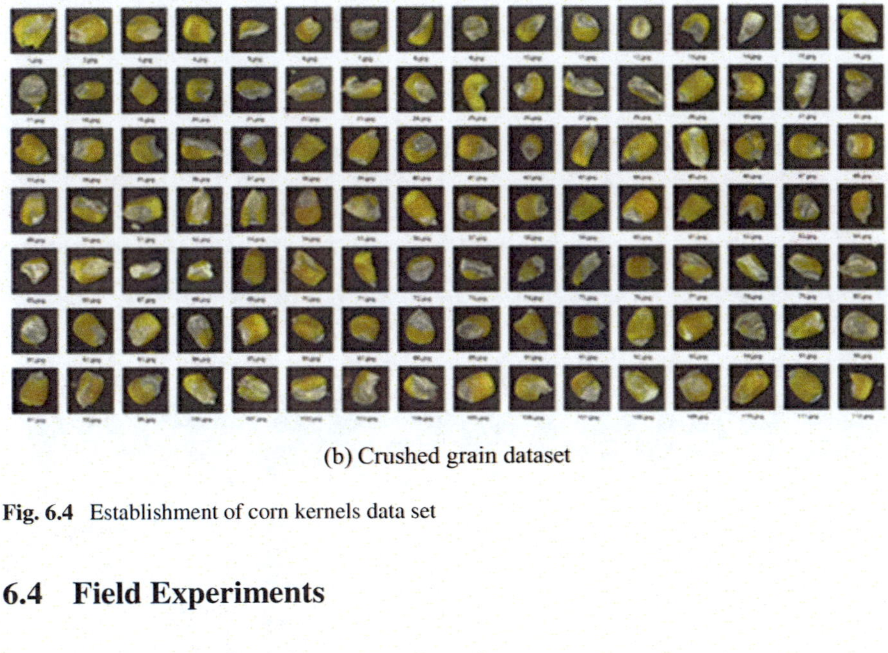

(b) Crushed grain dataset

Fig. 6.4 Establishment of corn kernels data set

6.4 Field Experiments

In order to verify the feasibility of online detection device and method for maize grain crushing rate, field experiments were carried out in Zouping County, Binzhou City, Shandong Province, as shown in Fig. 6.7.

At the field experiment site, the pictures of corn kernels collected by the corn grain crushing rate online detection device are shown in Fig. 6.8.

In the field experiment, the results of online detection and identification of corn kernels are shown in Fig. 6.9.

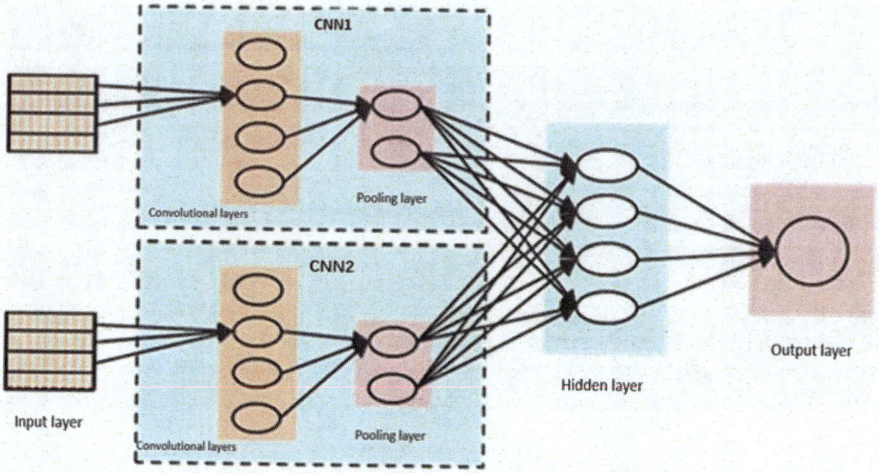

Fig. 6.5 Structural model of convolutional neural network

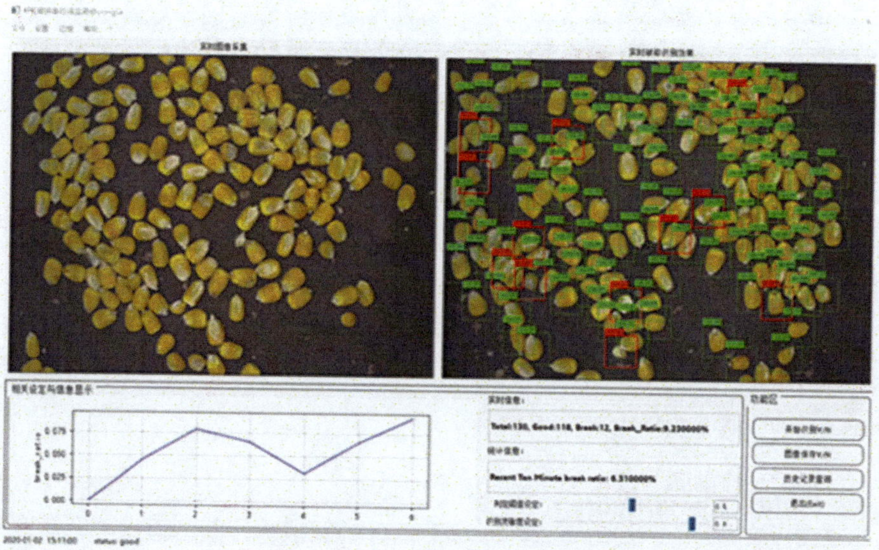

Fig. 6.6 Experimental results of online measuring of corn kernels

Fig. 6.7 Field experiment site in Zouping, Shandong Province

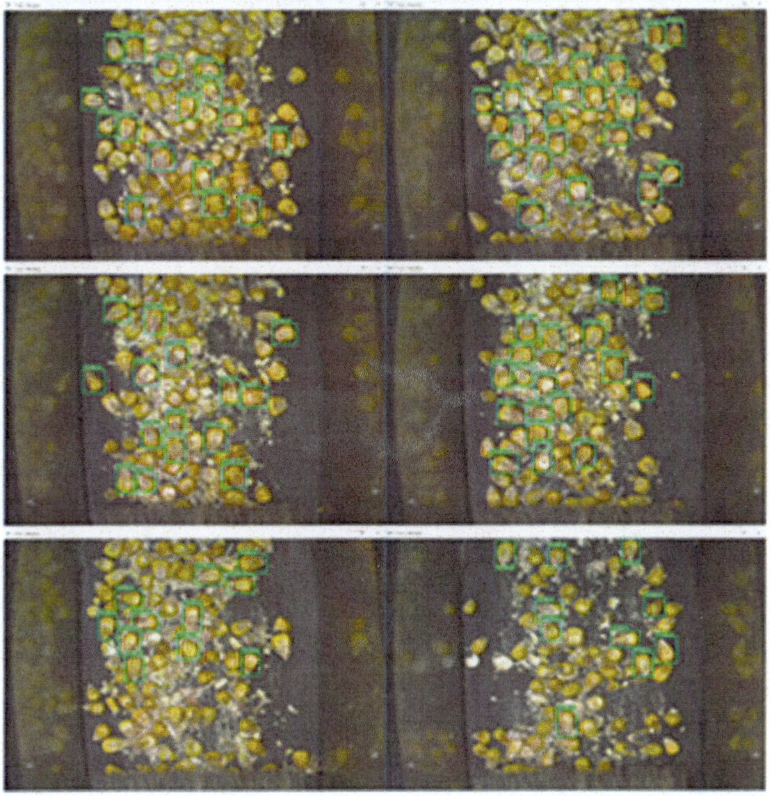

Fig. 6.8 Pictures collected by corn kernels measuring device in field experiments

Fig. 6.9 Results of online detection and identification of maize kernels in field experiments

6.5 Conclusion

Through field experiments, the online detection device and method of corn grain crushing rate were actually verified. The experimental results show that the online detection device for corn grain crushing rate can effectively realize sampling and reflux on the combined harvester of direct grain harvesting. It can effectively tile stacked corn kernels for subsequent image acquisition and method analysis.

The online detection method of corn grain crushing rate analyzes the real-time collection of corn grain pictures, and the analysis results show that it can effectively identify intact corn kernels and crushed corn kernels.

At present, the online detection efficiency of corn grain crushing rate is 3–4 sheets/second, and with the optimization of the algorithm and the improvement of the mechanism, the recognition efficiency and accuracy will be further improved.

Acknowledgements This work was financially supported by the Key Research and Development Program of LiaoNing Province(2020JH2/10200026), the National Key Research and Development Program of China(2020YFB1709600) and the Resarch and Development of Self-Adaptive Spraying Control System for Unmanned Vehicles (XLYC1902006).

References

1. Diao, Z.H., Huidan Wang, H.D., Wei, W.: Summary of research on machine vision application in agricultural production. J. Agric. Mechanization Res. **36**(03), 206–211 (2014)
2. Hai, T., Fu, G.Y.: Analysis of the research status of machine vision technology in grain appearance quality inspections. Agric. Equip. Technol. **39**(04), 25–28 (2013)
3. Zhao, J., Peng, Y.K., Dhakal, S., Zhang, L.L.: On-line detection of apple surface defect based on image processing method. Trans. Chin. Soc. Agric. Mach. **44**(S1), 260–263 (2013)
4. Li, X.Y., Tao, H,L., Gao, H.L., Li, P., Huang, T., Ren, J.P.: Nondestructive detection method of potato scab based on multi-sensor information fusion technology. Trans. Chin. Soc. Agric. Eng. **29**(19), 277–284 (2013)
5. Xiong, J.T., Zou, X.J., Wang, H.J., Peng, H.X., Zhu, M.S., Lin, G.C.: Recognition of ripe litchi in different illumination conditions based on Retinex image enhancement. Trans. Chin. Soc. Agric. Eng. **29**(12), 170–178 (2013)
6. Zapotoczny, P.: Computer vision algorithm for barley kernels identification, orientation estimation and surface structure assessment. Elsevier Science Publishers B. V(2012)
7. Li, X.P., Gao, L.X., Ma, F.L.: Analysis of finite element method on mechanical properties of corn seed. Trans. Chin. Soc. Agric. Mach. (**10**), 64–67+72 (2007)
8. Li, X.P., Zhang, F., Gao, L.X.: Analysis of construction and technique of threshing equipment on seed corn. J. Agric. Mechaniz. Res. (**6**), 24–26+30 (2008)
9. Li, X.P., Ma, F.L., Gao, L.X.: Impact of mechanical damage of corn seed to it's germination percentage. J. Agric. Mechaniz. Res. **31**(3), 34–35+39 (2009)
10. Yang, L., Wang, Z., Gao, L., et al.: Measuring of maize kernels breakage rate based on K-means clustering. Amer. Institute of Physics Conference Series. American Institute of Physics Conference Series, pp: 90–92 (2017)
11. Chen, J., Zhang, S., Li, Y.M., Zhu, L.J., Xia, H., Zhu, Y.H.: Reaserch on online identification system of rice broken impurities in combine harvester. J. Chin. Agric. Mechaniz. **42**(6) (2021)
12. Chen, J., Gu, Y., Lian, Y., Han, M.N.: Online recognition method of impurities and broken paddy kernels based on machine vision. Trans. Chin. Soc. Agric. Eng. **34**(13), 187–194 (2018)
13. Chen, J., Han, M.N., Lian, Y., Zhang, S.: Segmentation of impurity rice grain images based on U-Net model. Trans. Chin. Soc. Agric. Eng. **36**(10), 174–180 (2020)
14. Ding, K., Gunasekaran, S.: Shape feature extraction and classification of food material using computer vision. Trans. ASAE **37**(5), 1537–1545 (1994)
15. Reid, J.R., Kim, C., Paulsen, M.R.: Computer vision sensing of stress cracks in corn kernels. Trans. ASAE **34**(5), 2236–2244 (1991)
16. Valiente-González, J.M., Andreu-García, G., Potter, P., et al.: Automatic corn (Zea mays) kernels inspection system using novelty detection based on principal component analysis. Biosys. Eng. **117**(1), 94–103 (2014)

Chapter 7
Maize Kernels Endosperm Feature Extraction Based on Genetic Algorithm Image Enhancement

Liang Yang, Zhuo Wang, Xiaoping Bai, and Leyu Hao

Abstract Maize kernels endosperm is an important part of maize and has a close relationship with the physical characteristics of maize kernels, nutritional status, and post-breeding of maize. It also has important reference value in maize genetic analysis and variety identification. Therefore, this paper presents the endosperm character extraction of maize kernels based on genetic algorithm image enhancement. The experimental results show that the enhancement effect of the image based on the genetic algorithm is better than that of histogram equalization, which is propitious to the extraction of endosperm characteristics of maize kernels.

7.1 Introduction

In recent years, with the rapid development of computer software and hardware technology, computer vision technology has been widely used in face recognition, path planning, target tracking, intelligent sorting systems, and other [1] fields, which has brought many conveniences to people's production and manufacturing as well as improved people's quality of life.

Computer vision technology and processing methods have also been applied in agricultural intelligence detection and other fields. Liu [2], applied computer graphics

L. Yang · Z. Wang (✉) · X. Bai
Shenyang Institute of Automation Chinese Academy of Sciences, Shenyang 110016, China
e-mail: yangliangsia@sina.cn

Institutes of Robotics and Intelligent Manufacturing, Chinese Academy of Sciences, Shenyang 100019, China

Key Laboratory of Agricultural Equipment Intelligent Technology, Liaoning Province, Shenyang 110016, China

L. Yang
e-mail: yangliangsia@sina.cn

L. Hao
College of Engineering, Shenyang Agricultural University, Shenyang 110866, China

R. Kountchev et al. (eds.), *Multidimensional Signals, Augmented Reality and Information Technologies*, Smart Innovation, Systems and Technologies 374,
https://doi.org/10.1007/978-981-99-7011-7_7

theory and pattern recognition technology to study grape diseases, and the Bagged trees classifier was used to evaluate and recognize six grape diseases with a high recognition rate of 86.4%; Huang et al. [3, 4], the main research crop is sugarcane. Through computer vision technology, it can identify the stem nodes of sugarcane and detect the cutting height of sugarcane during harvest; Xiong [5], using bilateral filtering Retinex image enhancement algorithm to highlight the litchi fruits and fruit stalks in the image, in HSI color space, OTSU automatic threshold segmentation is carried out for the H component to remove the image background, and clustering segmentation is carried out for the image in the knot bicubic interpolation method and fuzzy C-means algorithm to realize the recognition of litchi fruits and fruit stalks; Li [6], in order to improve the recognition accuracy of potato scab disease, median filtering was performed on the original image collected, and binary processing was performed on the image after median filtering. In combination with the difference method and Markov random field model, the diseased area of the potato image was segmented. According to 12 color characteristic parameters, the method of support vector machine was used to establish the pattern of potato scab disease recognition, The image recognition information and spectral information are fused to improve the efficiency of pathological feature recognition; Yang [7], using machine vision technology, uses the color difference that the broken part of the broken maize grain is milky white and the whole maize grain is yellow, and combines K-Means color image segmentation algorithm to distinguish the complete maize grain from the broken maize grain.

Szczypiński [8], got the digital image of barley kernels through the optical scanner, and used the digital image as the input data of the algorithm to analyze the smooth and wrinkled areas indicated by the grain and the location of the grain germ. The results show that the accuracy of the method is 93%; Liang [9] et al., using image processing technology in computer vision, proposed a new detection method for maize grain traits, which automatically detected 12 traits such as length, width, and weight of maize grains. The experimental results show that this method can effectively evaluate maize grain traits, and the similar error rate with manual measurement is less than 5%; Chen [10], combined discriminant analysis with BP neural network method to identify five maize varieties, including Nongda 86, Nongda 108 and Nongda 4967, with a recognition rate of up to 90%; Valiente Gonz á lez [11], based on machine vision technology and the principal component analysis method, set up a maize grain separation system to distinguish complete maize grains from those with surface defects. Under HSV color space, the correct recognition rate can reach 92%; Li [12], used the linear transformation method to process the collected maize grain image, and converted the uneven intensity distribution on the object to the uniform intensity distribution, so that the gray level of the maize grain wormhole area is lower than the gray level of other areas, so as to distinguish the maize grains in the wormhole; Yan [13] realized the projection of color images from high-dimensional data points to one-dimensional space by using Fisher discriminant theory on the collected color images, combined with K-means clustering method to cluster the one-dimensional data points, so as to effectively identify the purity of maize seeds. Maize kernels endosperm is an important part of maize, and the quality of maize kernels endosperm determines the physical characteristics and nutritional quality of maize kernels, as well as the starch yield [14, 15]. At the same time, different maize varieties have certain differences in the shape and absorption peak of endosperm, which can be used to distinguish different maize varieties [16]. Therefore, this paper uses the image enhancement method based on a genetic algorithm to extract the characteristics of maize kernels endosperm, so as to provide a reference for researchers specializing in maize genetic traits and maize breeding.

7.2 Histogram Equalization Image Enhancement

In the process of image analysis and processing, image enhancement is one of the main means, which mainly includes: grayscale stretching, image sharpening, and denoising. The image enhancement of the initial image can improve the contrast between the overall and local details of the image, so as to highlight the details of the image, making the details more consistent with the characteristics of the human eye and facilitating machine recognition [17].

Image enhancement is divided into spatial domain image enhancement and frequency domain image enhancement. Spatial domain image enhancement is a direct operation on the gray level pixel value of the image, mainly represented by histogram equalization, linear transformation, etc. [18]. Frequency domain image enhancement is to transform the spatial image into a spatial image in the frequency domain after Fourier transformation. After filtering and other operations in frequency domain space, the image is transformed from frequency domain space to spatial space after inverse Fourier transformation, so as to realize image enhancement. Mainly represented by: wavelet transform, and super wavelet transform.

A Gray histogram is the statistics of the distribution of gray levels in an image. It counts the frequency of its occurrence according to the size of the pixel values in the gray image. Histogram equalization is the approximate uniform distribution of the gray levels of the gray image, so as to improve the contrast of the image [19, 20]. The histogram equalization algorithm is as follows:

Use x to represent the pixel value of the image to be enhanced. The allowable range of pixel values is $[0, L-1]$, for L - give the value (256), for a 8-bit grayscale image. Each pixel value in the initial image corresponds to a gray level y through the conversion function, namely:

$$y = f(x) \tag{7.1}$$

$$x = f^{-1}(y) \tag{7.2}$$

The transformation function $f(x)$ satisfies:

(1) It is a monotone-increasing function in the interval of $[0, L-1]$.

(2) $f(x)$ is it between $[0, L-1]$.

After $f(x)$ transformation, the gray histogram of the image is uniformly distributed, which maximizes the entropy of the image. In image processing, the function:

$$y = (L-1)\int Px(x)dx \quad (7.3)$$

The above conditions are met.

In the formula: $Px(t)$ represents the probability density of pixels.

Discrete time:

$$Px(t) = \frac{nk}{n} \quad (7.4)$$

In the formula: nk is the number of pixels corresponding to gray level y, and n is the total number of pixels of the image.

If the image belongs to discrete distribution, the histogram equalization expression is:

$$y(k) = (L-1)\sum\nolimits_{i=0}^{k} \frac{nk}{n} \quad (7.5)$$

As discrete is a special case of continuity, the continuous function is analyzed, and it is proved that the histogram becomes uniform after $f(x)$ transformation: by taking the derivative of both sides of Eq. (7.3), we can get:

$$f\prime(x) = (L-1)Px(x) \quad (7.6)$$

After $f(x)$ transformation, according to probability theory, its probability density is:

$$Py(y) = Px(x)\left|(f^{-1}(y))\prime\right| \quad (7.7)$$

Since the derivative of the inverse function is the reciprocal of the derivative of the original function, there are:

$$\left(f^{-1}(y)\right)' = \frac{1}{(L-1)P_x(x)} \quad (7.8)$$

Substituting Eq. (7.8) into (7.6), we can get:

$$Py(y) = \frac{1}{L-1} \quad (7.9)$$

The above formula shows that after $f(x)$ transformation, the probability of each gray level is equal, indicating that the histogram is evenly distributed.

The histogram equalization algorithm is simple, but under uneven lighting, using the equalization algorithm to enhance the image can make the lighting more rugged, make the image enhancement effect poor, and easily lose many details.

7.3 Genetic Algorithm Image Enhancement

Genetic algorithm is an adaptive global probability search algorithm, which simulates the genetic and evolutionary laws of natural organisms. At present, the main applications of genetic algorithms in the image field include image enhancement, image restoration and image reconstruction [21, 22]. The genetic algorithm process is as follows (Fig. 7.1).

The essence of image enhancement based on genetic algorithm is to transform the image information to be enhanced into the process of finding the optimal approximate solution, that is, the process of obtaining the optimal parameter value of the *Beta* function to maximize the value of the *Beta* function, and automatically fit the image transformation curve to achieve the image enhancement effect.

The *Beta* function is as follows:

$$F(u) = B^{-1}(\alpha, \beta) \times \int_0^u t^{\alpha-1} \times (1-t)^{\beta-1} dt \tag{7.10}$$

In the formula,α, β is the solution parameter

$$B(\alpha, \beta) = \int_0^u t^{\alpha-1} \times (1-t)^{\beta-1} dt \tag{7.11}$$

Set the initial image matrix as $f(x, y)$, find out the maximum pixel value $L\max$ and minimum pixel value $L\min$ in the initial image matrix, and use them to normalize the image:

$$g(x, y) = \frac{f(x, y) - L\min}{L\max - L\min} \tag{7.12}$$

Take $g(x, y)$ as the initial value of the algorithm after rounding. Genetic algorithm is used to find the best value α, β, to maximize $F(f(x, y))$, and the best value α, β, is used as the image enhancement parameter to automatically fit the image curve, so as to achieve the image enhancement effect (Fig. 7.2).

The following formula is used as the evaluation function:

$$EV = \frac{\mu 1}{m \times n - \mu_2^2} \tag{7.13}$$

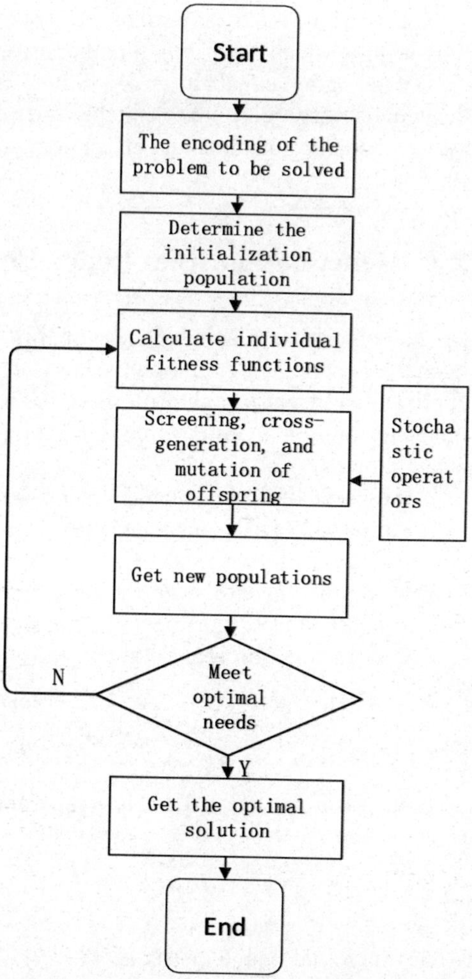

Fig. 7.1 Process of genetic algorithm

In the formula, $\mu 1 = \sum_{i=1}^{m\times n} f(x, y)^2$,$\mu 2 = \sum_{i=1}^{m\times n} \frac{f(x,y)}{m\times n}$,$m$ is the number of initial image rows, n is the number of initial image columns. The larger the EV value, the more uniform the image gray distribution, and the image quality is better.

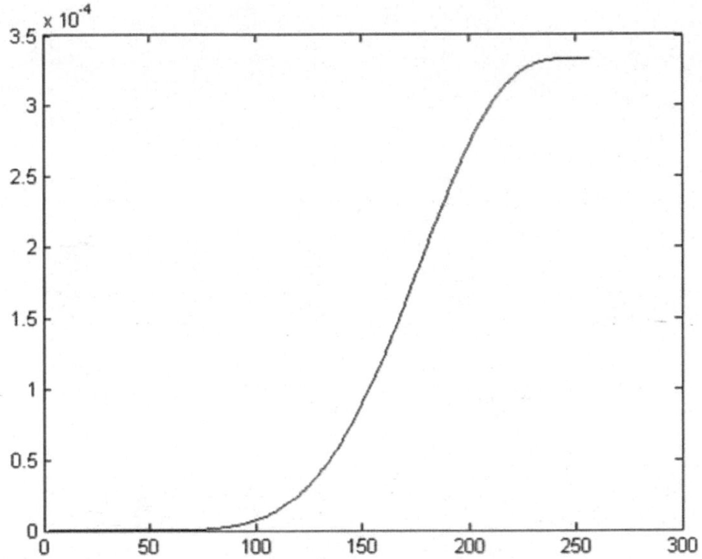

Fig. 7.2 Automatic fitting of curve

7.4 Test Design and Result Analysis

7.4.1 Test Design

The genetic algorithm initializes the population, including two chromosomes $x1$ and $x2$. The number of individuals in the population is set as 10. The initial gray level image of maize grain is shown in Fig. 7.4. The maize grain is Zhengdan 958 planted in northeast China.

The best parameter values obtained by using the genetic algorithm are $\alpha = 8.6719$, $\beta = 4.4531$. Image curve automatic fitting is shown in Fig. 7.3, image enhancement effect based on genetic algorithm is shown in Fig. 7.4, and feature extraction of maize endosperm is shown in Fig. 7.5.

The results of the initial gray level image after histogram equalization and enhancement processing are shown in Fig. 7.6, and the effect of maize endosperm extraction is shown in Fig. 7.7.

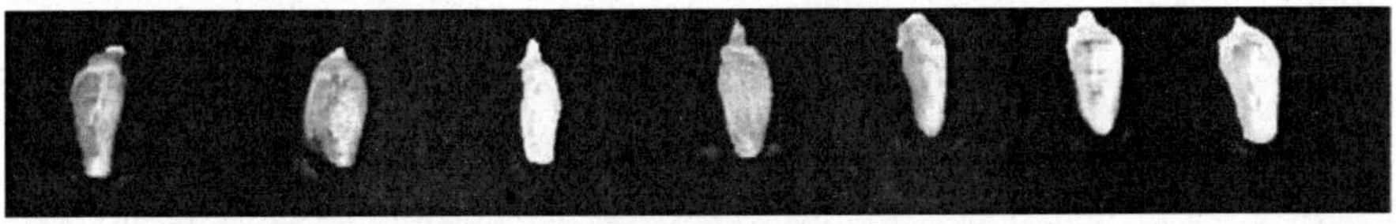

Fig. 7.3 Image enhancement based on genetic algorithm

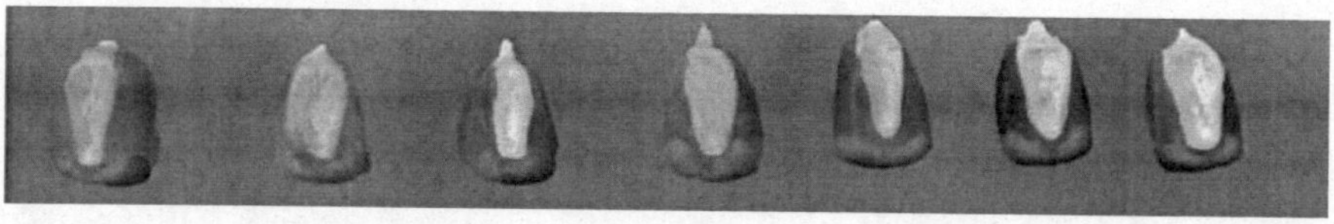

Fig. 7.4 Initial grayscale image

Fig. 7.5 Feature extraction of maize kernels endosperm based on genetic algorithm

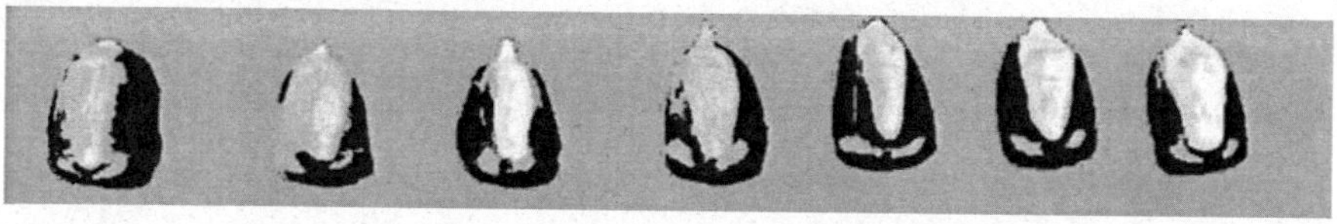

Fig. 7.6 Image enhancement based on histogram equalization

Fig. 7.7 Feature extraction of maize kernels endosperm based on histogram equalization

7.4.2 Analysis of Test Results

The evaluation value of the initial gray image is $EV = 723.15$, the evaluation value after image enhancement by genetic algorithm is $EV = 2708.7$, and the evaluation value after histogram equalization is $EV = -34773$. From the EV value results, it can be seen that the image processed by genetic algorithm is better than the initial image and the image is enhanced by histogram equalization. From the analysis of the results of maize endosperm feature extraction, the image enhanced by genetic algorithm is more conducive to extracting maize endosperm features than the image enhanced by histogram equalization. After analysis, the reason why the image evaluation value of histogram equalization enhancement is far lower than the evaluation value of the initial gray image is that the acquisition is under non-uniform lighting. After histogram enhancement, many details are lost and noise points are introduced.

7.5 Conclusion

In this paper, genetic algorithm is applied to the extraction of maize endosperm features. Genetic algorithm is better than histogram equalization in enhancing the initial gray level image, which is more conducive to the extraction of maize endosperm features. It has certain reference and practical value for genetic analysis and automatic variety identification based on maize endosperm.

Acknowledgements This work was financially supported by the Key Research and Development Program of LiaoNing Province(2020JH2/10200026), the National Key Research and Development Program of China(2020YFB1709600) and the Research and Development of Self-Adaptive Spraying Control System for Unmanned Vehicles (XLC1902006).

References

1. Ni, H.P., Liu, Y.N., Zhang, C.R., et al.: Sorting system algorithms based on machine vision for delta robot. Robot **38**(01), 49–55 (2016)
2. Liu, Y., Feng, Q.: Identification method of grape diseases based on computer technology. J. Chin. Agric. Mechanization **38**(4), 99–104 (2017)
3. Huang, Y.Q., Huang, T.S., Huang, M.Z., et al.: Recognition of sugarcane nodes based on local mean. J. Chin. Agric. Mechanization **38**(2), 76–80 (2017)
4. Huang, Y.Q., Huang, T.S., Yang, R., et al.: Detection and test of sugarcane cutting height based on machine vision. J. Chin. Agric. Mechanization **38**(9), 81–87 (2017)
5. Xiong, J.T., Zou, X.J., Wang, H.J., et al.: Recognition of ripe litchi in different illumination conditions based on Retinex image enhancement. Trans. Chin. Soc. Agric. Eng. (Trans. CSAE) **29**(12), 170–178 (2013). (in Chinese with English abstract)
6. Li, X.Y., Tao, H.L., Gao, H.L., et al.: Nondestructive detection method of potato scab based on multi-sensor information fusion technology. Trans. Chin. Soc. Agric. Eng. (Trans. CSAE) **29**(19), 277–284 (2013). (in Chinese with English abstract)
7. Yang, L., Wang, Z., Gao, L., et al.: Detection of maize kernelss breakage rate based on K-means clustering. American Institute of Physics Conference Series, American Institute of Physics Conference Series, pp. 90–92 (2017)
8. Szczypiński, P.M., Zapotoczny, P.: Computer vision algorithm for barley kernels identification, orientation estimation and surface structure assessment. Comput. Electron. Agric. **87**(3), 32–38 (2012)
9. Liang, X., Wang, K., Huang, C., et al.: A high-throughput maize kernels traits scorer based on line-scan imaging. Measurement **90**, 453–460 (2016)
10. Xiao, C., Yi, X., Wei, L., et al.: Combining discriminant analysis and neural networks for maize variety identification. Comput. Electron. Agric. **71**(1), S48–S53 (2010)
11. Valiente-González, J.M., Andreu-García, G., Potter, P., et al.: Automatic maize (Zea mays) kernels inspection system using novelty detection based on principal component analysis. Biosys. Eng. **117**(1), 94–103 (2014)
12. Li, J., Huang, W., Zhang, C.: Detection of wormhole in maize based on kernels RGB image and lighting transformation method. Adv. J. Food Sci. Technol. **5**(1), 63–67 (2013)
13. Yan, X., Wang, J., Liu, S., et al.: Purity identification of maize seed based on color characteristics. International Conference on Computer and Computing Technologies in Agriculture. Springer Berlin Heidelberg, pp. 620–628 (2010)
14. Yi, Z.T., Zhang, H.Y.: Starch granule size distribution of horny and silty endosperm in different maize varieties. J. Jilin Agric. Sci. **40**(4), 108–112 (2015)

15. Yan, C.Q., Shi, Y.M., Li, D.Y., et al.: FTIR Characteristics of maize endosperm and embryo. Chin. J. Spectroscopy Lab. **29**(6), 003525–003528 (2012)
16. Cheng, H., Shi, Z.X., Yin, H.J., et al.: Detection of multi-maize kernels embryos characteristic using machine vision. Trans. Chin. Soc. Agric. Eng. (Trans. CSAE) **29**(19), 145–151 (2013). (in Chinese with English abstract)
17. Wang, H., Zhang, Y., Shen, H.H., et al.: Review of image enhancement algorithms. Chin. Opt. **10**(4), 438–448 (2017)
18. Gu, L.L., Wang, Y.H.: Image enhancement based on simulated annealing algorithm and genetic algorithm. Laser J. **2**, 19–22 (2015)
19. Jiang, B.J., Zhong, M.X.: Improved histogram equalization algorithm in the image enhancement. Laser & Infrared **6**, 702–706 (2014)
20. Yang, W.Z., Xu, Y.L., Qiao, X., Rao, W., Li, D.L., Li, Z.B.: Method for image intensification of underwater sea cucumber based on contrast-limited adaptive histogram equalization. Trans. Chin. Soc. Agric. Eng. **32**(6), 197–203 (2016)
21. Zhu, C.R.L., Zhang, D.M., Zhang, M.X., et al.: Application of genetic algorithm in image processing. Commun. Technol. **50**(7), 1401–1406 (2017)
22. Huang, N.: Application research of genetic algorithm image enhancement. Comput. Simul. **29**(8), 261–264 (2012)

Chapter 8
Data Balancing of Ceph Distributed Storage System Based on Deep Reinforcement Learning

Di Yang, Yiyi Liu, Jiahao Chen, and Zhikang Jiang

Abstract Aiming at the problem of unbalanced data distribution caused by the Controlled Replication Under Scalable Hashing (CRUSH) algorithm used in the Ceph distributed storage system, this paper proposes a data balance algorithm based on deep reinforcement learning for Ceph system. When the traditional method optimizes the uneven distribution of the data of the CRUSH algorithm, it destroys the advantages of the small data movement of the CRUSH algorithm, and the effect is not satisfactory when the number of data copies is large. In this paper, the data balancing problem is transformed into a weight optimization problem of Object Storage Device (OSD) in the Ceph system, and a deep reinforcement learning algorithm based on TD3 (Twin Delayed Deep Deterministic Policy gradient algorithm) is constructed. Experimental results show that the proposed method can improve the balance of data distribution without changing the logic of the CRUSH algorithm and multiple copies.

8.1 Introduction

In recent years, with the development and widespread application of internet technology, life and production have gradually transformed into digitalization, the global data volume will reach 175 ZB [1] by 2025, and will continue to grow. Ceph [2], as one of the most popular distributed storage systems [3, 4], has the advantages of high

---

D. Yang (✉) · Y. Liu
State Grid Shanghai Communication Company, Shanghai, China
e-mail: ydtina@foxmail.com

Y. Liu
e-mail: liuyiyi@sh.sgcc.com.cn

J. Chen · Z. Jiang
Information Science and Engineering School, Southeast University, Nanjing, China
e-mail: 220210840@seu.edu.cn

Z. Jiang
e-mail: 220210841@seu.edu.cn

R. Kountchev et al. (eds.), *Multidimensional Signals, Augmented Reality and Information Technologies*, Smart Innovation, Systems and Technologies 374,
https://doi.org/10.1007/978-981-99-7011-7_8

performance, scalability, and reliability. It uses the Controlled Replication Under Scalable Hashing (CRUSH) algorithm for data storage, but suffers from the problem of unbalanced data distribution. To address this problem, literature [5] proposed the B-CRUSH algorithm, a new hashing algorithm that can achieve good data balancing effects. However, its underlying logic is significantly different from the original CRUSH algorithm, so when storage devices are expanded, the data migration volume of the B-CRUSH algorithm is much larger than that of the CRUSH algorithm. Literature [6] proposed a CRUSH algorithm based on temperature factors, which changes the weight of object storage device (OSD) nodes to make low-load OSD nodes serve as storage nodes. However, the effect of this algorithm is not ideal when there are multiple replicas. Therefore, improving the data distribution balance for CRUSH algorithm is still necessary [7]. This paper proposes a TD3 deep reinforcement learning algorithm-based data balancing method for the Ceph distributed storage system.

8.2 Data Distribution Problem in Ceph System

The CRUSH algorithm first selects storage devices based on the weight value of each device, and then maps the data to OSD through the set replica rules to maximize the read and write performance of the distributed system [8]. However, because the mapping process of CRUSH is pseudo-random, it may lead to an uneven distribution of user data on various device nodes. To verify this conclusion, we simulated the Ceph cluster architecture shown in Fig. 8.1.

This paper simulates that the number of computer rooms is 2, each computer room corresponds to 3 racks, each rack corresponds to 1 host, each host deploys 10 OSDs

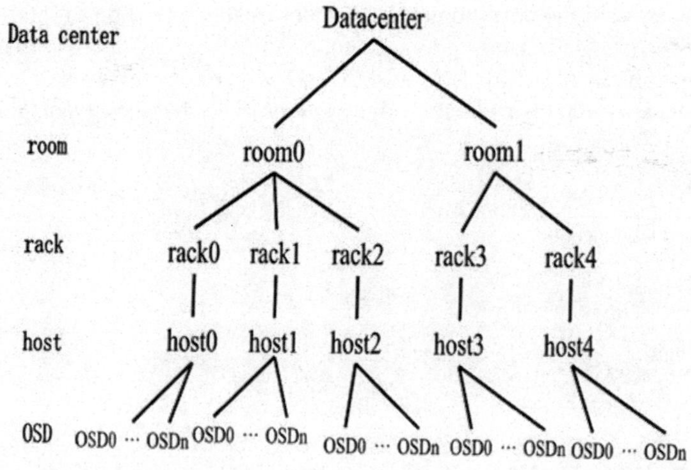

Fig. 8.1 Ceph simulation environment architecture diagram

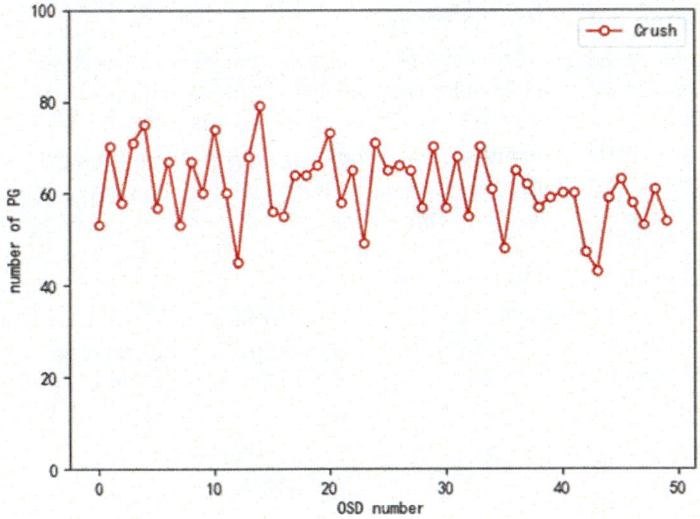

Fig. 8.2 PG distribution in OSD

with the same capacity, and 1024 PG passes through CRUSH algorithm is mapped to OSD [9–11], the final distribution of PG in OSD is shown in Fig. 8.2. The maximum number of PG in OSD is 79, the minimum is 43, and the relative theoretical value ranges from −29.9% to 28.6%, the standard deviation of the number of PG in OSD is 7.89. It can be seen that when the CRUSH algorithm is used for mapping, the distribution of PG on OSD will be unbalanced, which will further affect the read and write performance of the distributed storage system.

8.3 TD3-Based Data Balance Algorithm in Ceph System

The data balancing problem in the Ceph storage system can be considered as an optimization problem for OSD weights [12, 13]. Deep reinforcement learning (DRL) methods are suitable for solving optimization problems because they can interact with the current environment, learn autonomously, and flexibly adjust to the current environment to select the optimal solution. The problem studied in this article has a large state space and action space, so we proposed a data balancing algorithm based on DRL to solve the problem.

8.3.1 Markov Decision Process

To solve the data balancing problem in the Ceph distributed storage system using the TD3 algorithm, it is first necessary to convert the problem into a Markov decision process [14, 15]. This paper abstracts the optimization problem from the perspective of reinforcement learning, converts the data balancing problem into an OSD weight optimization problem, and thus achieves the optimization problem with the highest total profit value. The Markov decision process is described by the basic elements of state, action, and immediate reward.

State design: The state at time t is determined by the number of placement groups (N_i) in the current time of the i-th OSD, the number of placement groups (N_s) in each object storage device when the data is completely evenly distributed, and the maximum weight value (W_{max}) of the current time OSD, as shown in Eq. (8.1).

$$S_t = (N_i, N_s, W_{max}) \tag{8.1}$$

Action design: The action at time t is determined by the change in weight value of the i-th OSD at the current time, as shown in Eq. (8.2).

$$a = \left(W_c^i\right) \tag{8.2}$$

Immediate reward design: The evaluation index is the product of the difference between the number of placement groups in a certain object storage device at a certain time and the number of placement groups in each object storage device when the data is completely evenly distributed, and the weight of the action taken at that time. Combined with the reward obtained when the next state is reached, a reward function is formed. The expression of the reward function is $r(S_t, a_t, S_{t+1})$, where S_t is the state in the state space S at time t, a_t is the action in the action space A at time t, and S_{t+1} is the state in the state space S at time $t + 1$. The evaluation index at time t is the product of the action weight taken at time t and the difference between the number of placement groups in the i-th OSD at time t and the number of placement groups in each OSD when the data is completely evenly distributed. The reward obtained when reaching state S_{t+1} is: if the product at time t is positive, $r = -1$, otherwise $r = 1$. If the action weight value is too large and causes the number of PGs in the OSD to exceed a certain threshold ρ ($\rho = 0.25 * N_s$), $r = -2$. The reward function is shown in Eq. (8.3).

$$r = \begin{cases} -1, a_t \cdot (N_i - N_s) > 0 and |N_i - N_s| < 0.25 * N_s \\ 1, a_t \cdot (N_i - N_s) \leq 0 and |N_i - N_s| < 0.25 * N_s \\ -2, else \end{cases} \tag{8.3}$$

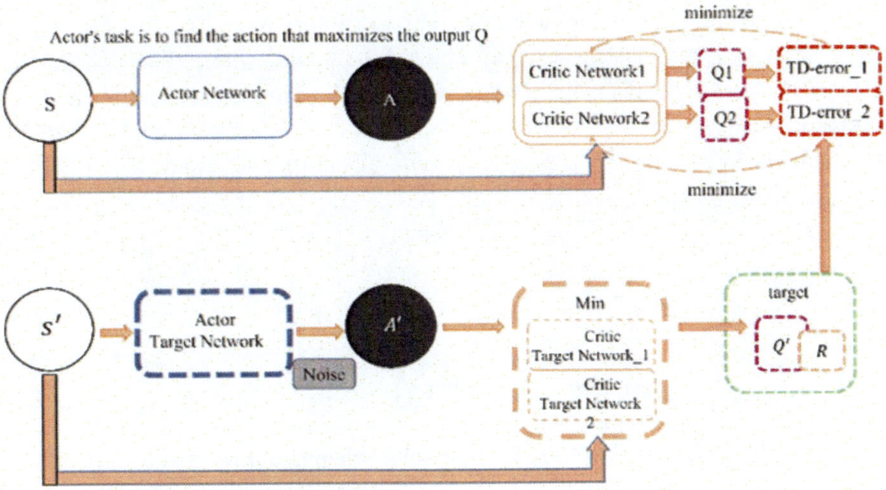

Fig. 8.3 TD3 algorithm frame diagram

8.3.2 Algorithm

In this paper, the TD3 deep reinforcement learning algorithm is used to optimize the weights of OSDs, and it is named the TD3_CRUSH algorithm [16, 17]. Figure 8.3 shows the framework of the TD3 algorithm. The Actor network in TD3 takes the state as input and outputs continuous actions, which are input together with the state into the Critic network. There are two target networks and one training network in the Critic network, so there are a total of six networks in the TD3 algorithm. Since the network parameters are randomly initialized, the output Q values must vary. We update the network parameters using the smaller Q value. The two training networks of the Critic are updated according to Eqs. (8.4) and (8.5).

$$\theta_c^{(t+1)} = \theta_c^{(t)} - \mu_c \Delta_{\theta_c^{(train)}} l(\theta_c^{(train)}) \tag{8.4}$$

$$l(\theta_c^{(train)}) = (r^{(t)} + \gamma \min_{i=1,2} q_i(\theta_c^{(target)} | s^{(t+1)}, a') - q(\theta_c^{(train)} | s^{(t)}, a^{(t)}))^2 \tag{8.5}$$

where μ_c is the learning rate of the trained network of Critic. a' is the action output from the Target network of Actor, $\Delta_{\theta_c^{(train)}} l(\theta_c^{(train)})$ represents the gradient of training network $\theta_c^{(train)}$. $\theta_c^{(train)}$ and $\theta_c^{(target)}$ represent the training network for Critic and the network for Target.

The update of Actor's training network is

$$\theta_a^{(t+1)} = \theta_a^{(t)} - \mu_a \Delta_a \min_{i=1,2} Q(\theta_c^{(target)} | s^{(t)}, a) \Delta_{\theta_a^{(train)}} \pi(\theta_a^{(train)} | s^{(t)}) \tag{8.6}$$

Among them, μ_a is the learning rate of the training network of Actor, $\pi(\theta_a^{(train)}|s^{(t)})$ represents the training network of Actor, where $\theta_a^{(train)}$ is DNN's parameter, $\Delta_a \underset{i=1,2}{min} Q(\theta_c^{(target)}|s^{(t)}, a)$ is the Critic training network relative to the action gradient,

The update of the trained network for Actor and Critic is as follows

$$\theta_{c1}^{(target)} = \tau_{c1}\theta_{c1}^{(train)} + (1 - \tau_{c1})\theta_{c1}^{(target)} \tag{8.7}$$

$$\theta_{c2}^{(target)} = \tau_{c2}\theta_{c2}^{(train)} + (1 - \tau_{c2})\theta_{c2}^{(target)} \tag{8.8}$$

$$\theta_{a}^{(target)} = \tau_{a}\theta_{a}^{(train)} + (1 - \tau_{a})\theta_{a}^{(target)} \tag{8.9}$$

Among them, τ_a and τ_c are the learning rates when updating the network.

Both the Critic network and the deep neural network in the Actor network are fully connected, and both have the same structure, including an input layer, an output layer, and two hidden layers, and the number of neurons in the hidden layer is 128.

The training flow chart of the entire network is shown in Fig. 8.4. The agent initializes the state $S_0$to(N_0, N_s, W_{max}) and use actor network to select the action a_t with random noise. Then the agent will update the state and get a reward, after that, the experience (S_t, a_t, r_t, S_{t+1}) will be added to the experience pool. If the number of experiences in the experience pool exceeds 200, 32 historical experience data will be extracted from the experience pool to update the network. The state at the next moment S_{t+1}equalsto(N_i, N_s, W_{max}), if $N_i^{t+1} - N_s < 0.1 * N_s$ is satisfied, then set $i = (i + 1)modN\_OSD$, where N_OSD is the number of OSD, otherwise i remains unchanged. After each round reaches the specified step size, the initialization state is $S_t = (N_0, N_s, W_{max})$. Then repeat the training multiple times and save the weight that minimizes the OSD standard deviation. The specific algorithm steps are as follows:

Step 1: Initializing state $S_0 = (N_0, N_s, W_{max})$;

Step 2: Selecting action a_t using the actor network with random noise.

Step 3: Updating the status, getting rewards, and adding experience (S_t, a_t, r_t, S_{t+1}) to the experience pool.

Step 4: Extracting 32 historical experience data from the experience pool if the number of experiences in the experience pool exceeds 200. Updating network according to the update strategy, otherwise proceeding to step 2.

Step 5: Updating $S_{t+1} = (N_i, N_s, W_{max})$, if $N_i^{t+1} - N_s < 0.1 * N_s$ is satisfied, setting $i = (i + 1)modN\_OSD$, where N_OSD is the OSD quantity, otherwise i remains unchanged. After each round reaches the specified step size, the initialization state is $S_t = (N_0, N_s, W_{max})$, and the training is repeated multiple times to save the weight that minimize the OSD standard deviation. Algorithm ends.

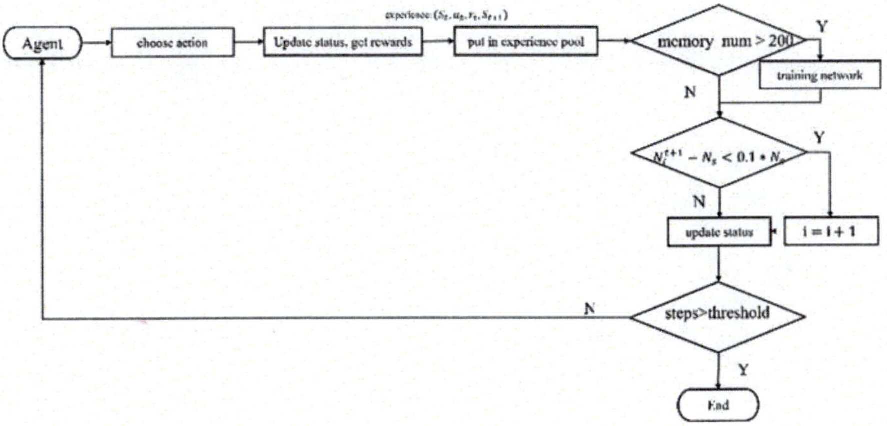

Fig. 8.4 Flow chart of training

8.4 Simulations

In the experiment, we used a computer configured with Intel CPU i5-9400F @ 2.90 GHz with six cores and 8 GB of running memory. A Ceph system simulation environment with 128, 256, 512, and 1024 PG, 10, 20, 50, and 100 OSD and a replica number of 3 was built using Python.

Figure 8.5 shows the comparison of PG distribution in 50 OSD before and after data equalization. It can be clearly seen that when the equalization algorithm is not used, the data distribution of the Ceph system is quite unbalanced. The number of PG on OSD has a large difference, which has a great impact on the performance of the system. Before the data is balanced, the range of change relative to the theoretical value is −29.9% to 28.6%, with a standard deviation of 7.89 for the number of PG in OSD. Comparing the TD3_CRUSH data equalization effect with DQN, after adopting DQN data equalization, the relative theoretical value ranges from −16.9% to 9.1%, the standard deviation of the number of PG in OSD is 3.59; after using TD3_CRUSH data equalization, the range of relative theoretical value is −16.9% to 9.1%, with a standard deviation of 3.29 for the number of PG in OSD.

In order to compare the data balancing effects of TD3_CRUSH and DQN algorithms in Ceph systems with different numbers of OSD and PG, simulations were performed on Ceph systems with 10, 20, 50, and 100 OSD, and with 128, 256, 512, and 1024 PG respectively, using TD3_CRUSH and DQN algorithms for data balancing. The simulation results are as follows:

Figure 8.6 shows the comparison of the equalization effect between the TD3_CRUSH and DQN algorithms and the traditional CRUSH algorithm at different OSD scales. It can be seen that after the TD3_CRUSH and DQN algorithms are balanced, in various scale scenarios, there are obvious optimizations compared with the traditional CRUSH algorithm, especially in small-scale scenarios, using TD3_

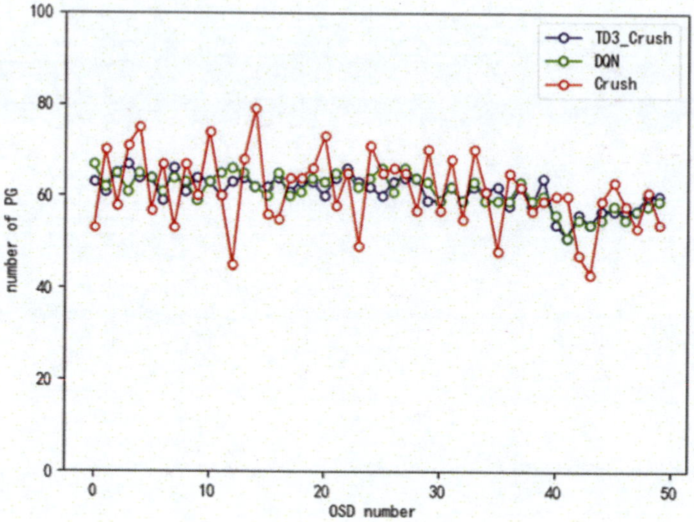

Fig. 8.5 Comparison of the number of PG in OSD before and after data equalization

CRUSH After balancing with the DQN algorithm, the balance of data distribution is significantly improved.

Figure 8.7 presents a comparison of the balancing effects of TD3_CRUSH and DQN algorithms with the traditional CRUSH algorithm in different cases where the number of PG increases from 128 to 1024 in a Ceph system with 50 OSD. It can be observed that the increase in the number of PGs leads to a more uneven data

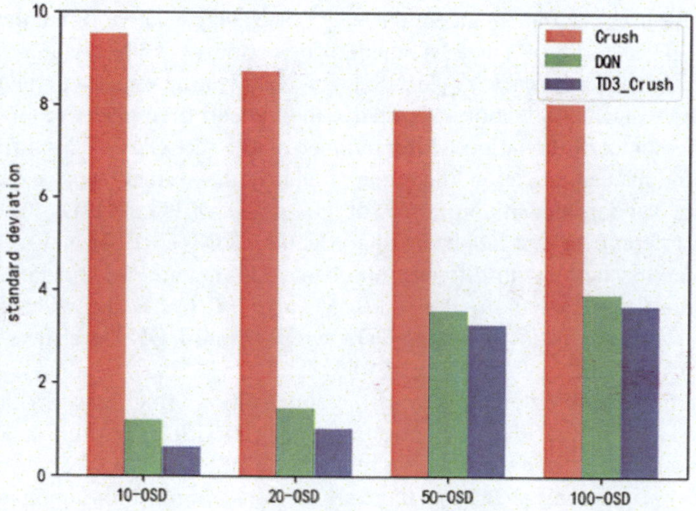

Fig. 8.6 Comparison chart of PG standard deviation under different scales of OSD

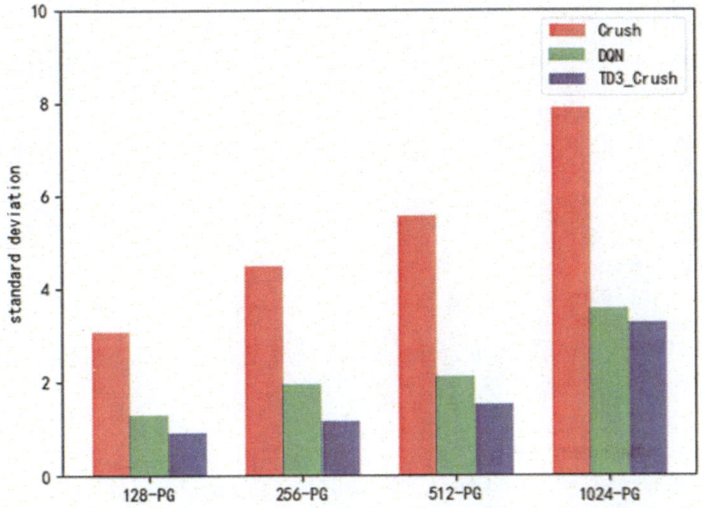

Fig. 8.7 Comparison of standard deviations of different PGs under the same scale OSD

distribution, and the standard deviation increases significantly with the increase of PGs. However, the data balancing effects of TD3_CRUSH and DQN algorithms are significant, which can effectively reduce the standard deviation and optimize the data distribution of the Ceph system. Meanwhile, when the number of PG changes, the optimization effect of TD3_CRUSH is slightly better than that of DQN algorithm, and it can obtain the optimal solution of the distribution of PG on different OSD faster.

8.5 Conclusion

For the Ceph distributed storage system, the traditional CRUSH distribution algorithm has insufficient data distribution balance. This paper presents a deep reinforcement learning-based data distribution algorithm which transforms the problem of data distribution balance into the weight optimization problem of OSD. The experiment proves that the proposed algorithm can well solve the problem of data distribution in the traditional CRUSH distribution algorithm. However, when the number of objects reaches about 200, the optimization effect will be not good enough, and at the same time, as the number of objects increases, the optimization time becomes longer and longer.

References

1. Reinsel, D., Gantz, J., Rydning, J.: Data age 2025: the evolution of data to life-critical. Don't Focus on Big Data 2 (2017)
2. Weil, S.A., Brandt, S.A., Miller, E.L., et al.: Ceph: a scalable, high-performance distributed file system. Proceedings of the 7th Symposium on Operating Systems Design and Implementation, pp. 307–320 (2006)
3. McKusick, M.K., Quinlan, S.: GFS: evolution on fast-forward: a discussion between Kirk McKusick and Sean Quinlan about the origin and evolution of the google file system. Queue **7**(7), 10–20 (2009)
4. Borthakur, D.: HDFS architecture. Document on Hadoop Wiki. URL http://hadoop.apache.org/common/docs/r0 (2010)
5. He, Y.J.: Research and Implementation of Big Data Distributed Storage Method for Load Balancing. Shanghai Jiao Tong University (2015)
6. Mu, Y.L.: Research and Improvement of CRUSH Algorithm in Ceph Storage Technology. Chengdu University of Information Technology (2016)
7. Yusuf, I., Mulyana, E., Hendrawan, H., et al.: Utilizing CRUSH algorithm on ceph to build a cluster of reliable data storage. 2019 IEEE 13th International Conference on Telecommunication Systems, Services, and Applications (TSSA), IEEE (2019)
8. Khan, Z., Singh, R., Kumar, J.A., et al.: Performance analysis of dynamic load balancing techniques for parallel and distributed systems. Int. J. Comput. Netw. Secur. **2**(2), 123–127 (2010)
9. Wang, Y., Ye, M., He, Q., Huan, Y.M., Kang, W.J.: A new node selecting approach in ceph storage system based on software defined network and multi-attributes decision-making model. Chinese J. Comput. (2019)
10. Wu, D., Wang, Y., Feng, H., Huan, Y.: Optimization design and realization of Ceph storage system based on software defined network. 2017 13th International Conference on Computational Intelligence and Security (CIS). IEEE Computer Society (2017)
11. Chen, H.M., Li, C.J., Ke, B.S.: Designing a simple storage services (S3) compatible system based on Ceph software-defined storage system. The 2017 2nd International Conference (2017)
12. Fei, L., Jun, J.D., Huan, Z., Jing, C., Jun, W., Jin, X.: Heterogeneous storage aware data placement of ceph storage system. Comput. Sci. (2017)
13. Wu, Z.Q., Wei, J., Zhang, F., Guo, W., Xie, G.W.: Mdlb: a metadata dynamic load balancing mechanism based on reinforcement learning. Frontiers Inf. Technol. Electron. Eng. **21**(7), 1034–1046 (2020)
14. Daoud, H., Dagenais, M.R.: Performance analysis of distributed storage clusters based on kernel and userspace traces. Software Pract. Exp. **51**(1) (2021)
15. Geng, W., Zhang, H., Liu, L., He, R., Tang, D.: Optimization of the recovery time of pyramid code in distributed storage system. J. Phys: Conf. Ser. **1634**, 012075 (2020)
16. Anschel, O., Baram, N., Shimkin, N.: Averaged-DQN: Variance Reduction and Stabilization for Deep Reinforcement Learning (2016)
17. Yang, Y., Juntao, L., Lingling, P.: Multi-robot path planning based on a deep reinforcement learning dqn algorithm. CAAI Trans. Intell. Technol. **5**(3) (2020)

Chapter 9
The Fault Detection Method of the Seismic Image Based on Semantic Segmentation

Jing Chen, Qingqing Chen, and Xin Wang

Abstract Seismic image fault detection is of practical significance in solid mineral resources investigation, petroleum geology investigation, and other multidisciplinary fields. Traditional fault detection requires manual interpretation, limited by inefficiency and subjectivity. Nowadays, deep learning methods have significantly improved segmentation tasks, but most of these methods were supported by multitudinous seismic data. However, the semantic segmentation method achieves enhanced performance in insufficient data. This paper proposed Channel-UNet, a novel U-Net architecture that integrates the channel attention module to recalibrate feature weight adaptively. Simultaneously, to obtain adequate seismic images, we synthesize a dataset of 2000 images containing apparent fault features and corresponding labels. The experimental results indicate that our network has the highest average IOU and Dice coefficient on the synthetic seismic datasets compared with other semantic segmentation networks. In addition, we further confirm that when applying the Hough Transform as a post-process, the results of Channel-UNet on the actual seismic images perfectly match the predicted results and labels given by the interpreter. The experimental results reveal that the Channel-UNet delineates the fault features in seismic sections more accurately.

9.1 Introduction

Faults are formed during crustal structural change. Faults are arranged under different values and directions of in situ stress, existing in the deep rock mass [1]. Moreover, the formation of many metals is related to natural amalgamation initiated by activating the platform and adjacent deep faults. Hence, fault detection is inextricably related to exploring stratigraphic structures and reservoir properties from seismic images.

J. Chen · Q. Chen · X. Wang (✉)
College of Computer Science and Technology, Jilin University, Changchun, China
e-mail: w_x@jlu.edu.cn

R. Kountchev et al. (eds.), *Multidimensional Signals, Augmented Reality and Information Technologies*, Smart Innovation, Systems and Technologies 374,
https://doi.org/10.1007/978-981-99-7011-7_9

Traditional fault detection uses a manual tracking method to select faults. Therefore, it is difficult to ensure efficiency due to enormous manual intervention. Hence, researchers have proposed various methods to improve the efficiency and accuracy of fault detection. Bahoric [2] proposed the coherence method. Because the curvature attribute was previously applied to fault identification, Robert [3] systematically classified the two-dimensional curvature properties and gave the calculation formula for each curvature. Thanks to Dorn and James, AFE automatic fault interpretation technology [4] was proposed. This method can self-identify and interpret faults in the seismic data volume. In the meantime, with the prevalence of deep learning, many researchers have developed deep-learning-based fault detection networks to shorten the interpretation cycle time. Zhang et al. [5] presented a technique in which a machine learning approach is used to directly synthesize the original input seismic data into 2D faults, which is beneficial to alleviate interpreters' work. By combining convolutional neural networks (CNNs) and image processing technology, Zhao et al. [6] investigated a fault detection workflow to improve the accuracy of CNNs fault detection. Lu et al. [7] improved the fault detection quality by using generative adversarial networks (GANs) to conduct upsampling on seismic data. Taking fault orientation estimation as an image classification problem, Wu et al. [8] proposed an automatic interpretation method that uses convolutional neural networks (CNN) to classify fault orientations (dips and strikes) by seismic image slices. However, most of these methods have been developed using synthetic data sets because the actual seismic data involves the confidentiality of land and resources.

To ensure that accurate fault detection results can be obtained despite the absence of actual seismic data, we design the Channel-UNet network based on the Channel attention module. In addition, to address the problem of scarce seismic data, we apply the synthetic seismic datasets platform to generate seismic images with random faults, define fault parameters, and generate fault labels. Furthermore, the semantic segmentation method is used to improve model performance on various datasets, and the Hough Transform method is used to post-process the detection results to enhance the quality of results. Eventually, the synthetic seismic images and the labeled datasets are used to train several semantic segmentation networks and the effects of multiple semantic segmentation networks in fault detection are compared.

9.2 Methodology

9.2.1 The Semantic Segmentation Method

The semantic segmentation method taken by labeling each pixel with the class of its enclosing object or region is now mainly applied to tumor image segmentation and other fields [9]. Compared with the deep learning method, it achieves better performance in insufficient data, so it is suitable for fault detection.

Long et al. [10] proposed the Fully Convolutional Networks (FCN), which became the basic network structure for many subsequent works. However, their success was limited due to the loss of information associated with discarding the final average pooling layer. Ronneberger et al. [11] modified and extended FCN architecture, naming it UNet to yield more precise segmentations. Additionally, UNet's skip connection concatenates the shallow feature map and the deep feature map together. Accordingly, the proposed network is built based on the U-Net structure.

9.2.2 Post-Processing of Fault Images

Regarding the segmentation results, a binary image is obtained where the likely fault points are highlighted as white pixels and non-fault points are highlighted as black pixels. In this paper, the Hough transform is applied as the post-processing of segmentation results for the following reasons:

1. The neural network will extract some subsets of fault separately that are ignored in the manual interpretation. Admittedly, these minor faults will interfere with the overall performance.
2. Under the influence of discontinuous reflectance properties in seismic profile, a long fault line in the seismic image is probably represented by minor likely fault points.

Because the fault curvature in the seismic section is slight, we employ the approach of two faults in the same direction with a close distance connected as post-processing. This paper discusses the results of utilizing different post-processing methods, including the Hough transform, Edge Detection, Line Segment Detector (LSD), and the Expansion/Corrosion method.

Obviously, Fig. 9.1b shows the six white fault lines detected by the Hough transform. In a word, the Hough transform can connect successive fault lines and remove minor likely fault points, significantly improving the detection of fault features. Hence, it is natural that the Hough transform be chosen as post-processing.

9.2.3 The Adam Optimization Algorithm

The Adam optimization algorithm was initially proposed by Diederik Kingma and Jimmy Ba [12]. The Adam algorithm assigns an individual adaptive learning rate for different parameters. Therefore, the Channel-UNet adopts the Adam algorithm to improve the performance of the network.

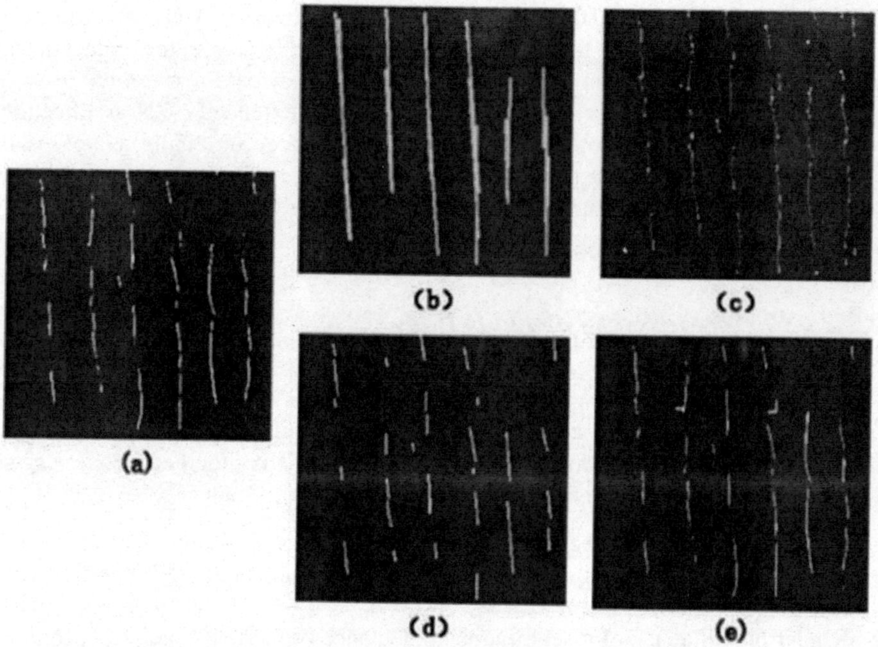

Fig. 9.1 Results of different methods. **a** The results of the Channel-UNet. **b** Extracted faults using the Hough transform on (**a**). **c** Extracted faults using Edge Detection on (**a**). **d** Extracted faults using LSD on (a). **e** Extracted faults using the Expansion/Corrosion method on (a)

9.2.4 Network Architecture

Since it was proposed, UNet [11] has gradually become the most popular image segmentation network. Squeeze-and-Excitation Networks (SENet) [13] were built based on the channel attention module to emphasize the relationship between channels. Afterward, research shows that the channel attention module improves segmentation quality significantly [14, 15]. Therefore, Channel-UNet is built based on the U-Net structure and integrates the channel attention module of SENet.

Compared with the original UNet, the Channel-UNet has a similar architecture: the downsampling, upsampling, and skip connection parts. The downsampling module consists of two stacked 3×3 convolutions and 2×2 max-pooling for feature extraction, the following describes the composition of the three parts:

- Deconvolution module: It is obtained by 2×2 deconvolution of the upper layer that aims to halve the number of channels and double the size of the feature picture.
- The channel attention module: It borrows squeeze operation from SENet [13], named SE block. SE Block converts the feature map of upper 1024-channel into features map with 1-channel through mean-pooling. Figure 9.2 shows the architecture for the SE block.

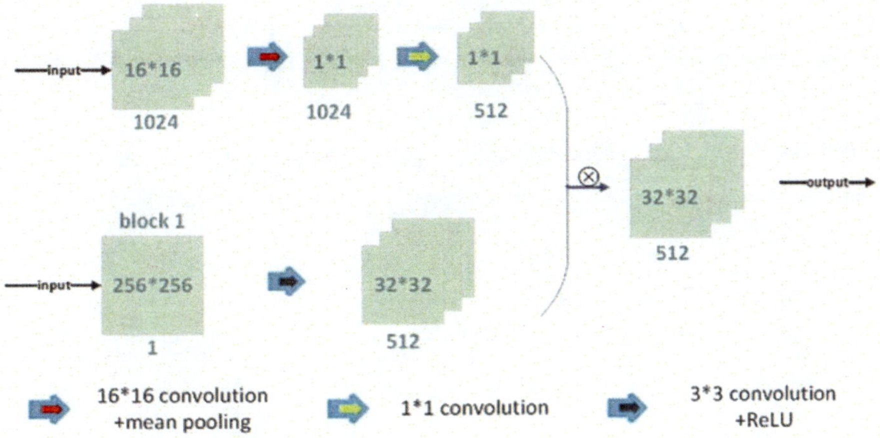

Fig. 9.2 SE block architecture

- The skip connection module: The skip connection is used for feature fusion, directly connecting the feature map with the same number of channels in downsampling. Remarkably, information loss will not occur in this process.

Because the feature maps obtained by the above three modules have the same size, we can apply the Concat operation to perform feature fusion. The network structure is shown in Fig. 9.3. The feature map combined with three modules passes through a 1×1 convolution and ReLU. Finally, through the Sigmoid activation function, the results for Channel-UNet become the same size as the input seismic image. Briefly, the feature map of the downsampling and upsampling are fused, which can separate closely interacting segmentation targets and backgrounds.

The essential of fault detection is a binary classification task, and BCE suitable for binary classification is chosen as the loss function. It is calculated as follows:

$$BCE = -y\log(\hat{y}) - (1 - y)\log(1 - \hat{y})(1) \quad (9.1)$$

where y represents the ground truth and $\hat{y}$ represents the predicted value. In addition, the input value must be between [0,1]. To ensure that the result for the Channel-UNet is between [0, 1], the Sigmoid function is utilized in this paper. Overall, the Channel-UNet finally obtains a segmented binary mask image as output.

9.3 Synthetic Data Set Generation

To maximize Channel-UNet effectiveness, we require multitudinous seismic data and corresponding labels to train the network. Adopting synthetic data sets or actual data is the primary method to obtain a training set. Unfortunately, when using original

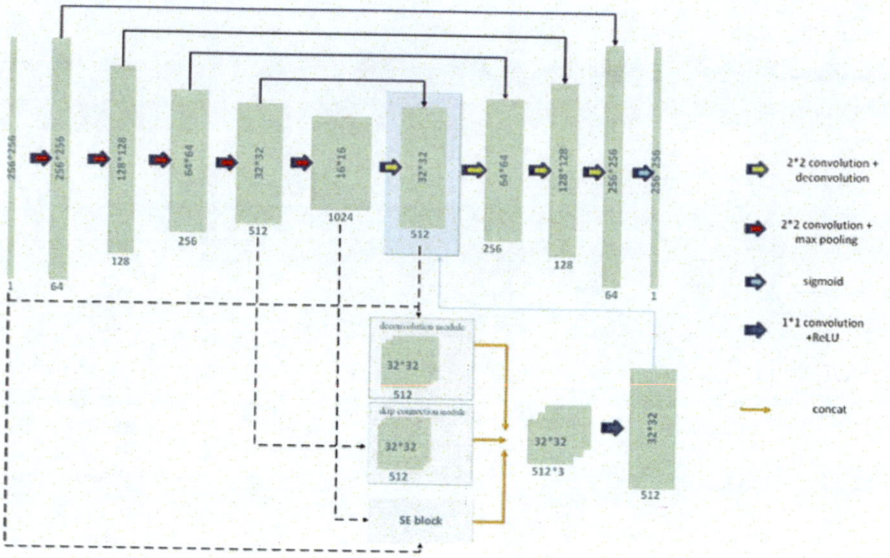

Fig. 9.3 Channel-UNet architecture

seismic data, manual interpretation of fault directly influences the performance of the network. Particularly, the spread of actual seismic data increases the risk of land and resource exposure, which directly brings obstacles to obtaining adequate seismic data containing vivid faults. Hence, we apply synthetic seismic data in the network training stage.

This paper proposes an efficient method for constructing synthetic seismic images and dependent labels. The following assumptions are made in the process of synthesizing seismic images:

1. A local fault is a uniquely defined line by a single fault orientation;
2. The dip angles of the two-dimensional fault range from $-63°$ to $-86°$ and from $86°$ to $63°$;
3. Fault parameters: fault dip and offset. The redefined fault dip angle is formed by the intersection of the fault line and the vertical line in the image. Simultaneously, the offset is the percentage of the point where the fault line and the upper edge intersect on the upper edge ruler.

Fault dips and offsets can determine a unique line running through the whole image in the synthetic seismic image, which is equivalent to the fault in the actual seismic image.

The open-source code from Hale allowed us to reproduce the process of synthesizing seismic datasets [16]. We built a dataset of 2000 images of 128×128 pixels as the training set, all containing one straight fault crossing the section entirely and corresponding fault labels by modifying the fault dip angle and offset randomly. Compared to accurate data, constructing a synthetic dataset presents the great advantage of

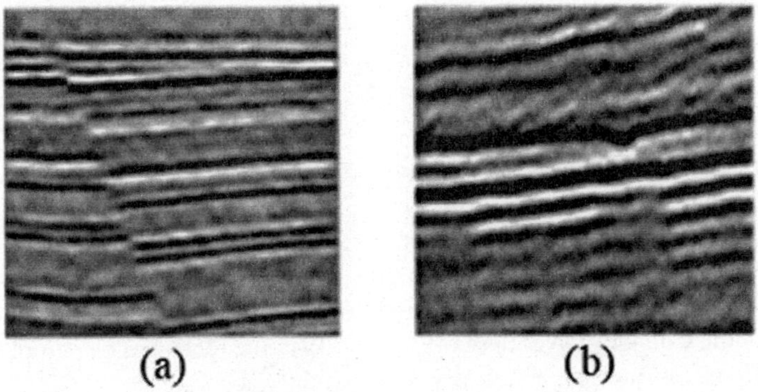

Fig. 9.4 **a** Synthetic seismic image. **b** Actual seismic image

providing a simple extension of the amount of training while allowing automatic label synthesis and auto-labeling to avoid mismatch errors.

As shown in Fig. 9.4, the generated seismic image also simulates topographic relief and structure remarkably. Meanwhile, the synthetic image's distinct or blurry area represents the real-world stratum's depth. Therefore, the synthetic seismic image is suitable for fault detection.

9.4 Experiments

We trained several semantic segmentation networks, including UNet, ResNet34-UNet, R2U-Net, Attention-UNet, and Channel-UNet with the channel attention mechanism proposed in this paper, using the synthetic seismic datasets generated autonomously and the corresponding labels used in the previous section. Channel-UNet is implemented based on PyTorch. The experimental environment for training and testing is shown in Table 9.1.

Table 9.1 Experimental Environment

Category	Configuration
OS	Windows10
CPU	Intel(R) Core(TM) i5-10,300
GPU	NVIDIA GeForce GTX 1650
PyTorch	1.1.0
Python	3.6.5

9.4.1 *Experimental Results on Synthetic Datasets*

Three synthetic seismic images generated by the seismic data platform are chosen at random and used as input to a network trained for detection in this paper. The comparison of the fault labels from seismic images with the network detection results is shown in Fig. 9.5.

According to the comparison results, it can be seen that the results of Channel-UNet are closest to the actual label. Similarly, the UNet, Attention-UNet, and ResNet34-Unet can predict the fault direction and location, but the accuracy is less than that of the Channel-UNet network. In contrast, the R2U-Net can crudely predict fault locations, but fault detection is not sufficiently accurate.

To objectively evaluate the detection accuracy of the model, 100 synthetic seismic images generated by the seismic data platform were acquired in this study, and five semantic segmentation networks were used to detect faults on them. The detection results are compared to the labels to compute the mean IOU and Dice coefficient. The results are given in Table 9.2. It shows that our method has the highest IOU and Dice for fault detection for the seismic image, followed by UNet and ResNet34-UNet methods. However, Attention-UNet and R2U-Net have relatively low accuracy. In particular, ResNet34-UNet, R2U-Net, and Attention-UNet all add structures to UNet, but their detection results are not as good as those of UNet, indicating that not all structure is conducive to fault detection in seismic images. The Channel-UNet outperforms UNet due to the addition of SE Block to the UNet, which enables the network to have a feature recalibration mechanism. Therefore, in summary, Channel-UNet achieves the best performance than any other network on the synthetic datasets.

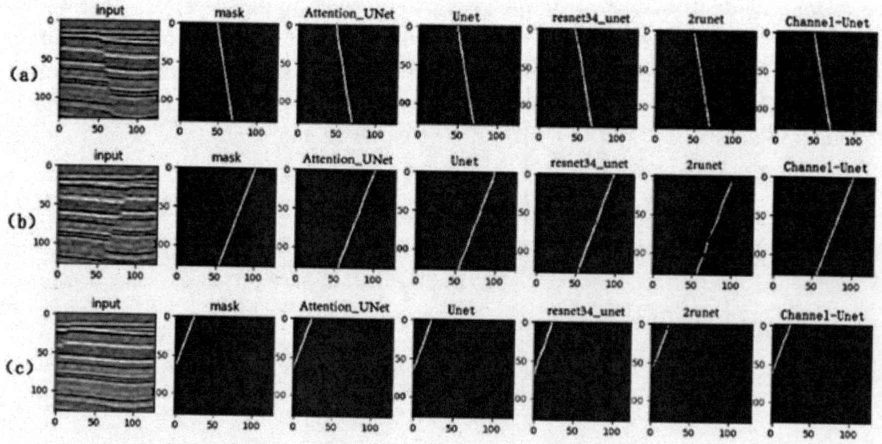

Fig. 9.5 Comparison of the fault labels with the fault detection results

Table 9.2 Model Accuracy Evaluation

Performance metrics	Attention-UNet	UNet	ResNet34-UNet	R2U-Net	Channel-UNet
IOU	0.68	0.76	0.7	0.65	**0.82**
Dice	0.84	0.82	0.76	0.64	**0.90**

9.4.2 Experimental Results on Real-World Data from Great South Basin (GBS) in New Zealand.

In order to verify the accuracy of the Channel-UNet network for fault detection on real-world datasets, in this study, we use seismic data and labels from GSB, New Zealand, as used by Babc et al. [17].

It can be seen from the comparison that the Channel-UNet network detects faults with greater clarity and accuracy than the other networks in Fig. 9.6.

(a) The fault label provided by Haibin and used by Babc. (b) The fault detection result of the Channel-UNet, where blue is the same artificial label as (a), and white is the output of the network prediction. (c)–(f) The prediction results of the network obtained by Babc. The blue lines are the fault labels provided by Haibin. The red areas are the prediction of the faults.

We conclude that the average FDA of the 100 post-processed seismic images is 0.78, which is highly precise. The comparison results are presented in Fig. 9.7. These experiments demonstrate that using Hough transform makes important contributions to fault detection.

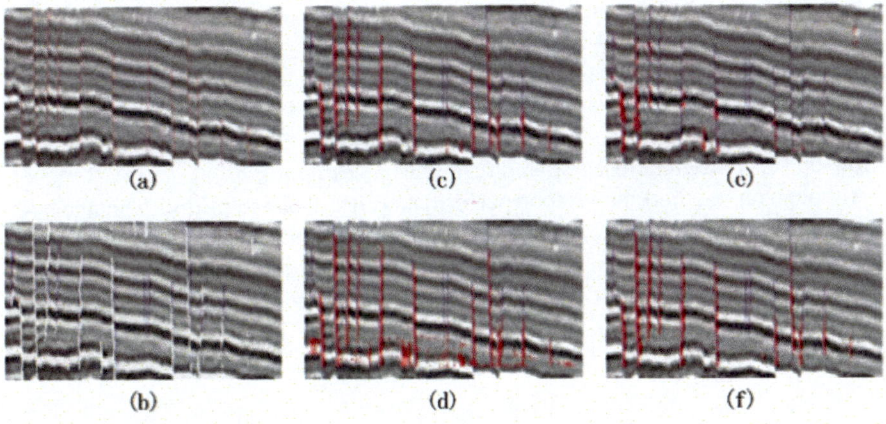

Fig. 9.6 Experimental results of different methods on data from GBS

(a)
(b)
(c)

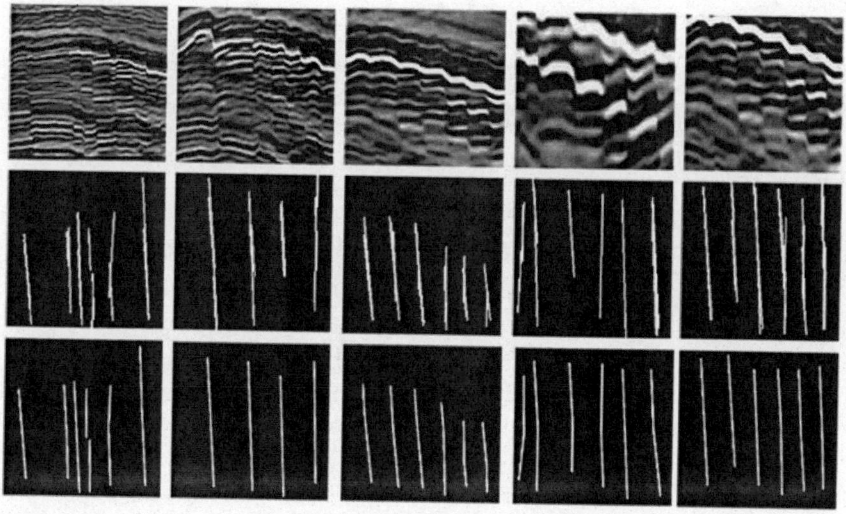

Fig. 9.7 Fault detection results of the chosen F3 seismic image: **a** A randomly selected F3 seismic image processed with noise reduction and contrast enhancement. **b** Channel-UNet's detection results. **c** The detection image after using Hough transform

9.5 Conclusion

This paper proposes the Channel-UNet network for fault detection based on U-Net architecture. SE block is applied to adaptively learn the weight of each feature channel to emphasize informative features and suppress less useful ones. Simultaneously, we redefine the fault dip and define another parameter, named offset, to try to generate a data set containing faults and common geological structures. Notably, the Channel-UNet is trained with synthetic data only and yields good performance when applied to novel synthetic data. The experimental results indicate that our network has the highest IOU and Dice average index compared with other semantic segmentation networks. After applying the Hough Transform as a post-process in the seismic images of the F3 block in the Netherlands-North Sea region, the Channel-UNet has a perfect match of the predicted results and labels given by the interpreter. In short, the Channel-UNet network can accurately locate seismic faults, greatly assisting the follow-up seismological research and resources investigation.

Although our method shows good performance, it still has some disadvantages. As for future research directions, we plan to improve the generalization of the proposed method. Specifically, (1) combined with 3D seismic data as a training set, which can enhance the network's fault detection ability; (2) by further exploring the potential distinguishing features in seismic fault data under different geological conditions, a more general fault detection framework can be designed.

References

1. Li, S., Ma, Z., Shi, H., Gao, X.: New understanding of Anderson fault formation model based on butterfly plastic zone theory. Arabian J. Geosci. **14**(8), 1866–7511 (2021)
2. Mike, B., Steve, F.: 3-d seismic discontinuity for faults and stratigraphic features: the coherence cube. Leading Edge **14**(10), 1053–1058 (1995)
3. Andy, R.: Curvature attributes and their application to 3D interpreted horizons. First Break **19**(2), 85–100 (2001)
4. Dorn, G.A., James, E.H.: Automatic Fault Extraction of Faults and a Salt Body in a 3-D Survey from the Eugene Island Area, Gulf of Mexico. Search and Discovery (2005)
5. Chiyuan, Z., Charlie, F., Mauricio, A., Detlef, H.: Machine-learning based automated fault detection in seismic traces. 76th European Association of Geoscientists and Engineers Conference and Exhibition 2014: Experience the Energy—Incorporating SPE EUROPEC 2014, Vol. 2014, p. 1–5 (2014). https://doi.org/10.3997/2214-4609.20141500
6. Tao, Z., Pradip, M.: A fault detection workflow using deep learning and image processing. 2018 SEG International Exposition and Annual Meeting (2018)
7. Ping, L., Matt, M., Seth, B., Cody, C., Yuan, X.: Using generative adversarial networks to improve deep-learning fault interpretation networks. The Leading Edge **37**(8), 578–583 (2018)
8. Wu, X., Shi, Y., Fomel, S., Liang, L.: Convolutional neural networks for fault interpretation in seismic images. 2018 SEG International Exposition and Annual Meeting, pp. 1946–1950 (2018)
9. Savadikar, C., Kulhall, R., Garware, B.: Brain tumour segmentation using probabilistic U-Net. Lecture Notes in Computer Science (including subseries Lecture Notes in Artificial Intelligence and Lecture Notes in Bioinformatics), Vol.12659, pp. 255–264 (2021)
10. Shelhamer, E., Long, J., Darrell, T.: Fully convolutional networks for semantic segmentation. IEEE Trans. Pattern Anal. Mach. Intell. **39**(4), 640–651 (2017)
11. Ronneberger, O., Fischer, P., Brox, T.: U-Net: convolutional networks for biomedical image segmentation. Medical Image Computing and Computer-Assisted Intervention—MICCAI 2015, Vol.9351, pp. 234–241 (2015)
12. Kingma, D., Ba, J.: Adam: A method for stochastic optimization. Comput. Sci. (2014)
13. Hu, J., Shen, L., Albanie, S., Sun, G., Wu, E.: Squeeze-and-excitation networks. IEEE Transactions on Pattern Analysis and Machine Intelligence, Vol. 42(8), pp. 2011–2023 (2020)
14. Kushnure, D., Talbar, S.: HFRU-Net: high-level feature fusion and recalibration UNet for automatic liver and tumor segmentation in CT images. Computer Methods and Programs in Biomedicine, Vol. 213, pp. 106501–106506 (2022)
15. Soans, R.V., Fukumizu, Y.: Improved facial keypoint regression using attention modules. Front. Comput. Vis. **1578**, 182–196 (2022)
16. Sørensen, J.C., Gregersen, U., Breiner, M., Michelsen, O.: High-frequency sequence stratigraphy of upper Cenozoic deposits in the central and southeastern North Sea Areas. Marine Petroleum Geol. **14**(2), 99–123 (1997)
17. Cunha, A., Pochet, A., Lopes, H., Gattass, M.: Seismic fault detection in real data using transfer learning from a convolutional neural network pre-trained with synthetic seismic data. Comput. Geosci. **135**(9), 1–9 (2020)

Part II
Augmented Reality and Deep Learning

Chapter 10
Analysis on Grey Space Form and Simulation Evaluation in Landscape Design

Songlin Wu

Abstract In order to improve the landscape design effect, this paper combines intelligent grey space algorithm to improve the design algorithm, and uses Mean-shift algorithm to smooth the multi-focus image. Moreover, this paper considers the spatial information of multi-focus images as well as the grey information of the image, which effectively removes the image noise and retains more grey texture information, thus achieving a better smoothing effect. In addition, this paper uses the maximum inter-class difference algorithm as the grey histogram fitting function of multi-focus images, and completes the recognition of grey overlapping areas of multi-focus images. Finally, this paper uses DEM model to design the terrain, which can not only simulate the terrain of the site, but also complete the site analysis, site leveling and site reconstruction. Through experimental analysis, we can see that the grey space design form proposed in this paper has a good effect in landscape design, and the design scheme proposed in this paper can meet the actual design needs.

10.1 Introduction

"Grey space" is mainly reflected in the connection, transition, separation and foil of internal and external spaces of landscape. Through a series of free and flexible combinations and structures, it scientifically designs the original chaotic and irregular forms of expression, thus creating a more open, shared, ecological and socialized development space, endowing the landscape with more spirituality and creativity, and adding beautiful scenery to urban development. There are many forms of "grey space" in landscape design, which need to be endowed with more forms and forms according to the specific content of the landscape, so as to realize the coordination and unity of the whole and part, reach a deeper design height and embody greater value and significance.

S. Wu (✉)
Wuhan Qingchuan University, Wuhan, China
e-mail: wsl326390739@163.com

R. Kountchev et al. (eds.), *Multidimensional Signals, Augmented Reality and Information Technologies*, Smart Innovation, Systems and Technologies 374,
https://doi.org/10.1007/978-981-99-7011-7_10

From the perspective of spatial attributes, grey space has "fuzziness", which refers to the lack of clear attribution and division of this space, and the characteristics of adjacent spaces will be brought into grey space. This fuzziness determines the diversity of the grey space environment and can more effectively utilize environmental resources. With the acceleration of urban modernization, there is an increasing demand and expectation from the public for landscape design. The excessive pursuit of formal methods in design results in a lack of integrity, continuity, and poor hierarchy in urban landscape design. In order to pursue the perfection of form, many designers learn from each other and neglect the organic unity and coordination with the overall development of the city [1]. The spatial differences are becoming increasingly large, and many places are difficult to effectively integrate, which also brings great difficulties to urban construction and development. To better meet the public's high-quality requirements for urban construction, many designers gradually introduce and create the "grey space" form of garden landscape design, Thus improving the overall quality of the urban landscape [2].

With the continuous development of the industry, the theory of grey space is also constantly improving and improving. The research on grey space has gradually expanded from a single field of the construction industry to urban landscape design, and has given more connotation and significance [3]. By introducing the form of grey space in landscape design, the public can experience more changes and innovations in the spatial sequence of landscape design, giving more sense of hierarchy and richness. The form of grey space is gradually becoming more and more widely used and embodied in landscape design, giving urban landscapes greater vitality and scalability [4].

In the process of architectural art design, the application of grey space design techniques is often limited due to the carrier. However, in fact, grey space carriers are very rich, which requires architectural art design personnel to understand and master the basic concepts of grey space, and through reasonable application, make grey space more concrete to display. Vegetation and water bodies are common grey space carriers [5]. Vegetation, as a vibrant landscape, can effectively create a dynamic and vibrant space in the application process. When architectural art and design personnel use vegetation grey space carriers, they should reasonably match different color combinations and pay attention to the combination forms between the same vegetation, providing users with a visual and auditory stimulus, With seasonal changes, these grey spaces, mainly carried by vegetation and water bodies, have to some extent undergone changes in form and color, making the grey space more abundant. Corridors are also the main carrier for the application of grey space in architectural art design [6]. As a corridor with transportation characteristics, in the process of serving as a transitional space carrier, it often combines with corridors with different height differences and conversion angles due to different ground heights, forming a space shape with excellent artistic charm, making the grey space of the building more rhythmic. The porch is the preferred grey space carrier for most residential architectural art designs. In residential buildings, the porch has a closer relationship with users. Although it may sometimes create a certain degree of spatial ambiguity with guests, restaurants, etc., in the process of blurring spatial boundaries, it also buffers the visual experience of residential space and plays a certain finishing touch [7].

In modern society, people pay high attention to their own privacy, especially in buildings closely related to people's daily life and work. They have high requirements for their spatial privacy and spatial domain. Enclosing and limiting the space creates, varying degrees of enclosure or limitation can determine the privacy and domain of the building space. Although this meets some people's expectations for living and living [8]. The privacy protection requirements of the workspace also affect the relationship between spatial communication to a certain extent. In most cases, the internal space of a building is composed of enclosed enclosures, so open enclosures or semi-open enclosures are used. This can create a richer interior and exterior space of the building, providing richer possibilities for transitional spaces. In the process of compensating for the insufficient functionality and span of the internal and external spaces of buildings, extending a certain element of modern buildings to the outside reasonably to form a distinctive grey space can not only meet people's actual needs but also make up for people's psychological deficiencies [9].

Grey space can enrich the transition forms and increase the proximity of transition forms. In the internal and external spaces of buildings, multiple nearby landscapes can be selected for fusion to achieve transition effects; At the same time, it can also increase the form of distant transition, selecting semi-enclosed building spaces and utilizing pavilions to overlook the relatively open water surface and vegetation in the distance; And it is also possible to choose a timely transition to integrate meteorological, astronomical, and vegetation landscapes. The application of grey space can enrich the intervention content, making the internal and external spaces of buildings more abundant. In the interior of buildings, plants, water surfaces, and other architectural landscapes outside the building can be utilized to gradually enrich the indoor landscape. At the same time, meteorological landscapes such as sunrise and sunset, spring breeze, and snow scenery can also be utilized [10].

In the process of development, the concept of grey space is no longer just an individual level of buildings, but gradually develops towards the direction of building ethnic and regional levels. In addition, in the process of modern architectural art design, the application of grey space design techniques is more extensive, and its role cannot be underestimated. It can already be fully displayed as a unique characteristic of the building, and can express the unique artistic style of the building, with a certain degree of decoration. In the process of applying grey space to architectural design art, it transforms the color hierarchy of the building into richness and diversity, and has excellent coordination, making the use of the entire space or the color tones reflected in the building more harmonious [11].

The "grey space" in landscape design is embodied in all aspects. In the specific design and combination, it is necessary to make a concrete and comprehensive analysis according to the requirements of urban development design and overall function. On the basis of ensuring and embodying ecology, it gives more layers, richness and extensibility to roads, plants, surfaces and environmental facilities in the landscape, so as to enhance the image of the city and create a more livable environment.

10.2 Adaptive Image Recognition Based on Grey Overlapping

In this paper, Mean-shift algorithm is used to smooth multi-focus images, which can consider the spatial information of multi-focus images and the grey information of images, effectively removing image noise and retaining more grey texture information, so as to achieve a better smoothing effect.

10.2.1 Grey Value Enhancement of Image Grey Overlapping Area

We formulate n model points $x_i, i = 1, 2, ..., n$ in d-dimensional space. The form of Mean-shift at point x is defined as formula (10.1):

$$M_k(x) = \frac{1}{k} \sum_{x_i \in S_h} (x_i - x) \tag{10.1}$$

In the formula, S_h represents the grey area with radius h, which is a set satisfying formula (10.2) [11].

$$S_h(\mathrm{x}) = \left\{y : (y - x)(y - x) \leq h^2\right\} \tag{10.2}$$

In the formula, k means that among the n sample points, k points fall in the g region, x represents the center point of the region, y represents the radius centered on x, and h is any point in the grey region.

The multi-focus image can be expressed as a two-dimensional grid P-dimension, each point represents a pixel, P = 1 represents a grey image, and the coordinates of grid points are regarded as spatial information of the multi-focus image. Considering the spatial information and grey information of multi-focus image, a P + 2-dimensional vector $X = (\mathrm{x}^s, x^y)$ is created. Among them, x^s and x^y represent the abscissa and ordinate on the grid, that is, the spatial description of grey feature vectors, and x^r represents the description of P-dimensional vectors on the image grid, which can also be regarded as the spatial description of feature vectors. The kernel function $K(h_s, h_r)$ is used to predict the x distribution, and $K(h_s, h_r)$ has the following form [12]:

$$K(h_s, h_r) = \frac{C}{h_s^2 h_r^2} k\left(\left\|\frac{x^s}{h_s}\right\|^2\right) k\left(\left\|\frac{x^r}{h_r}\right\|^2\right) \tag{10.3}$$

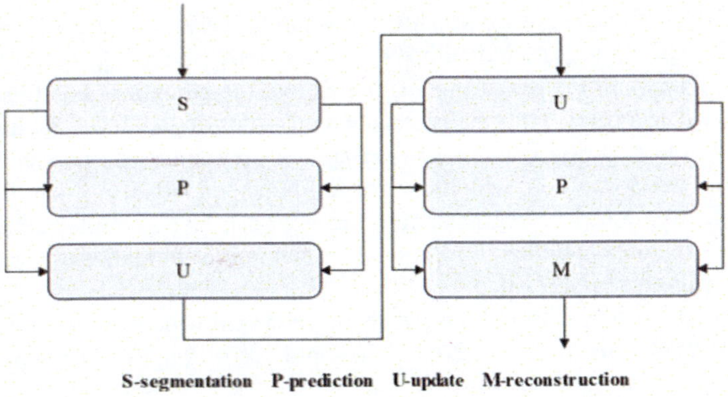

Fig. 10.1 Schematic diagram of lifting wavelet transform

Among them, h_s, h_r represents the bandwidth of the kernel used in the grey area and chrominance area of the image. By using the two bandwidth parameters of the kernel function, the kernel size can be controlled and the resolution of the multi-focus image smoothing stage can be adjusted. C is the unified constant.

Lifting wavelet transform is applied to the smoothed multi-focus image to enhance the grey overlapping area of the multi-focus image. Wavelet Transform algorithm is directly used to enhance the multi-focus image, which will lead to the loss of some details of the enhanced image. Therefore, a wavelet transform lifting scheme is proposed.

Figure 10.1 is the schematic diagram of decomposition and reconstruction of lifting wavelets. The signal s_{i} is first split into two signals, and then predicted to obtain the prediction error. Finally, the prediction error is used to update the signal to obtain a new signal $s_{\mathrm{i}-1}$. The reconstruction process is the opposite.

That is, an operator U is constructed to enhance the grey value of the grey overlapping area of the image to:

$$s_{\mathrm{i}-1} == s_{\mathrm{i}} - U(d_{\mathrm{i}-1}) \tag{10.4}$$

10.2.2 *Classification of Grey Overlapping Areas*

In the past, the segmentation of grey overlapping region of multi-focus image is mostly based on a single image target to complete single threshold image segmentation. In practical application, the texture information of the image background region is extremely complex, the grey features of a single target are few, and there are many peaks in the histogram. At this time, the recognition result obtained by using single threshold segmentation method is not accurate. In this paper, the multi-threshold image segmentation method is used to identify the features of overlapping grey areas. The main feature of this method is that the potential basis function can be established according to the statistical characteristics of grey levels in multi-focus images, and the partition features of overlapping grey areas can be calculated by the potential basis function, which can reflect the characteristics of key peaks and valleys in histograms. The segmentation threshold can be determined according to the partition features and the intersection points of curves to realize the recognition of overlapping grey areas [13].

$H_N(k)$ is the probability function of the unified multi-focus image between the intervals [0, 1], and the function $P(\mathrm{k})$ of the multi-focus image can be expressed as:

$$P(\mathrm{k}) = \sum_{g=0}^{G-1} \frac{H_N(k)}{1 + \mathrm{a}(g-k)^2} \tag{10.5}$$

g is the residual information of the multi-focus image, a is the constant control factor in the basis function $\mathrm{C}(x) = \frac{1}{1+ax^2}$, , and the function is unified so that the value is between [0, 1], and the unified multi-focus function is:

$$P(\mathrm{k}) = \frac{1}{\max P(k)} P(k), k = 0, 1, ..., G \tag{10.6}$$

It can be seen that $P(\mathrm{k})$ is the interpolation result of $H_N(k)$ to function $\mathrm{C}(x)$, and when using a suitable constant control factor a, the peak characteristics of $P(\mathrm{k})$ are similar to those of $H_N(k)$.

According to the above procedure, if $P_0(\mathrm{k}) = P_N(\mathrm{I})$, the remaining functions of the histogram are set to:

$$P_k(i) = P_{k-1}(i) - P_k^* \frac{1}{1 + f_d(i - x_d)^4} \tag{10.7}$$

$$P_k^* = \max\{P_{k-1}(\mathrm{i}), i = 0, 1, ..., G-1\} \tag{10.8}$$

$$\mathrm{x_k} = \left\{i | P_{k-1}(\mathrm{i}) = P_{\mathrm{k}}^*\right\} \tag{10.9}$$

Among them, i = 0, 1, ..., $G - 1$ is the grey level of multi-focus image. K = $1, 2, ..., C$, and C is the number of classes to be segmented in the image, which can segment the multi-focus image according to the requirements. f_d is the attenuation radius of the control factor, which is to remove the influence of the previous peak on the subsequent peak. Usually, $f_d = \beta \frac{C-1}{5D_H^2}$ is taken, D_H is the size of the dynamic region of multi-focus image, β is a constant, and $\beta = 12$ is taken in this paper. By virtue of the residual potential function of the grey histogram of the multi-focus image obtained above, the partition potential function group of the grey histogram of the multi-focus image is set as follows:

$$F_k(i) = P_k^*(i)\frac{1}{1 + f_d\|i - x_k\|^4} \tag{10.10}$$

In the formula, F_k is represented as the k-th cluster subset with x_k as the center point and P_k^* as the length among C function groups. The characteristic of each peak in the grey histogram of multi-focus image is well fitted, and the intersection point of adjacent partition potential function curves is the segmentation threshold.

From another point of view, the threshold segmentation method is applied to the segmentation of grey overlapping areas, which can be described as:

We set the grey overlapping image as $f(x, y)$ has N pixels and L grey levels. The grey histogram is made, and the probability p_i corresponding to each grey level is obtained by normalization.

$$\sum_{i=1}^{L} p_i = 1(i = 1, 2, ..., L) \tag{10.11}$$

We select the threshold value T to divide $f(x, y)$ into two categories, namely R_1 and R_2. R_1 consists of pixels of $f(x, y)$ in grey level $[0, T]$, and R_2 consists of pixels of $f(x, y)$ in grey level $[T + 1, L]$.

The grey overlapping areas of the whole multi-focus image are classified as:

$$u_T = \sum_{i=1}^{L} ip_i \tag{10.12}$$

10.2.3 *Recognition of Overlapping Areas of Image Grey Scale*

Maximum inter-class difference algorithm is used as the fitting function of grey histogram of multi-focus image, which is an infinitely smooth function, so it can accurately approach the edge of grey histogram of multi-focus image and complete the recognition of grey overlapping areas of multi-focus image. The grey values of class R_1 are all:

$$u_1 = \frac{\sum_{i=0}^{T_i} ip_i}{\sum_{i=0}^{T_i} p_i} \tag{10.13}$$

The grey values of class R_2 are:

$$u_2 = \frac{\sum_{i=T_l+1}^{L} i p_i}{\sum_{i=T_l+1}^{L} p_i} \tag{10.14}$$

The recognition of grey overlapping areas of images is as follows:

$$\sigma_B^2 = p_i (u_1 - u_2)^2 \tag{10.15}$$

We use T to traverse $[0, L]$, and then find the maximum corresponding T value of σ_B^2. Then, at this time, the T value is the best threshold of the maximum inter-class difference method.

10.3 Model Simulation and Analysis

10.3.1 Landscape Design Simulation

Due to the restriction of weather, architecture and other factors, it is difficult to collect garden spatial characteristic data, so three-dimensional images are used to extract garden spatial data, as shown in Fig. 10.2.

In order to make clear the accuracy of spatial feature data collection, three-dimensional image simulation judgment is used to analyze the function, and the process of judging whether the collected information is standard is the accurate guarantee and the key to simulate the spatial feature distribution of gardens. Function resolution is shown in Fig. 10.3.

The three-dimensional landscape design simulation system is firstly a computer-aided design system, which has comprehensive functions and can help designers quickly complete the early design and later results display. The system takes landscape design as its main purpose, terrain design, planning design and planting design as its main contents, and integrates architectural modeling and road design. It can dynamically browse the scene in OpenGL (Open Graphics Library) while completing graphic design, and can instantly make realistic rendering drawings and three-dimensional simulation cartoons to display the scene in multiple directions and from all angles. The system can be roughly divided into modeling and editing, terrain design, planning and design, planting design, sprinkler irrigation design, data statistics, construction drawing production, rendering and animation production. The overall architecture of the system is shown in Fig. 10.4.

The accuracy of terrain simulation depends on the sampling accuracy of natural terrain. Reducing the original terrain to three-dimensional digital model accurately

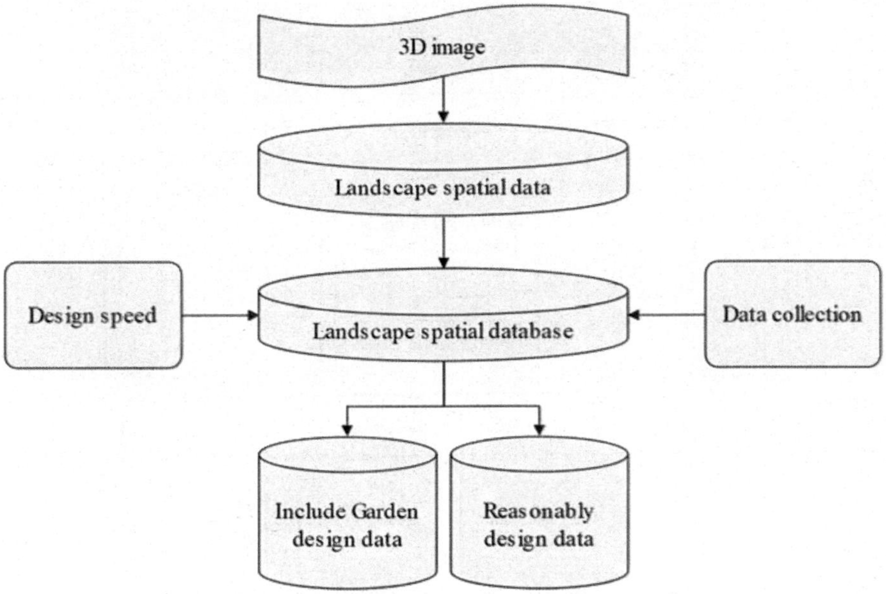

Fig. 10.2 Acquisition diagram of landscape data of three-dimensional image

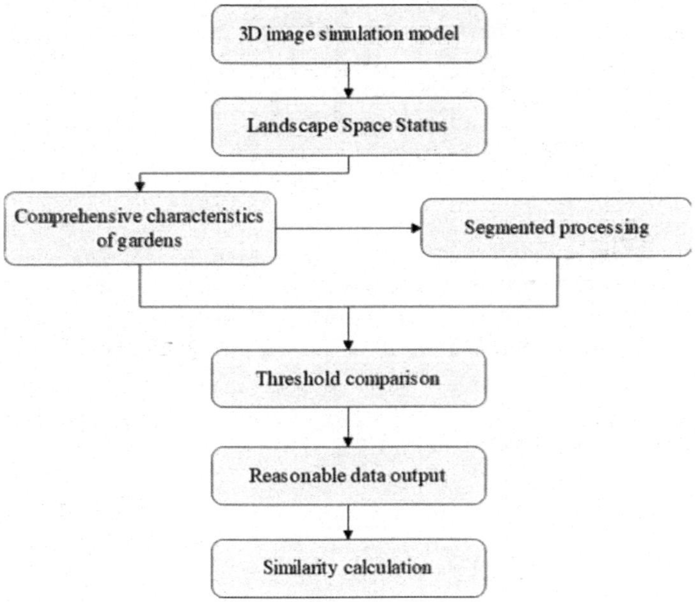

Fig. 10.3 Analytical diagram of garden state characteristics

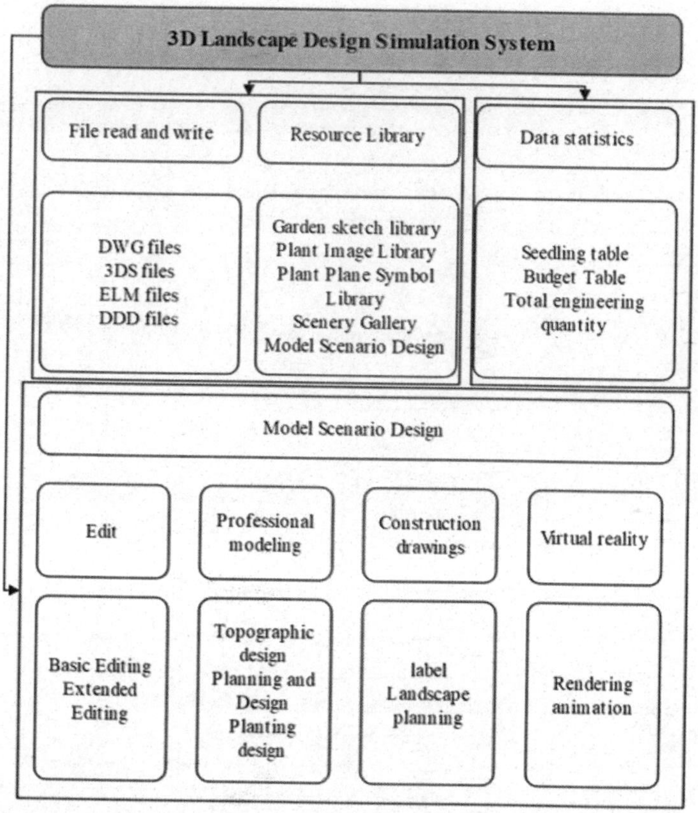

Fig. 10.4 System architecture

according to terrain data information can effectively provide a data analysis basis for designers. In order to avoid the influence of sampling accuracy and range on the simulation results, the system adopts manual intervention sampling accuracy to calculate and control, and achieve the expected effect. In addition, using DEM (Digital Elevation Model) model for terrain design can not only simulate site terrain but also complete site analysis, site leveling and site reconstruction. The specific terrain design process is shown in Fig. 10.5.

10.3.2 Simulation Analysis

This paper uses the model proposed in this paper to design and analyze the grey space form of the landscape. Figure 10.6a below shows the grey space example of landscape design in this paper, Fig. 10.6b shows the rendered design drawing, and Fig. 10.6c shows the actual construction drawing. From this figure, it can be seen

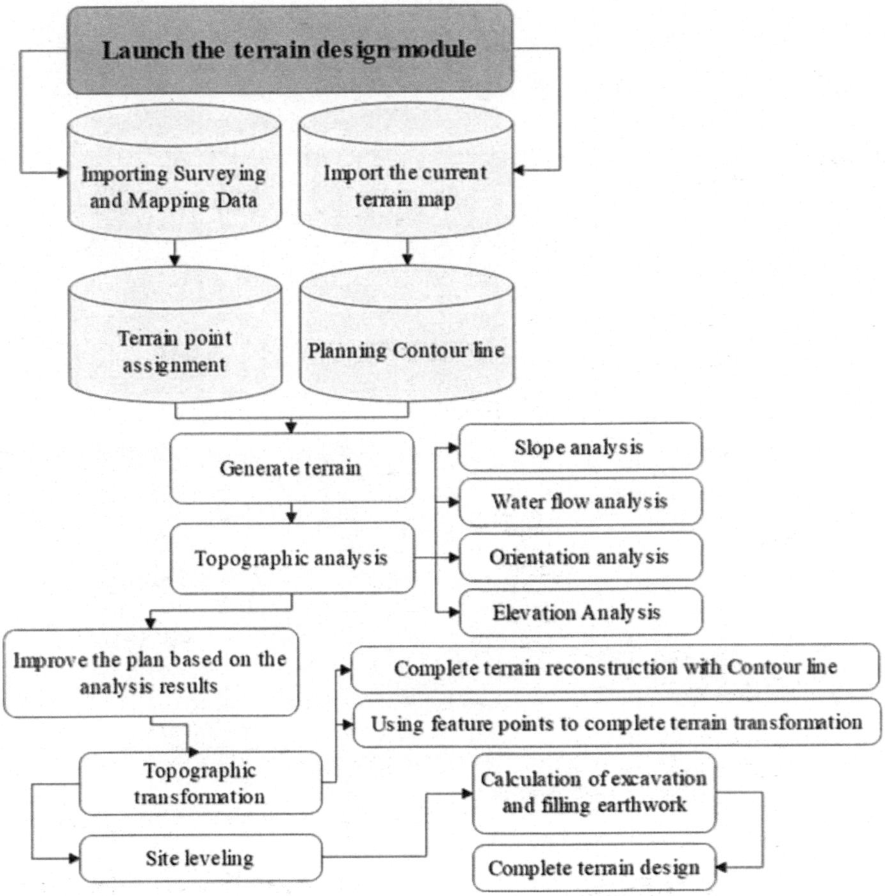

Fig. 10.5 The flow of terrain design process

that the design effect of landscape can be intuitively felt in the design stage through grey space simulation analysis.

Next, through several groups of simulation experiments, the effect of grey space form in landscape design is evaluated.

In order to verify the overall effectiveness of the proposed method, a correlation experiment needs to be conducted. The experimental platform is built on RTLS019AS and simulated based on the existing garden information in the network, and finally the experimental results shown in Fig. 10.7 are obtained.

As shown in Fig. 10.7, the final evaluation result shows that the lowest score exceeds 74 points, and the overall score is between 74 and 88, indicating a higher evaluation.

Fig. 10.6 Example of grey space design of landscape

(a) Garden grey space

(b) Design Image of Landscape Simulation

(c) Image of actual landscape construction

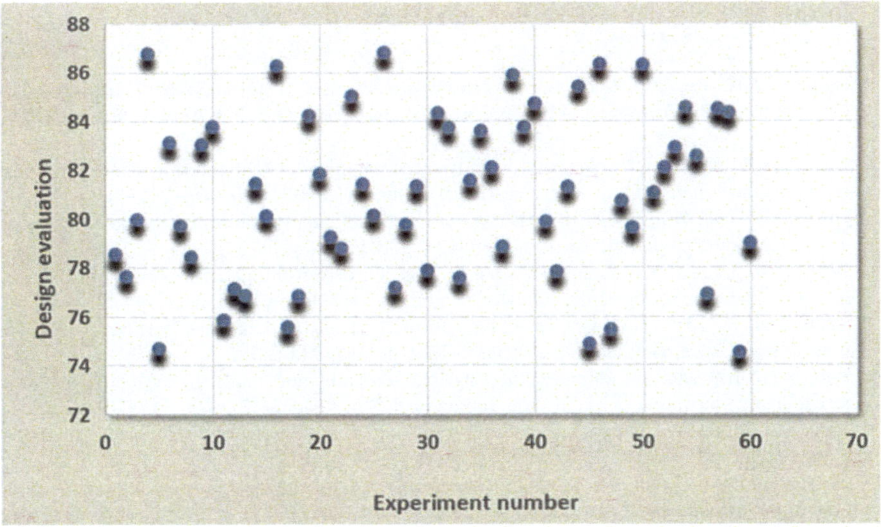

Fig. 10.7 Effect of grey space form in landscape design

From the above analysis, we can see that the grey space design form proposed in this paper has a good effect in landscape design, and the design scheme proposed in this paper can meet the actual design needs.

10.4 Conclusion

This paper discusses the development background, innovation and important technical means of landscape design simulation system based on virtual reality, deeply analyzes the calculation principle and realization method of professional modules, deeply discusses the important algorithms and technologies of simulation effect, and realizes the effective integration of various functional modules such as landscape modeling, landscape terrain design, landscape planning and design, landscape planting design, landscape professional construction drawing, real-time simulation of landscape scene, later rendering and simulation animation. In addition, this paper uses the model proposed in this paper to analyze the design of grey space form of landscape. Through analysis, we can see that the grey space design form proposed in this paper has a good effect in landscape design, and the design scheme proposed in this paper can meet the actual design needs.

The introduction of grey space evaluation method to analyze the layout of urban landscape architecture has made some progress. However, there is still room for improvement: the scope of future research may become increasingly broad. The limitations will also increase, which will have a certain impact on the accuracy of landscape information extraction. Further research should be conducted in this area.

References

1. La Sorte, F.A., Lepczyk, C.A., Burnett, J.L., Hurlbert, A.H., Tingley, M.W., Zuckerberg, B.: Opportunities and challenges for big data ornithology. Condor Ornithol. Appl. **120**(2), 414–426 (2018)
2. Gamache, R., Kharrazi, H., Weiner, J.P.: Public and population health informatics: the bridging of big data to benefit communities. Yearb. Med. Inform. **27**(01), 199–206 (2018)
3. Gupta, S., Altay, N., Luo, Z.: Big data in humanitarian supply chain management: a review and further research directions. Ann. Oper. Res. **283**(1), 1153–1173 (2019)
4. Wen, J., Yang, J., Jiang, B., Song, H., Wang, H.: Big data driven marine environment information forecasting: a time series prediction network. IEEE Trans. Fuzzy Syst. **29**(1), 4–18 (2020)
5. Manogaran, G., Lopez, D.: A survey of big data architectures and machine learning algorithms in healthcare. Int. J. Biomed. Eng. Technol. **25**(2–4), 182–211 (2017)
6. Rao, N.H.: Big data and climate smart agriculture-status and implications for agricultural research and innovation in India. Proc. Indian Natl. Sci. Acad. **84**(3), 625–640 (2018)
7. Kang, G.K., Gao, J.Z., Chiao, S., Lu, S., Xie, G.: Air quality prediction: Big data and machine learning approaches. Int. J. Environ. Sci. Dev. **9**(1), 8–16 (2018)
8. Ghernaout, D., Aichouni, M., Alghamdi, A.: Applying big data in water treatment industry: a new era of advance. Int. J. Adv. Appl. Sci. **5**(3), 89–97 (2018)
9. Thakur, S., Dharavath, R.: Artificial neural network based prediction of malaria abundances using big data: a knowledge capturing approach. Clinical Epidemiology Global Health **7**(1), 121–126 (2019)
10. Maya-Gopal, P.S., Chintala, B.R.: Big data challenges and opportunities in agriculture. Int. J. Agric. Environ. Inf. Syst. **11**(1), 48–66 (2020)
11. Lv, Z., Song, H., Basanta-Val, P., Steed, A., Jo, M.: Next-generation big data analytics: state of the art, challenges, and future research topics. IEEE Trans. Industr. Inf. **13**(4), 1891–1899 (2017)
12. Pencheva, I., Esteve, M., Mikhaylov, S.J.: Big data and AI–A transformational shift for government: so, what next for research? Public Policy Administr. **35**(1), 24–44 (2020)
13. Mavragani, A., Tsagarakis, K.P.: Predicting referendum results in the Big Data Era. J. Big Data **6**(1), 1–20 (2019)

Chapter 11
3D Animation Design and Production Based on Intelligent Algorithm and Virtual Reality

Honglin Huang

Abstract Three-dimensional animation production is an important technology in the animation industry at present. The traditional three-dimensional modeling of animation production is mainly carried out through the computer's early creative design and post production. VR (Virtual Reality) technology is a multi-viewpoint, real-time, and dynamic three-dimensional environment based on computer graphics, which can describe the real world and build a fictional world beyond reality. Three-dimensional animation can express the three-dimensional effect in virtual space without the limitation of place and time, which can not be expressed by two-dimensional animation. In the 3D animation production stage, using different forms of VR glasses in network systems or mobile screens, presenting works can achieve immersive and interactive characteristics, which can bring a stronger visual impact to people and enhance the infectivity of the screen. After further analysis and summary, a more perfect animation design is constructed. Under the influence of technology, users can interact at any time and watch the works from different angles, which will enhance the user's visual experience and improve the effect of animation production.

11.1 Introduction

The creative constraints on three-dimensional animation are becoming smaller and smaller. People can draw three-dimensional animation using design software to fully express their thoughts. Most 3D animation modeling methods acquire partial information about the surface of an animation scene. The traditional 3D modeling of animation production is mainly carried out through early creative design and post production through computers [1]. Three-dimensional animation can express three-dimensional effects in virtual space that cannot be expressed by two-dimensional animation, regardless of the constraints of location and time. It is greatly different

H. Huang (✉)
Guangzhou Institute of Foreign Economics and Technology, Guangzhou, China
e-mail: 574788437@qq.com

R. Kountchev et al. (eds.), *Multidimensional Signals, Augmented Reality and Information Technologies*, Smart Innovation, Systems and Technologies 374,
https://doi.org/10.1007/978-981-99-7011-7_11

from techniques such as shading and position processing in two-dimensional images. This three-dimensional effect is very realistic and can reflect the characteristics of animation from multiple angles. This approach is not only time-consuming, inefficient, and costly, but also difficult to achieve the desired quality and requirements of the completed animation [2].

Applying intelligent algorithms to group animation can remedy some shortcomings of keyframe technology. Therefore, this article integrates intelligent algorithms and VR into the design and production of three-dimensional animation to achieve human–computer interaction to meet the relevant needs of people's daily life and work learning [3]. After multiple technical processing, it can ensure the application value of data, in-depth analysis of the construction of adjustment mechanisms in the 3D animation production stage, and further analysis and summary to build a more complete animation production design. By shooting some still but continuous images at a certain speed, they are made into animated images. This method requires a lot of time and manpower to produce animation. Keyframe technology has improved the limitations of traditional animation production to some extent, but with the increasing demands of modern society on the autonomy and authenticity of animated characters, animation production based on keyframe technology cannot meet the expected requirements [4].

11.2 Application of VR in 3D Animation Production

The application scope of designing virtual scenes through VR is very wide, and the AR artificial intelligence technology that realizes scene transformation through artificial intelligence scene design is its main use [5]. Character animation is the core of three-dimensional production, and the production of character animation should pay attention to its motion state, as well as its movement and deformation in the process of movement. VR is not only a technical means, but also something deeper. Therefore, users always keep a passive state in the process of watching the works, and then obtain the corresponding information content in the works, which is the main difference between 3D animation technology and VR. Figure 11.1 shows the different actions and pose transitions of 3D animated characters.

The 3D animation modeling method based on VR can ensure good continuity in the case of more or less 3D animation scenes. In the process of applying technology, VR glasses are used to watch works, and the physical space is combined with projection equipment to enhance the user's perception experience effect in the process of watching movies. It is the use of modern computer equipment to mobilize human senses in many aspects, realize the effective cooperation of network equipment, and give people an immersive feeling [6]. At the same time, it is an exchange function and an interactive function in three-dimensional animation technology to deal with the reaction of participants to visual effects by applying computer perception technology. This is mainly because more 3D scene information can be obtained in 3D animation scene modeling with VR.

Fig. 11.1 Posture conversion of animated characters

11.3 Application of Intelligent Algorithms and VR in 3D Animation Production

11.3.1 Animation Production Concept Adjustment

People can draw three-dimensional animation using design software to fully express their thoughts. Most 3D animation modeling methods acquire partial information about the surface of an animation scene. The traditional 3D modeling of animation production is mainly carried out through early creative design and post production through computers [7]. In the process of animation production, technical personnel pay more attention to the realization of artists' ideas, control the process of image movement, complete the construction of rendering ideas, and ultimately express personal ideas on the quality of the image. The construction of a 3D scene generally involves several steps, such as model creation, material mapping, light settings, and rendering. The construction process is shown in Fig. 11.2.

3D animation design based on intelligent algorithms is closely related to VR itself. Three-dimensional animation can express three-dimensional effects in virtual space that cannot be expressed by two-dimensional animation, regardless of the constraints of location and time. It is greatly different from techniques such as shading and position processing in two-dimensional images. This three-dimensional effect is very realistic and can reflect the characteristics of animation from multiple angles of virtual worlds, and the process of using computers to create 3D software is the same

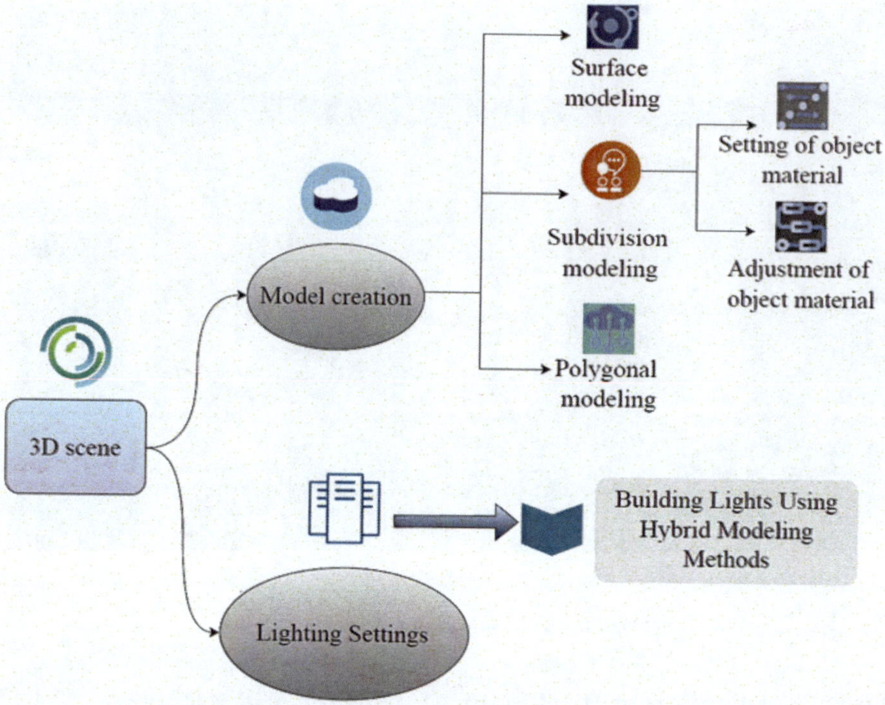

Fig. 11.2. 3D scene construction process

as the process of making 3D animation by computers [8]. Group all nodes in a 3D animation scene and describe the 3D animation scene as follows:

$$\begin{bmatrix} X_a \\ Y_a \\ Z_a \end{bmatrix} = \begin{bmatrix} a & x & u \\ 0 & 0 & 1 \\ 0 & a_y & u_0 \end{bmatrix} \tag{11.1}$$

In the formula, a, x, u, 0, u_0, and a_y all represent the coordinate points of the three-dimensional animation scene image.

Use intelligent algorithms and VR to obtain the scene scale of node groups in 3D animation scenes, with the formula as follows:

$$C_x = \sum_i a(i, k) \tag{11.2}$$

where C_x is the scene scale of the axis of x in the three-dimensional animation scene; $a(i, k)$ is the central pixel of the three-dimensional animation scene.

Three-dimensional animation can express the three-dimensional effect in virtual space without the limitation of place and time, which can not be expressed by two-dimensional animation. Therefore, this paper designs and produces three-dimensional animation based on intelligent algorithms and VR pictures more realistic and the environment more virtual. In the process, the visual communication changes from plane to three-dimensional, and the original creative concept of animation is adjusted.

11.3.2 Application of VR Products

With the continuous development of science and technology, and the increasing improvement of human living standards, some novel and magical film and television scenes, vivid television advertisements, vivid cartoons, and convenient and flexible product displays continue to pour into people's production and life, bringing different visual enjoyment and experience to people [9]. In the 3D animation production stage, using different forms of VR glasses in network systems or mobile screens, presenting works can achieve immersive and interactive characteristics, which can bring a stronger visual impact to people and enhance the infectivity of the screen.

By shooting some still but continuous images at a certain speed, they are made into animated images. This method requires a lot of time and manpower to produce animation. Keyframe technology has improved the limitations of traditional animation production to some extent, but with the increasing demands of modern society on the autonomy and authenticity of animated characters, animation production based on keyframe technology cannot meet the expected requirements. Using VR glasses can project content onto the screen, making the screen more realistic and effective, and also achieving the goal of alleviating user eye fatigue [10]. Using VR to represent the geometric information of 3D animation, define the surface features of objects in 3D animation, take the average value of image pixel points in the 3D animation scene, and obtain an image $I(x, y)$ composed of the same pixel points of the 3D animation scene image:

$$I(x, y) = \text{avg}(i, k) \tag{11.3}$$

where avg is the neighborhood of the animation scene image. Searching for the optimal transformation relationship between image point sets and point sets of three-dimensional animation scenes, and obtaining depth information f of all scenes in three-dimensional animation;

$$f = \frac{1}{N}\|y_i - (r + t)\| \tag{11.4}$$

where N is the total number of points in the neighborhood of the animation scene image table; r is the feature point of the animation scene image; y_i is a three-dimensional vector sample point; t is the point cloud data merging factor of the 3D animation scene.

With the rapid development of computer hardware, the creative constraints on three-dimensional animation are becoming smaller and smaller. People can draw three-dimensional animation using design software to fully express their thoughts. Most 3D animation modeling methods acquire partial information about the surface of an animation scene. The traditional 3D modeling of animation production is mainly carried out through early creative design and post production through computers, they have many complementary places. Motion capture technology mainly includes distributed network virtual platform, stereo projection system, motion capture system, and interactive input device. This technology has strong participation so that the experiencer can see the specific scene during roaming. Character animation is the core of three-dimensional production, and the production of character animation should pay attention to its motion state, as well as its movement and deformation in the process of movement. For the animation that can't be shot by an ordinary camera, we can fully coordinate the relationship between directional dynamics and pre-dynamics by using motion capture technology and a three-dimensional modeling system.

11.3.3 The Formation of Three-Dimensional Visual Elements

Under the background of modern technology, the number of animation presentation forms has increased, especially after the formation of three-dimensional spatial planning, and the use of VR to complete real spatial construction. In the process of using VR for three-dimensional animation production, attention should be paid to effectively combine different production methods, and at the same time, attention should be paid to reasonable analysis and application of data information obtained through the system. In the process of animation production, VR is introduced, and object tracking systems are used to achieve real-time tracking of object motion status. Effective data information is tracked according to actual work needs. Transform the visual expression of animation spatial planes into three-dimensional expression, and present three-dimensional animation in a comprehensive manner. Build an independent virtual space, utilize the perspective difference between the two eyes, leave rich imagination space to the brain, and create a brand new virtual space with far and near dimensions. The difference in perspective between the two eyes is shown in Fig. 11.3.

In order to ensure that the data can meet the production needs, the adjustment techniques in VR should be reasonably used to enhance the authenticity of the scene. Users can interact at any time under the influence of technology, view the work from different angles, and enhance the user's visual experience, ultimately significantly improving the effectiveness of animation production. To ensure that the audience can effectively participate in the interactive process and obtain a good sensory experience, the effective combination of personal feelings and environmental atmosphere, and the analysis of the user's real experience, is a link that cannot be ignored in the practical work stage.

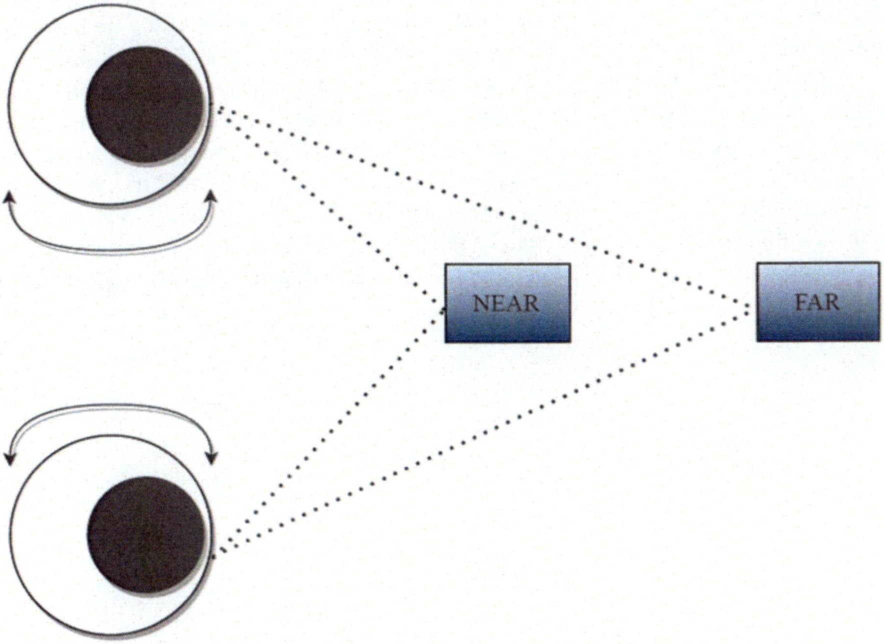

Fig. 11.3 Difference in visual angle between two eyes

In addition to the similarities between VR and 3D animation design, there are many complementary aspects between the two. Motion capture technology mainly includes distributed network virtual platforms, stereo projection systems, motion capture systems, and interactive input devices. This technology has strong participation, allowing users to see specific scenes during the roaming process, which is mainly manifested in the level of interactivity, VR includes various technologies such as internet technology, 3D graphics technology, and hardware sensors, among which people have a unique sense of realism and immersion. And I feel a stronger sense of participation, greatly narrowing the distance between the experimenter and the screen, effectively narrowing the distance between the experimenter and the screen.

11.4 Simulation Experiment

Character animation is the core of three-dimensional production, and the production of character animation should pay attention to its motion state, as well as its movement and deformation in the process of movement. VR is not only a technical means, but also something deeper. By shooting some still but continuous images at a certain speed, they are made into animated images. This method requires a lot of time and manpower to produce animation. In order to verify the reliability of this method in

3D animation design, the accuracy of coordinate values between this method and traditional methods is compared. The experimental results are shown in Table 11.1.

From the data in Table 11.1, it can be seen that the method in this paper has a high accuracy in obtaining three-dimensional animation coordinate values, with a maximum accuracy of 94.4%. The traditional method has a maximum accuracy of 88.7% in obtaining three-dimensional animation coordinate values. Through comparison, it can be seen that the method in this paper is superior to the traditional method, with certain feasibility and significance in this field. To verify the effectiveness of the algorithm in this article, first select three functions for testing, and the experimental functions are as follows:

$$f_1 : \sum_{i=1}^{n} x_i^2 - 10\cos \tag{11.5}$$

The search range is [−10,10], and the minimum value of 0 is obtained at (0, …, 0).

$$f_2 : \sum_{i=1}^{n} x_i^2 - \prod_{i=1}^{n} \cos \tag{11.6}$$

The search range is [−600,600], and the minimum value of 0 is obtained at (0, …, 0).

$$f_3 : \sum_{i=1}^{n-1} x_{i+1}^2 - x_i \tag{11.7}$$

The search range is [−100,100], and the minimum value is 0 at (1, …, 1).

The evolutionary comparison diagrams of the experiments on three functions by the algorithm in reference [4], the algorithm in reference [6] and the algorithm in this paper, respectively, where the abscissa is evolutionary algebra and the ordinate is the common logarithm of the average value of 50 experimental functions. The evolutionary comparison diagram is shown in Figs. 11.4, 11.5, and 11.6.

By comparing the experiments in Figs. 11.2, 11.3 and 11.4, it can be seen that this method not only effectively solves the problem, but also effectively improves the solving speed. After various technical processing, it is necessary to ensure that

Table 11.1 Precision comparison of coordinate values obtained by different methods

Number of experiments	Method in this article	Traditional method
1	90.6	84.2
2	94.4	88.4
3	88.6	85.6
4	93.5	88.7

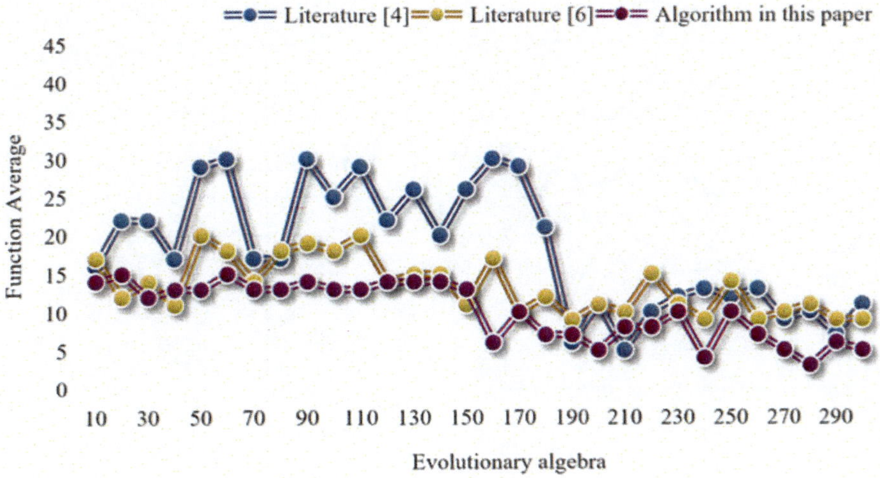

Fig. 11.4 Evolution of the average optimal value of function f_1

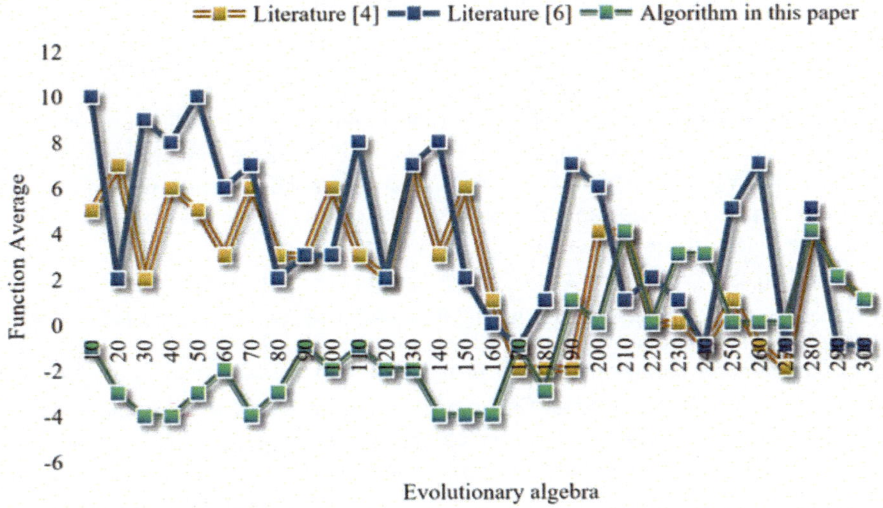

Fig. 11.5 Evolution of the average optimal value of function f_2

these data have practical application value. The adjustment mechanism during the 3D animation production stage has been thoroughly analyzed and summarized, thus establishing a more comprehensive animation production design. Applying intelligent algorithms to group animation can remedy some shortcomings of keyframe technology. Therefore, this article integrates intelligent algorithms and VR into the design and production of three-dimensional animation to achieve human–computer interaction to meet the relevant needs of people's daily lives and work learning.

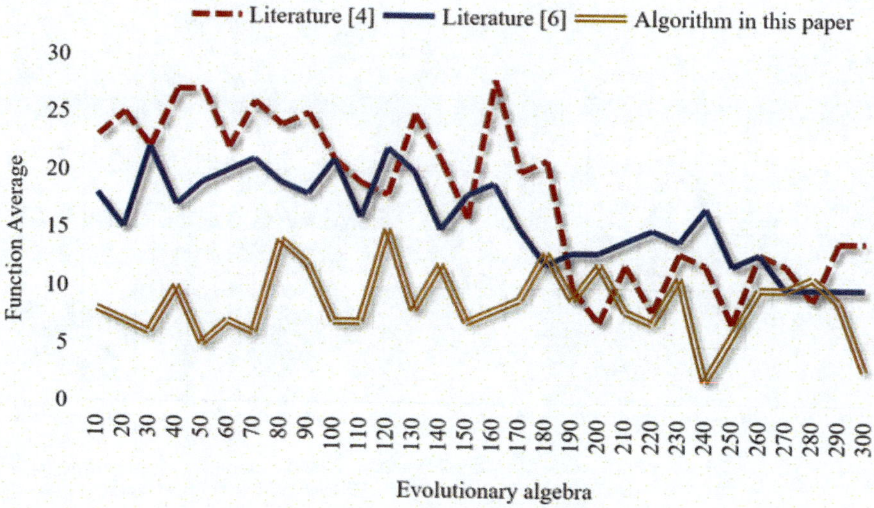

Fig. 11.6 Evolution of the average optimal value of function f_3

11.5 Conclusions

In practice, we should combine the actual development of the industry, and do a good job of international exchanges and cooperation. The three-dimensional animation made by VR has realistic images and scenes and has strong real-time and interactive. Apart from the similarity between VR and three-dimensional animation design, they have many complementary places. Motion capture technology mainly includes distributed network virtual platform, stereo projection system, motion capture system, and interactive input device. This technology has strong participation so that the experiencer can see the specific scene during roaming. Motion capture technology mainly includes a distributed network virtual platform, stereo projection system, motion capture system, and interactive input equipment. This technology has strong participation, which enables the experiencer to see specific scenes during roaming, mainly at the interactive level. VR includes Internet technology, three-dimensional graphics technology, and hardware sensors, which bring people a unique sense of realism and immersion, and at the same time, the optimization speed has been greatly improved. After many times of technical processing, the application value of data can be guaranteed, and the adjustment mechanism construction of the 3D animation production stage is deeply analyzed. After further analysis and summary, a more perfect animation production design is constructed.

11.6 Fund Project

2021 Guangdong Higher Vocational Computer Education Instruction Committee Education and Teaching Reform Research and Practice Project "Innovative Practice Research on the Hybrid Teaching Mode Based on" Internet plus Dichotomy "in Animation Teaching" (Project No.: JSJJZW2021053).

References

1. Wang, H.: Landscape design of coastal area based on virtual reality technology and intelligent algorithm. J. Intell. Fuzzy Syst.: Appl. Eng. Technol. **37**(5), 49–62 (2022)
2. Bao, W.: The application of intelligent algorithms in the animation design of 3D graphics engines. Int. J. Gaming Comput.-Mediated Simul. **13**(2), 26–37 (2021)
3. Liu, X.: Three-dimensional visualized urban landscape planning and design based on virtual reality technology. IEEE Access **58**(99), 27–56 (2022)
4. You, Z., Huang, J., Xue, J., et al.: A multiplayer virtual intelligent system based on distributed virtual reality. Int. J. Pattern Recognit. Artif. Intell. **64**(14), 35–47 (2021)
5. Yi, D., Wang, Q.: Intelligent optimization algorithm in virtual design of landscape VR. Springer, Cham, vol. 11, no. 27, pp. 11–24 (2022)
6. Cao, X.: Three-dimensional image art design based on dynamic image detection and genetic algorithm. J. Intell. Fuzzy Syst.: Appl. Eng. Technol. **42**(4), 37–48 (2021)
7. Kang, C., Wang, S., Ren, W., et al.: Optimization design and application of active disturbance rejection controller based on intelligent algorithm. IEEE Access **37**(18), 21–48 (2022)
8. Gan, B., Zhang, C., Chen, Y., et al.: Research on role modeling and behavior control of virtual reality animation interactive system in Internet of Things. J. Real-Time Image Process. **18**(1), 1–15 (2020)
9. Zou, S., Chen, H., Zhou, H., et al.: An intelligent image feature recognition algorithm with hierarchical attribute constraints based on weak supervision and label correlation. IEEE Access **42**(99), 12–27 (2020)
10. Xu, L.: Fast modelling algorithm for realistic three-dimensional human face for film and television animation. Complexity **21**(2), 1–10 (2021)

Chapter 12
4DFA: Four-Dimensional Full-Anatomy Reconstruction of Individualized Digital Human Models Based on Motion Videos

Rui Zhao, Jiachen Mi, Yanxin Jiang, Zhefu Chen, and Hongkai Wang

Abstract Reconstructing personalized models of the human body is a key step for digital twin modeling of sports motion. There have been many studies reconstructing the body surface models based on video sequences, but these algorithms lack the modeling of internal structures (bones, muscles, blood vessels, etc.). Some researchers map internal organ structures into the 3D point cloud scanning of the body surface, but such mapping does not guarantee realistic tissue deformation and is not developed for video processing. This paper focuses on the video-based reconstruction of a personalized full-anatomy digital model in sports motion. Our method creates a four-dimensional (4D) model of the moving process composed of the 3D models of each time moment. We first use a deep network to regress the body surface of each time point and then estimate the internal structures by registering a previously constructed deformable human anatomy atlas to the body surfaces. To mimic realistic internal structure deformation, we applied articulated linear transform to the bony structures to avoid bone shape distortion and applied smooth nonlinear transform to the soft tissue to mimic the motion deformation. We also developed an intersection removal method to prevent the possible intersection between bones and soft tissues during the deformation. Our method is tested with athletes' sports videos and automatically generates full-anatomy motion models which cannot be achieved by previous methods. Our method also surpasses the existing video-based body

R. Zhao · J. Mi · Y. Jiang · H. Wang (✉)
School of Biomedical Engineering, Faculty of Medicine, Dalian University of Technology, Dalian 116024, China
e-mail: wang.hongkai@dlut.edu.cn

J. Mi
INTESIM(Dalian)co, LTD, Dalian 116023, China

Z. Chen (✉)
School of Kinesiology and Health Promotion, Dalian University of Technology, Dalian 116024, China
e-mail: chenzhefu@dlut.edu.cn

H. Wang
Liaoning Key Laboratory of Integrated Circuit and Biomedical Electronic System, Liaoning, China

R. Kountchev et al. (eds.), *Multidimensional Signals, Augmented Reality and Information Technologies*, Smart Innovation, Systems and Technologies 374,
https://doi.org/10.1007/978-981-99-7011-7_12

surface modeling method by providing additional internal structures. The 4D human motion model constructed by our method is useful in sports modeling, biomechanics simulation, wearable device design, etc.

12.1 Introduction

In the field of kinesiology and health promotion, digital human models are widely used for biomechanical simulation of body motion, with the purpose of sports injury prediction, muscle load assessment, etc. Although simulation based on a generic human model has been studied for decades, personalized simulation using a digital twin model is just an emerging field. The simulation of individual motion not only requires accurate modeling of personal anatomy, but also demands precise capturing of the body movement. Thanks to the fast development of deep learning techniques in recent years, individualized motion modeling based on personal sports videos has become feasible.

So far, a series of deep learning algorithms have been proposed to address the above challenges. Most of these methods register deformable human surface models (such as SPACE [1] and SMPL [2]) to the sports photo or video sequence. Early methods in this field first infer bone joint positions from the 2D images and then use the joints to predict the 3D body surface model. Later, Kanazawa et al. [3] proposed an end-to-end method that realizes direct image-to-model mapping. There were also methods focusing on body part pose estimation such as the trunk and limbs [4, 5], hand posture [6, 7], and facial expression [8, 9]. Rong et al. [10] proposed a modular system that integrates the regression outputs of three parts to unify the output of whole-body pose estimation. Xiu et al. [11] introduced the normal graph and made geometric constraints on the body to improve the pose-level modeling accuracy.

Despite the fast progress in deep learning-based motion modeling, the current methods only generate body surfaces. However, biomechanical simulation requires not only the external surface, but also the internal structures like bones and muscles. To tackle this problem, some studies register a human atlas with internal anatomy to the body surface scan data or volumetric tomographic images. For example, Keller et al. [12] inferred the skeleton from a 3D surface model in an arbitrary pose, Ali-Hamadi et al. [13] combined the fat distribution and other information inferred from MRI data, and Sugar et al. [14] calculated the deformation of anatomical structures based on the displacement transformation of the predefined virtual skeleton in different poses [15–17]. However, these methods require complex data acquisition (surface scanning or tomographic imaging) which is not suitable for fast motion capturing. Moreover, these approaches use nonlinear spatial transform to map the internal structures into the individual body, resulting in unrealistic tissue deformation for large posture changes.

In summary, state-of-the-art (SOTA) deep learning methods only model the body surface without internal structures, while the current internal anatomy modeling approaches are not developed for video data and are prone to unrealistic tissue deformation. Our study aims to address these problems with the following key features:

(i) Our method combines the deep learning-based body surface model with the internal structure mapping of a whole-body anatomy atlas. We for the first time realizes full-anatomy modeling based on simple sports video.
(ii) We use a previously constructed deformable human atlas trained with thousands of computed tomographic images for internal structure modeling. The *prior* knowledge of internal anatomy from a large population dataset ensures reasonable estimation of unseen organs from the external surface.
(iii) To simulate the body motion, we apply articulated rigid transform and smooth nonrigid deformation to the bones and soft tissues, respectively. Unlike the existing methods which apply nonrigid deformation to all the structures, our method avoids anatomically implausible distortion of the bones and realizes anatomically realistic motion deformation of the whole-body anatomy.
(iv) We also develop a method to remove the possible intersections between the bones and the soft tissue during large pose changes. The generation of an intersection-free model is crucial for subsequent finite element (FEM) simulation of the motion kinetics.

12.2 Materials and Methods

12.2.1 3D Digital Human Anatomical Model

In this study, we used a deformable digital human atlas [18] constructed by our group based on 1063 whole-body CT images. Anatomical structures including the skin, bones, muscles, torso organs, and great vessels were segmented from the CT images, and the inter-subject anatomical variations were learned via the statistical shape model (SSM) method. With *prior* anatomical knowledge learned from large population data, this atlas can model the shape correspondence between the external skin and internal structures, facilitating the estimation of internal anatomy in the subsequent steps.

12.2.2 Motion Armature

To endow our human atlas with posture-changing ability, this study defines an articulated motion armature with the armature joints located at the skeleton joint centers. For the balance between the modeling reality and the computation complexity, the motion armature is defined with 26 individual segments. The control range of each armature segment is defined via the closest point searching between the model vertices and the armature segments. Figure 12.1 illustrates the deformable human

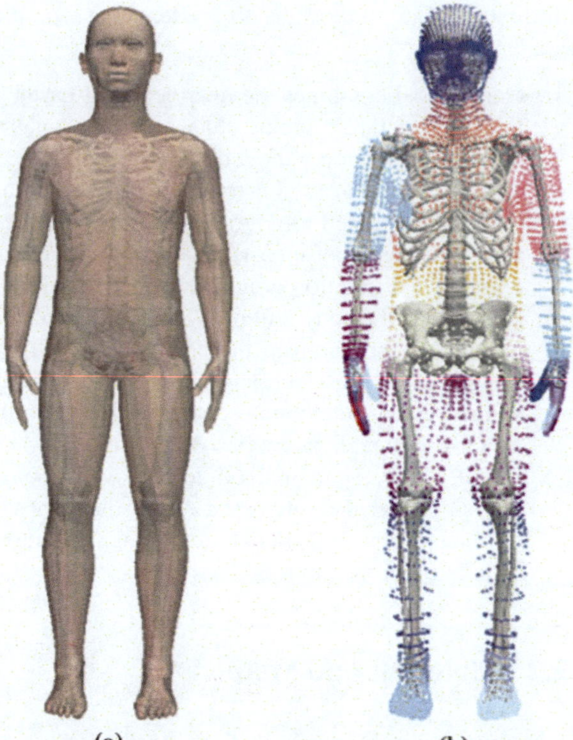

Fig. 12.1 Deformable human anatomy atlas with articulated motion armature. (a) The atlas in standing pose with transparent skin rendering to see the internal structures. (b) Control range of each armature segment, rendered with different pseudo colors for different segments

atlas with internal organs and the defined articulated armature with the individual segment control ranges.

12.2.3 Personalized Pose Deformation

Figure 12.2 shows the workflow of video-based full-anatomy motion modeling. First, the video sequence is input to a deep neural network for body surface pose estimation of every single moment. Then, for every single moment, the articulated bone rotation of every armature segment is obtained by computing the rigid spatial transform (including translation and rotation) from the standing pose to the motion pose of this segment. Based on the obtained transform, articulated rigid and nonrigid deformations are applied to the bones and soft tissues respectively. Since the bones and the soft tissues are deformed using different types of spatial transforms, the possible intersection between them may occur, thus the intersection removal step is finally performed to correct possible intersections.

For video-based body surface estimation, we adopt the deep neural network model used for 3D human pose modeling [19]. This network uses the Resnet50 [20] structure

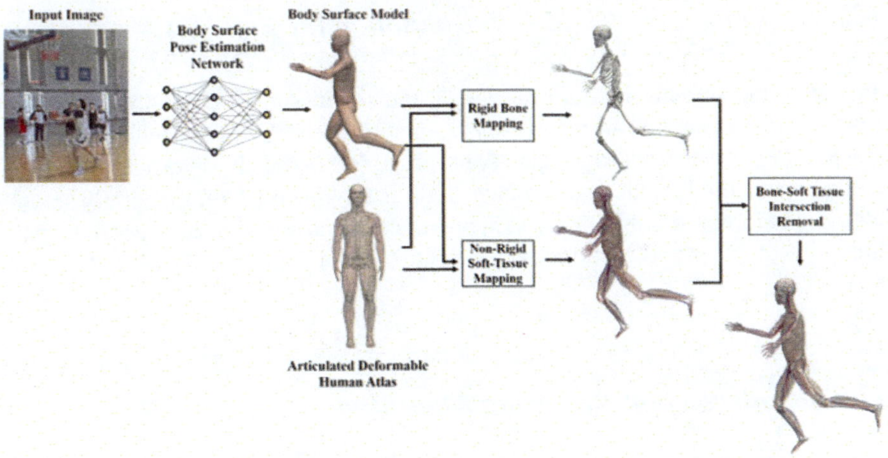

Fig. 12.2 Workflow of video-based full-anatomy motion modeling

as the image encoder to compute the spatial features of each time frame and then uses a two-layer gated recurrent unit with a hidden size of 1024 to compute the temporal features, followed by two fully connected layers with 1024 neurons to regress the pose and body shape parameters of the SMPL model. The reconstructed SMPL model is used to get the 3D joint points and 2D projection coordinate points. The loss function used for training is

$$L_g = L_{3D} + L_{2D} + L_{SMPL} + L_{adv} \tag{12.1}$$

where L_{3D}, L_{2D}, L_{SMPL}, and L_{adv} are the losses of 3D joint points, 2D projection coordinate points, and the regressed SMPL's pose and shape parameters losses, respectively.

12.2.4 Articulated Rigid Bone Motion

The articulated rigid bone transform is applied in a bottom-to-top manner following the parent–child joint inheritance relationship in the moving armature. The child joints with the lowest generation are transformed first, and then its relative parent joints are transformed, with all the children joints of this parent joints transformed simultaneously. This process is recurrently applied to all the armature segments and the articulated posing of the entire body is realized.

12.2.5 Smooth Nonrigid Deformation of the Soft Tissues

To mimic the smooth nonrigid motion of the soft tissues, we use the Centers of Rotation (CoR) method instead of the commonly used Linear Blending Skinning (LBS). The CoR method was originally proposed for solving the deformation artifacts of the LBS method for the body surface. In this study, we use CoR to produce smooth nonrigid deformation of both the skin and the internal soft tissues. Due to the page limitation of this paper, we refer the readers to [16] for detailed principles of the CoR method.

12.2.6 Intersection Removing Algorithm

In the process of digital human pose deformation, the intersection between the soft tissue and bones may be caused because the bones and soft tissues are deformed using different types of transforms. It is necessary to take intersection removal operations after the above deformation process to eliminate the possible surface crossing and generate a model available for FEM simulation.

Let the two surfaces that need to do intersection removal be S1 and S2, and both must ensure that they are closed mesh surfaces in 3D space. The intersection-removing process we developed is as follows:

(i) Iterate through all vertices on S_1 and detect whether each vertex is inside S_2. To do this, let any vertex $v \in S_1$, make n rays from v, and check the number of intersection points of each ray with S_2. If the number of intersection points between all n rays and S_2 is odd, it means that v is inside S_2, otherwise, it is outside. In this project, $n = 10$ can satisfy all test cases.

(ii) Find the point to be adjusted. If the object to be detected is between soft tissue and bone or between bone and bone, the point to be adjusted is the point in S_1 that is inside S_2; if the object to be detected is between soft tissue and skin, the point to be adjusted is the point in S_1 that is outside S_2.

(iii) For each point v^* to be adjusted, through v make vertical lines for all triangular faces in S_2 and record the distance from v^* to the vertical foot if the footfalls inside the triangle. Take the shortest one among all distances as the shortest distance from v^* to S_2, and then take the vertical foot o corresponding to the shortest distance as the target point of v^*, then the coordinates of v^* should be updated as

$$v^* \leftarrow 1.5 \bullet (o - v^*) \tag{12.2}$$

(iv) Repeat the process described in (i)–(iii) above until no point to be adjusted is detected.

(v) Based on the result of (iv), do the faces cross detection? Iterate through all the triangular faces on S_1 and detect their intersection relationship with all the faces on S_2. Specific method: set t_1 as any face of S_1, and detect the cross relationship between its three edges and t_2 of any face of S_2. As long as any one of the three edges intersects with t_2, t_1 is considered to intersect with t_2.

(vi) For the vertices of all faces on S_1 that intersect S_2, adjust their positions as in (ii), (iii), and (iv).

(vii) Repeat the process of (v)–(vi) until no intersecting faces are detected.

12.3 Result and Analysis

Table 12.1 compares our method with several existing methods, including modeling internal anatomical structures, a 3D human reconstruction based on videos or medical images, and complete automation of the reconstruction algorithm. In the table Framkmocap [10] and Vibe [19] are methods to reconstruct body surface models based on video sequences, where Framkmocap effectively integrates modules for hand pose and face expression estimation and outputs full body pose estimation results; Vibe proposes a grid structure based on time series, using an adversarial learning framework to generate realistic and reasonable human actions without real 3D labels. We used a deep learning method with high accuracy as the first step to reconstruct the human skin model. Anatomy transfer [14] proposed a semi-automatic method to estimate the internal anatomical structure by combining MRI images, while OSSO [12] combined the DXA detection results to generate a skin model that fits the parametric body model to achieve a 3D body skin to internal skeletal mapping.

Our proposed method is based on the digital atlas of personalized human whole-body anatomy, combined with the topological structure of human motion armature, to automatically rotate the joints and deform the posture of the anatomical structure of the whole body. At the same time, it solves the unrealistic distortion during large pose deformation and is more suitable for the reconstruction of 3D models with a large number of motion moments during the complete motion.

Table 12.1 Comparison between our method and existing methods

Method	Skin	Bone	Internal anatomy	Input data	Automation level
Framkmocap [10]	√	×	×	Video	Automatic
Vibe [19]	√	×	×	Video	Automatic
Anatomy transfer [13]	√	√	√	MRI	Semi-automatic
OSSO [12]	√	√	×	DXA	Semi-automatic
Ours	√	√	√	Video	Automatic

12.3.1 Realistic Anatomical Structure Deformation

In the process of digital human overall posture deformation, we use the Center of Rotation algorithm to simulate the real flexible soft tissue deformation, which solves the volume loss at the joints. At the same time, due to the inability to strictly simulate the complex rotation and sliding processes of human joints, human posture deformation may cause an intersection between adjacent anatomical structures. Among the currently available studies. Keller et al. [12] and Ali-Hamadi et al. [13] obtained visually plausible results for the postural transformation of human anatomical structures. However, the limitation is that the results still have some slight skin interpenetration or overlapping of internal tissues. For this reason, we developed a fully automated de-crossing algorithm to prevent possible crossings of bones and soft tissues during the deformation process.

Figure 12.3 compares the results after skin optimization, using the skin at the arm joint as an example. The left figure shows the results of the commonly used linear blending skinning (LBS), and the right figure shows the results of the COR skinning method we applied. The comparison shows that COR improves the volume loss problem when skinning the LBS, which has obvious artifact depressions in both wrist and shoulder, and COR solves most of the artifacts introduced in the LBS and improves the mesh quality. Our proposed method is based on the digital atlas of personalized human whole-body anatomy, combined with the topological structure of human motion armature, to automatically rotate the joints and deform the posture of the anatomical structure of the whole body. At the same time, it solves the unrealistic distortion during large pose deformation and is more suitable for the reconstruction of 3D models with a large number of motion moments during the complete motion.

Figure 12.4 compares the results of our method with those of existing studies in which 3D point cloud scans based on the body surface achieve mapping of internal structural structures. Our method uses both rigid and flexible spatial deformation for bones and soft tissues respectively, which avoids unrealistic distortion deformation during the simulation of postural deformation. The results show that our method

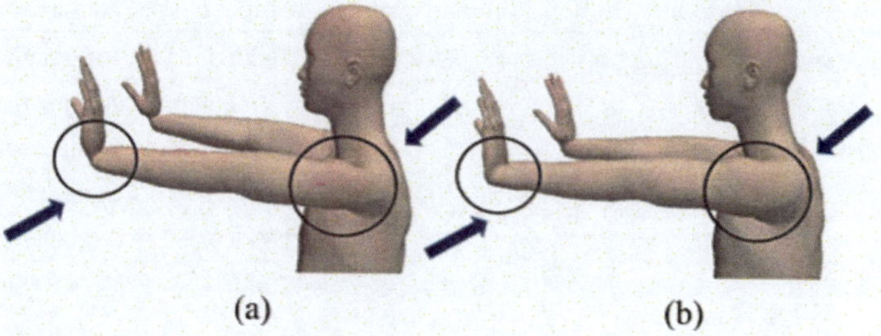

Fig. 12.3 The results of smooth deformation. **a** Linear blending skinning results with obvious artifacts at the joints; **b** Our realistic deformation result

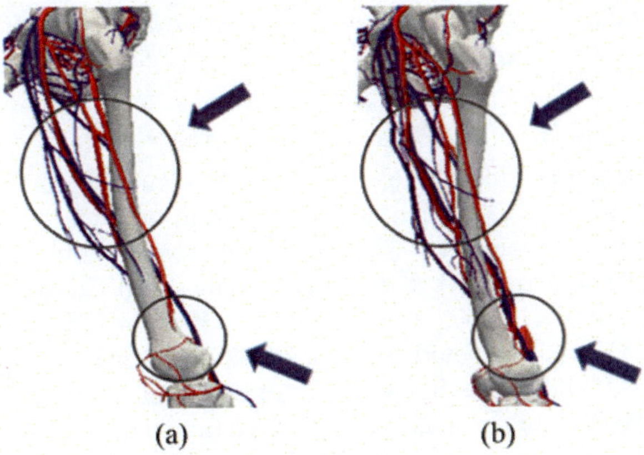

Fig. 12.4 The results of internal anatomical structure mapping. **a** Our method produces smooth soft tissue deformation and rigid bone transform; **b** The existing nonlinear transformation method may produce distortions of soft tissue and bones

produces more reasonable internal anatomical structure postural deformation results so that no unreasonable shape distortion occurs in bones and blood vessels.

We developed a fully automated de-crossing algorithm to prevent possible intersections between bones and soft tissues during deformation. Figure 12.5 shows the results of our fully automated de-crossing algorithm for an intersection between bones and soft tissues during deformation. Here the intersection between blood vessels and bones near the knee joint is used as an example, while the whole de-crossing process takes about 8 min to run on a laptop with an i7 CPU.

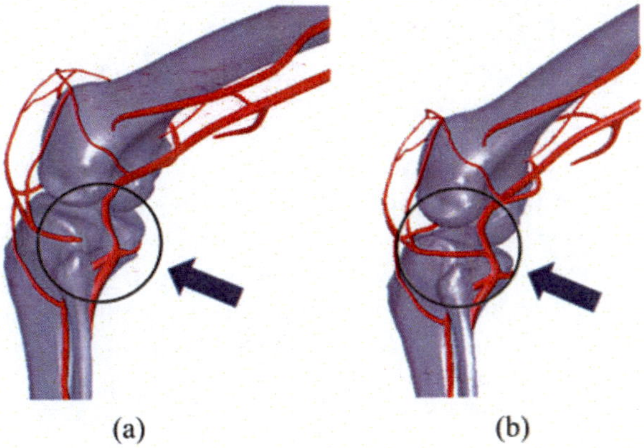

Fig. 12.5 The results of the intersection removal between the blood vessels and bones. **a** Before intersection removal; **b** After intersection removal

12.3.2 Adaptive Modeling Results of Whole-Body Anatomical Structures

Our method is based on the digital atlas of personalized human body anatomy structure, designing a whole set of human motion digital modeling methods. We reconstruct the 3D human skin model from the video, combine human motion armature topology for joint rotation and posture deformation of whole-body anatomical structures, where the skeleton and soft tissue are rigid and flexible deformation respectively, and introduce the de-crossing mechanism. Finally, reasonable internal anatomical structure and posture deformation structure are generated.

Figure 12.6a shows a series of action images acquired from a video sequence. (b) The 3D human model reconstructed from the motion images (c) The human anatomy model corresponding to each motion instant. Arteries and veins, soft tissues, and bones are shown in different colors, respectively. (d) Local zoomed-in view of the model. The human anatomy after postural transformation is accurate and without unreasonable distortion.

The limitation of the current study is that it is not possible to quantitatively assess the accuracy of the postural deformation results of the whole-body anatomical structure, but the obtained postural deformation results are visually plausible. These deformation results have no irrational shape distortion among bones, muscles, and internal organs, and no geometric intersection between adjacent anatomical structures, which can provide digital models for subsequent human finite element simulation or biomechanical simulation for simulation testing, thus effectively meeting the needs of different applications.

12.4 Conclusion

In this paper, we present an algorithm for video-based full-anatomy reconstruction of personalized digital twin human models in motion. We use a deep neural network to produce the body surface models of each time moment and develop an internal anatomy method by registering a deformable digital human anatomy atlas to the body surfaces. The integrated articulated deformation of bones and soft tissues produces anatomically realistic models of the moving anatomy, and the intersection removal method generates high-quality mesh for subsequent FEM analysis. The method in this paper provides a novel technical tool for digitally evaluating motion results and performing personalized motion mechanics simulations.

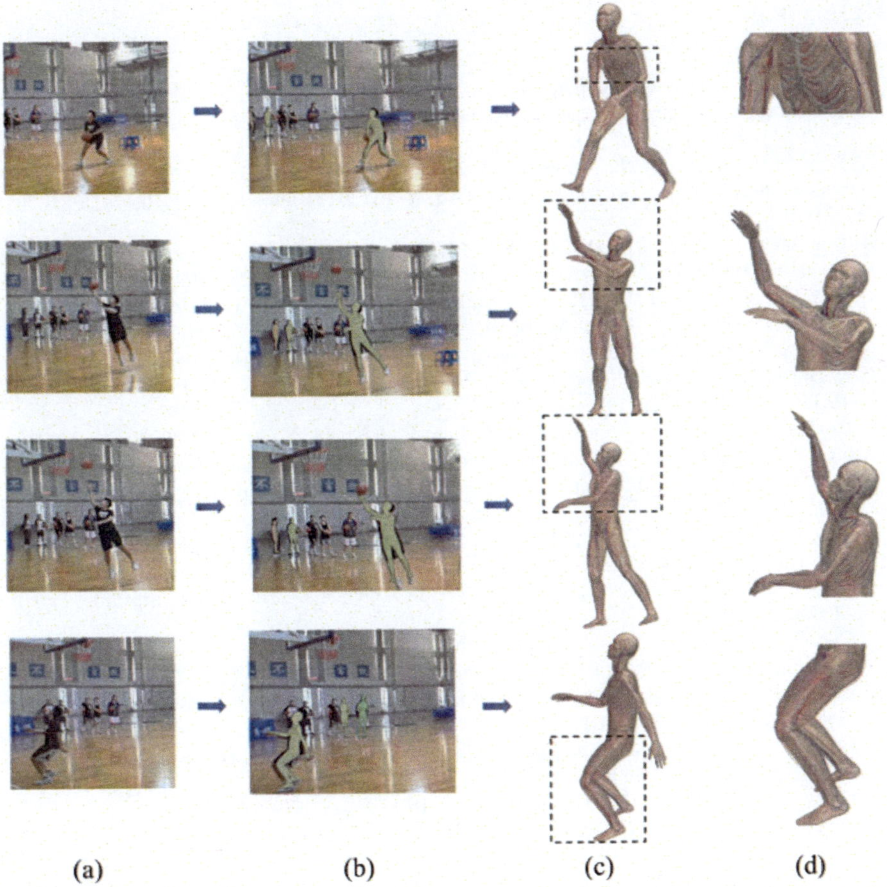

Fig. 12.6 Video sequence-based modeling of full body anatomy for each moment. **a** A series of action images acquired from the video sequence. b The body surface model reconstructed for each moment. **c** The human resultant anatomical model corresponding to each moment. Arteries and veins, soft tissues, and bones are shown in different colors. **d** Local zoomed-in view of the dashed boxes in (**c**)

Acknowledgements This work was supported in part by the National Key Research and Development Program No. 2020YFB1711500, 2020YFB1711501, and 2020YFB1711503, the general program of the National Natural Science Fund of China (No. 81971693, 61971445), the funding of Dalian Engineering Research Center for Artificial Intelligence in Medical Imaging, Hainan Province Key Research and Development Plan ZDYF2021SHFZ244, the Fundamental Research Funds for the Central Universities (No. DUT22YG229), the funding of Liaoning Key Lab of IC & BME System and Dalian Engineering Research Center for Artificial Intelligence in Medical Imaging.

References

1. Anguelov, D., Srinivasan, P., Koller, D., Thrun, S., Rodgers, J., Davis, J.: SCAPE: shape completion and animation of people. ACM Trans. Graph. **24**(3), Art. no. 3, (2005). https://doi.org/10.1145/1073204.1073207
2. Loper, M., Mahmood, N., Romero, J., Pons-Moll, G., Black, M.J.: SMPL: a skinned multi-person linear model. ACM Trans. Graph. **34**(6), 1–16 (2015). https://doi.org/10.1145/2816795.2818013
3. Kanazawa, A., Black, M.J., Jacobs, D.W., Malik, J.: End-to-End recovery of human shape and pose. In: 2018 IEEE/CVF Conference on Computer Vision and Pattern Recognition, Salt Lake City, UT, pp. 7122–7131 (2018). https://doi.org/10.1109/CVPR.2018.00744
4. Lassner, C., Romero, J., Kiefel, M., Bogo, F., Black, M.J., Gehler, P.V.: Unite the people: closing the loop between 3D and 2D human representations. In: 2017 IEEE Conference on Computer Vision and Pattern Recognition (CVPR), Honolulu, HI, pp. 4704–4713 (2017). https://doi.org/10.1109/CVPR.2017.500
5. Bogo, F., Kanazawa, A., Lassner, C., Gehler, P., Romero, J., Black, M.J.: Keep it SMPL: automatic estimation of 3D human pose and shape from a single image. In: Computer Vision – ECCV 2016, Springer International Publishing, pp. 561–578 (2016). https://doi.org/10.1007/978-3-319-46454-1_34
6. Kolotouros, N., Pavlakos, G., Black, M., Daniilidis, K.: Learning to reconstruct 3D human pose and shape via model-fitting in the loop. In: 2019 IEEE/CVF International Conference on Computer Vision (ICCV), Seoul, Korea (South), pp. 2252–2261 (2019). https://doi.org/10.1109/ICCV.2019.00234
7. Cai, Y., Ge, L., Cai, J., Yuan, J.: Weakly-supervised 3D hand pose estimation from monocular RGB images. In: Ferrari, V., Hebert, M., Sminchisescu, C., Weiss, Y. (eds.) Computer Vision—ECCV 2018, vol. 11210. Springer International Publishing, Cham, pp. 678–694 (2018). https://doi.org/10.1007/978-3-030-01231-1_41
8. Ge, L., et al.: 3D hand shape and pose estimation from a single RGB image. In: 2019 IEEE/CVF Conference on Computer Vision and Pattern Recognition (CVPR), Long Beach, CA, USA, pp. 10825–10834 (2019). https://doi.org/10.1109/CVPR.2019.01109
9. Feng, Y., Wu, F., Shao, X., Wang, Y., Zhou, X.: Joint 3D face reconstruction and dense alignment with position map regression network. In: Computer Vision—ECCV 2018. Springer International Publishing, pp. 557–574 (2018). https://doi.org/10.1007/978-3-030-01264-9_33
10. Rong, Y., Shiratori, T., Joo, H.: FrankMocap: a monocular 3D whole-body pose estimation system via regression and integration. In: 2021 IEEE/CVF International Conference on Computer Vision Workshops (ICCVW), Montreal, BC, Canada, Oct. 2021, pp. 1749–1759 (2021). https://doi.org/10.1109/ICCVW54120.2021.00201
11. Xiu, Y., Yang, J., Tzionas, D., Black, M.J.: ICON: implicit clothed humans obtained from normals. In: 2022 IEEE/CVF Conference on Computer Vision and Pattern Recognition (CVPR), New Orleans, LA, USA, pp. 13286–13296 (2022). https://doi.org/10.1109/CVPR52688.2022.01294
12. Keller, M., Zuffi, S., Black, M.J., Pujades, S.: OSSO: obtaining skeletal shape from outside. In: 2022 IEEE/CVF Conference on Computer Vision and Pattern Recognition (CVPR), New Orleans, LA, USA, pp. 20460–20469 (2022). https://doi.org/10.1109/CVPR52688.2022.01984
13. Ali-Hamadi, D., et al.: Anatomy transfer. ACM Trans. Graph. **32**(6), 1–8 (2013). https://doi.org/10.1145/2508363.2508415
14. Sujar, A., Casafranca, J.J., Serrurier, A., Garcia, M.: Real-time animation of human characters' anatomy. Comput. Graph. **74**, 268–277 (2018). https://doi.org/10.1016/j.cag.2018.05.025
15. Kavan, L., Collins, S., Žára, J., O'Sullivan, C.: Geometric skinning with approximate dual quaternion blending. ACM Trans. Graph. **27**(4), 1–23 (2008). https://doi.org/10.1145/1409625.1409627
16. Le, B.H., Hodgins, J.K.: Real-time skeletal skinning with optimized centers of rotation. ACM Trans. Graph. **35**(4), 1–10 (2016). https://doi.org/10.1145/2897824.2925959

17. Magnenat-Thalmann, N., Laperriere, R., Thalmann, D.: Joint-dependent local deformations for hand animation and object grasping. P~vc. Graphics Interface, pp. 26–33 (1988)
18. Wang, H., Sun, X., Huo, L., Tang, X., Liu, C.: Construction of deformable trunk atlas of chinese human based on multiple PET/CT images: preliminary results. In: Digital Human Modeling. Applications in Health, Safety, Ergonomics, and Risk Management: Ergonomics and Design, vol. 10286, Springer International Publishing, pp. 69–77 (2017). https://doi.org/10.1007/978-3-319-58463-8_7
19. Kocabas, M., Athanasiou, N., Black, M.J.: VIBE: video inference for human body pose and shape estimation. In: 2020 IEEE/CVF Conference on Computer Vision and Pattern Recognition (CVPR), Seattle, WA, USA, pp. 5252–5262 (2020). https://doi.org/10.1109/CVPR42600.2020.00530
20. He, K., Zhang, X., Ren, S., Sun, J.: Deep residual learning for image recognition. In: 2016 IEEE Conference on Computer Vision and Pattern Recognition (CVPR), Las Vegas, NV, USA, pp. 770–778 (2016). https://doi.org/10.1109/CVPR.2016.90

Chapter 13
Research on Fault Prediction Technology of Air Compressor Based on Wavelet Neural Network

Lin Shu, Tianxin Hu, and Yu Xia

Abstract The air compressor supply is the main gas source equipment for the water supply system of civil aviation aircraft. As a piece of complex electromechanical equipment, various faults become apparent with the increase in usage time. If the faults are not dealt with promptly, they will affect the equipment's performance and system function. In this paper, based on the working principle and structural characteristics of the air compressor, the main fault modes are analyzed, and the characteristic parameters are determined. The wavelet neural network algorithm is used to establish a fault prediction model. Based on the collected characteristic parameters, the service life of the air compressor is predicted, which improves the proactivity, guidance, and efficiency of maintenance support, enhances the availability of equipment, and provides technical support for airlines to reduce operating costs and improve customer satisfaction.

13.1 Introduction

The water supply system is an important component of civil aircraft and is an essential system for flight crew members' onboard life [1]. As the core gas source equipment of the water supply system, the air compressor compresses and boosts the air to provide pressure for water supply and drainage. The mechanical structure of the air compressor is complex, and various faults or symptoms of faults will inevitably occur during operation, affecting the comfort of onboard personnel and the efficiency of equipment operation. Currently, the maintenance support of electrical equipment

L. Shu (✉) · T. Hu
Wuhan Innovation Center, AVIC Aerospace Life-Support Industries LTD, Hubei, China
e-mail: shu52468@163.com

Y. Xia
The R&D Department of Electronic Control Technology, AVIC Aerospace Life-Support Industries LTD, Hubei, China

R. Kountchev et al. (eds.), *Multidimensional Signals, Augmented Reality and Information Technologies*, Smart Innovation, Systems and Technologies 374,
https://doi.org/10.1007/978-981-99-7011-7_13

such as air compressors usually adopts a combination of "after-the-fact maintenance" and "planned maintenance," which is prone to events of under-maintenance and over-maintenance, affecting the utilization rate of equipment and increasing the maintenance cost of the system's entire life cycle. Therefore, using fault prediction technology, based on the current health status and historical status of the equipment, adopting an appropriate prediction model, determining the remaining service life of components or systems, avoiding the occurrence of faults, and fully exerting the service life of the equipment can improve the accuracy and speed of maintenance.

13.2 Water Supply System

As an important component of the civil aircraft water and wastewater system, the water supply system's main function is to store sufficient drinking water in the onboard water tank and provide drinking water, washing water, and toilet cleaning water for the onboard kitchen and bathroom through pressure water supply during the flight to meet the water needs of passengers and crew members during the flight.

13.2.1 Working Principle

Before the aircraft takes off, it can be connected to the groundwater supply pipeline through the service panel to add water to the aircraft water tank and store it in the aircraft water tank. During the flight, the pressure water supply method is used to provide drinking water, hot and cold washing water, and toilet flushing water for the onboard kitchen and bathroom. After the aircraft lands, it can be connected to the ground gas supply pipeline through the service panel to pressurize and drain the water tank and water pipes. The air compressor compresses the air to provide sufficient pressure for water supply and drainage. The structure and composition of the water supply system are shown in Fig. 13.1.

13.2.2 Air Compressor Fault Modes

The fault-prone parts of reciprocating air compressors are concentrated in the power transmission part, gas inlet and outlet, sealing part, and auxiliary part [2]. The compressor fault characteristics are divided into two categories: one is with fluid properties, which belongs to the thermal performance failure of the machine. Its main characteristics are insufficient exhaust volume, abnormal pressure, and temperature difference, and the reasons are mainly due to failures in the gas valve, piston ring, and intake filter [3]. The other is with mechanical properties, which belong to the dynamic performance failure of the machine. The main characteristics are abnormal

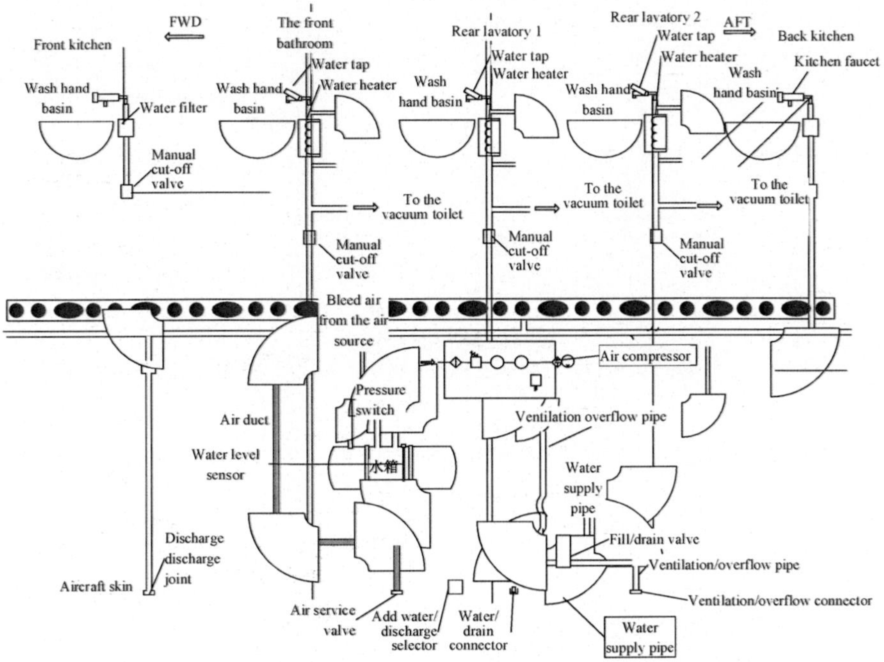

Fig. 13.1 Water supply system composition

noise, vibration, and overheating during machine operation, and the reasons are mainly due to changes in the clearance of moving parts and cracks in the structure [4]. The specific fault forms of the air compressor are shown in Table 13.1.

13.3 Fault Prediction Technology

The air compressor is a complex electromechanical product with functional mechanisms, system complexity, and continuous operation characteristics. It is very difficult to establish an accurate physical model. Therefore, data-driven prediction methods are adopted. By recording the important characteristic parameters of key components and analyzing their performance degradation curves, the time and probability of possible faults in the later stage are predicted to achieve fault prediction of key components. There are many data-driven fault prediction methods, including BP neural network, expert system, fuzzy comprehensive evaluation algorithm, etc. BP neural network is a relatively mature fault prediction algorithm, so this paper uses the BP neural network algorithm for fault prediction.

Table 13.1 Main fault list of air compressor functional components

Serial number	Functional unit	Major failure	Loss of function
1	Motor	Overburning of coil	Failure of function
2		Lack of phase	Failure of function
3		Bearing wear	Performance degradation
4		Rotor damage	Failure of function
5	Transmission assembly	Gear wear and tear	Performance degradation
6		Bearing wear	Performance degradation
7		Not enough lubrication	Performance degradation
8	Piston assembly	Piston seal damage	Performance degradation
9		Insufficient piston lubrication	Performance degradation
10	Trap door	Valve disc fracture	Failure of function
11		Break of spring	Failure of function
12		Valve disc wear	Performancedegradation

13.3.1 BP Neural Network

The BP neural network is a type of multi-layer feedforward network that uses error backpropagation as its learning and training rule [5]. Through learning and training, the BP neural network can describe the mapping relationship between input and output [6]. The topological structure of the BP neural network model is shown in Fig. 13.2.

In the network, the hidden layer nodes serve as the input and output of the network's internal, while the output layer neurons serve as the nodes outside the

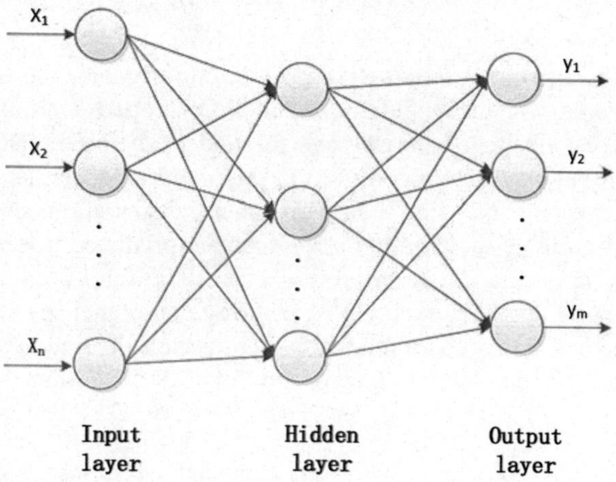

Fig. 13.2 BP neural network structure

network. The output of the previous layer node only affects the input of its lower layer node. Therefore, the BP neural network has a very good learning and approximation ability for expected mapping output.

Since the essence of the BP neural network algorithm is to solve the minimum value problem of the error function, it generally uses a search and solution algorithm along the gradient descent. Therefore, it may cause the network training to enter a local minimum state, resulting in problems such as low learning efficiency and slow convergence speed [7].

13.3.2 Wavelet Neural Network

Wavelet Neural Network combines the theories of neural network and wavelet analysis, which not only features pattern recognition, adaptive learning, and non-linear mapping capabilities, but also gains multi-resolution analysis ability from wavelet analysis. It has a wide range of application prospects.

The Wavelet Neural Network combines the theories of neural network and wavelet analysis, which not only features pattern recognition, adaptive learning, and non-linear mapping capabilities from the neural network theory, but also gains multi-resolution analysis ability from wavelet analysis. In this paper, we combine the BP neural network model and wavelet analysis, using the wavelet function as the hidden layer function [8]. The topological structure of the wavelet neural network model is shown in Fig. 13.3.

According to the network topology structure in Fig. 13.3, the calculation results of the hidden layer can be described by the following formula:

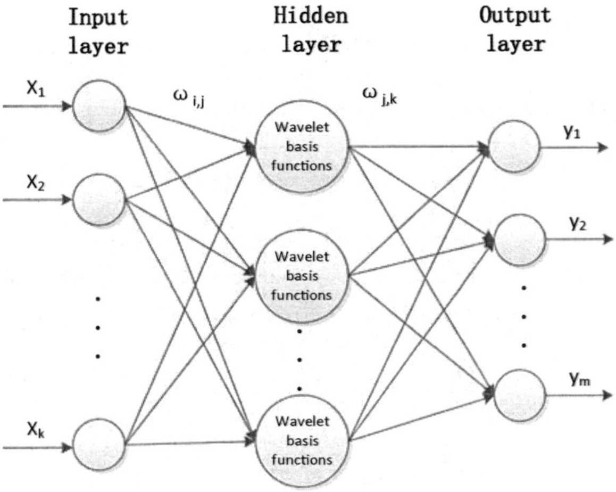

Fig. 13.3 Wavelet neural network structure

$$h(j) = \Psi_j \left(\frac{\sum_{i=1}^{k} \omega_{ij} x_i - b_j}{a_j} \right) j = 1, 2, \ldots, l \tag{13.1}$$

In the above formula, $h(j)$ represents the final result of the JTH node of the hidden layer, $\omega_{i,j}$ is the connection weight between the input layer and the hidden layer, and b_j bj is the translation factor of the wavelet basis function Ψ_j. a_j represents the scaling factor of the wavelet basis function Ψ_j. Ψ_j is represented as a wavelet basis function. The Morlet function is used as the wavelet basis function in the above mathematical expression, and the specific expression is shown below:

$$f(x) = \cos(1.75x)e^{-x^2/2} \tag{13.2}$$

The formula of the output for the wavelet network is as follows:

$$y(k) = \sum_{j=1}^{m} \omega_{jk} h(i) \quad k = 1, 2, \ldots, m \tag{13.3}$$

In the above formula, $\omega_{jk}\omega_{jk}$ represents the weight from the hidden layer to the output layer, and $h(i)$ is the output of the ith hidden layer node. l is the number of nodes in the hidden layer and m is the number of nodes in the output layer.

After constructing the wavelet neural network structure, to make the network more stable, it is necessary to train the network with relevant data. After the network is trained, some data needs to be inputted for verification to observe whether the network is successfully constructed. The training process of the wavelet neural network can be described as follows:

(1) Initialization of wavelet neural network parameters. Randomly initialize the scaling factors a_k and translation factors b_k of the wavelet function, as well as the connection weights ω_{ij} and ω_{jk} of the neural network. Set the learning rate of the network to η.
(2) Classification of sample data. Divide the collected data into the training set and test set, and input the training data into the network to train the model. The test set is used to test the accuracy of the trained network.
(3) Calculation Error. The training samples are input into the network, and the predicted values are output from the network. The error value e is obtained by performing a calculation between the predicted and expected output values.
(4) Weight Adjustment. Based on the error value e, the weights and coefficients of the network are adjusted to reduce the error between predicted values and expected values.
(5) Determine whether the algorithm output meets the conditions, otherwise continue the loop from steps (3) to (5). Figure 13.4 shows the flowchart of the training steps of the wavelet neural network.

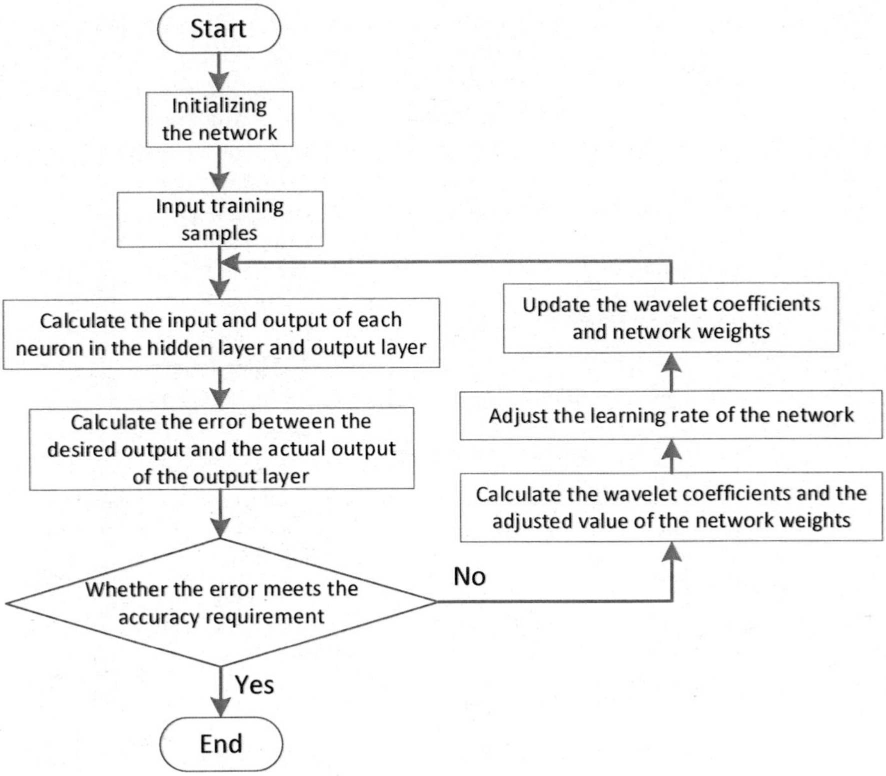

Fig. 13.4 Steps of wavelet neural network training

13.4 Simulation and Verification Data Acquisition

13.4.1 Data Acquisition

As a type of energy-consuming equipment, the efficiency of an air compressor will gradually decrease and its boosting capacity will weaken during continuous use. To obtain feature parameters, pressure sensors are installed at the inlet and outlet of the air compressor to collect real-time inlet and outlet pressures. As a mechanical and electrical device, the damage to the mechanical structure of the air compressor will also be reflected in the vibration of the equipment. Therefore, three-axis six-direction acceleration sensors are installed on both devices to obtain real-time acceleration signals during operation as feature parameters. In this study, the inlet and outlet pressure values and acceleration signals were selected as feature parameters based on the fault characteristics of the air compressor. After signal dimensionality reduction, denoising, and normalization, the wavelet neural network algorithm was used for air compressor fault prediction.

Two types of samples are required for wavelet neural network prediction: training samples and testing samples. The air compressor used in this study was a customized product developed based on functional requirements, and there was no historical data available for use as samples. Therefore, based on the functional structure of the water supply system and using AMESim software, a simulation model of the water supply system was established to simulate the working scenario of the water supply system and obtain the parameter indicators required for fault prediction as sample data.

13.4.2 Network Training

For the simulation data, its time-domain features were calculated, and PCA was used for dimensionality reduction. The data was divided into 15 groups of training data and 5 groups of testing data after PCA dimensionality reduction, which were used as input data for training and testing, respectively. In this test, a single hidden layer neural network model was used, with 4 nodes in the input layer and 3 nodes in the output layer (outputting three situations: fault, performance degradation, and normal). The transfer function of the hidden layer was selected as the Morlet function of the wavelet basis function, and the number of nodes in the hidden layer was determined to be 7 through experimental adjustment.

The creation and training results of the neural network are shown in Figs. 13.5, 13.6, and 13.7.

The output of the wavelet neural network test group is shown in Figs. 13.8 and 13.9.

From the output results shown in Figs. 13.8 and 13.9, it can be seen that the fault diagnosis of the BP neural network for the 5 groups of testing samples has achieved a 100% accuracy rate, indicating that the network can meet the requirements of fault diagnosis.

13.4.3 Fault Prediction

The feature parameters of the air compressor were collected and recorded through the comprehensive control platform, and the trained wavelet neural network algorithm was used for fault prediction.

According to system requirements, a work safety threshold is established, and when it falls below this value, the product is considered to have reached the end of its lifespan and needs to be replaced. Figure 13.10 shows the remaining life prediction of the air compressor. It can be seen from the figure that the reliability of the air compressor will decrease to 75% after 4800 working times, reaching the set safety threshold, and replacement should be considered.

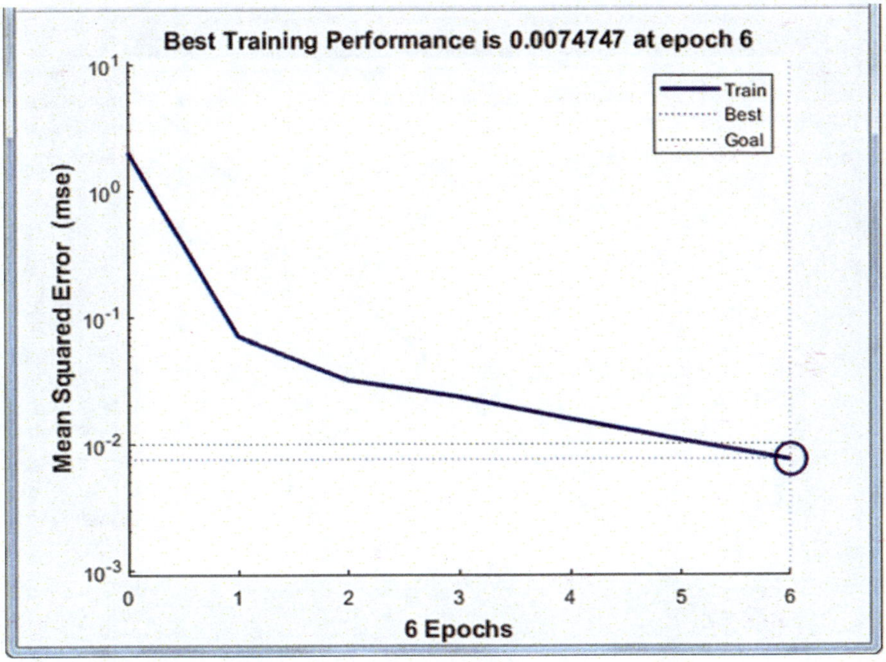

Fig. 13.5 Result analysis of wavelet neural network

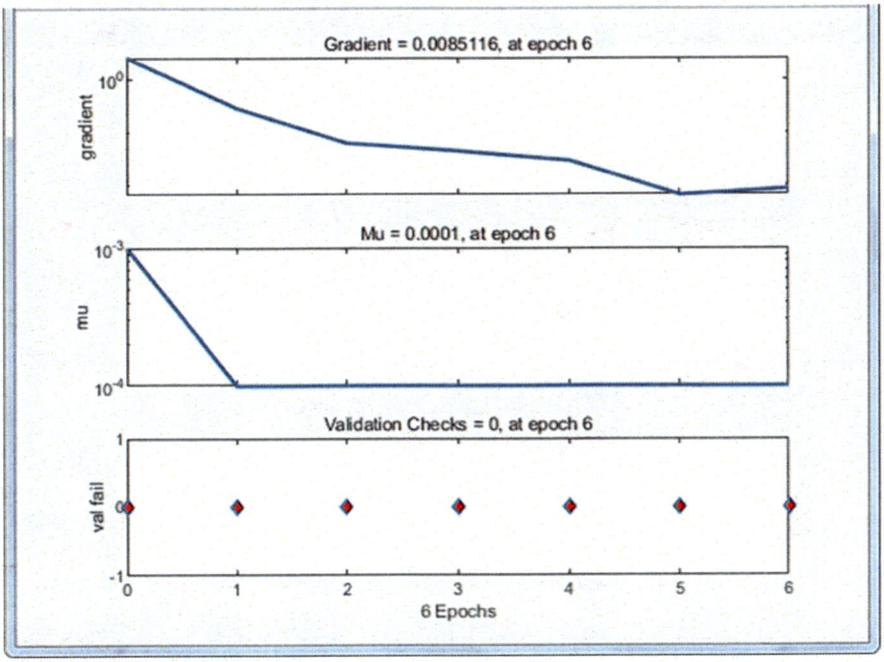

Fig. 13.6 Gradient and mean square error graph of training data

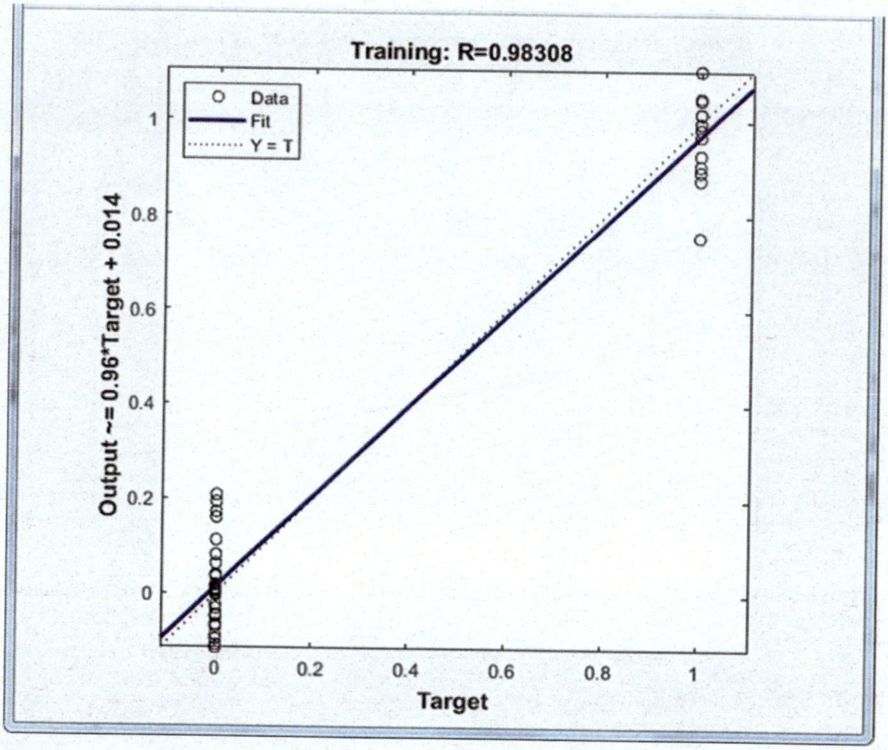

Fig. 13.7 Correlation coefficient R

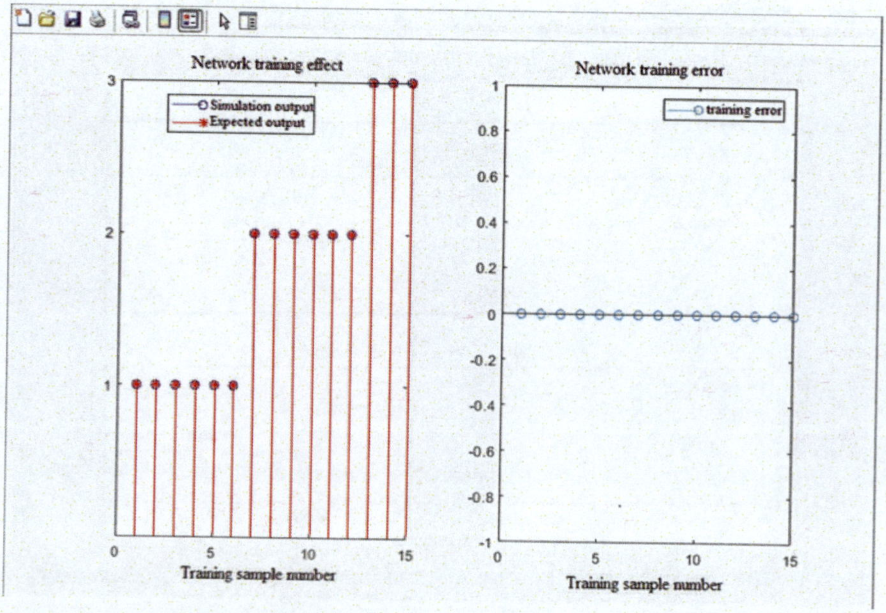

Fig. 13.8 Training results and errors of wavelet neural network

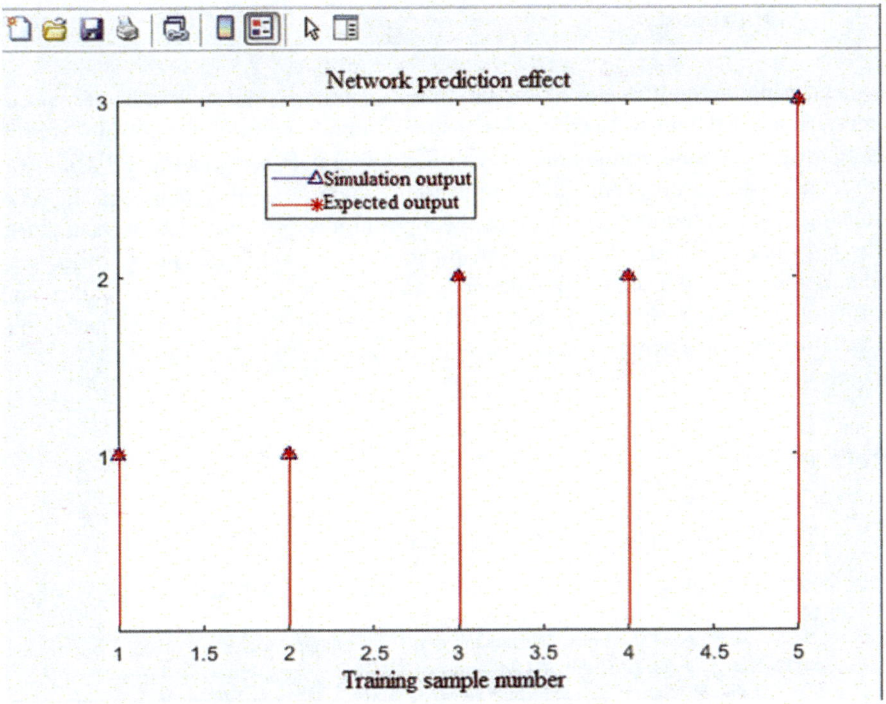

Fig. 13.9 Wavelet neural network test group output results

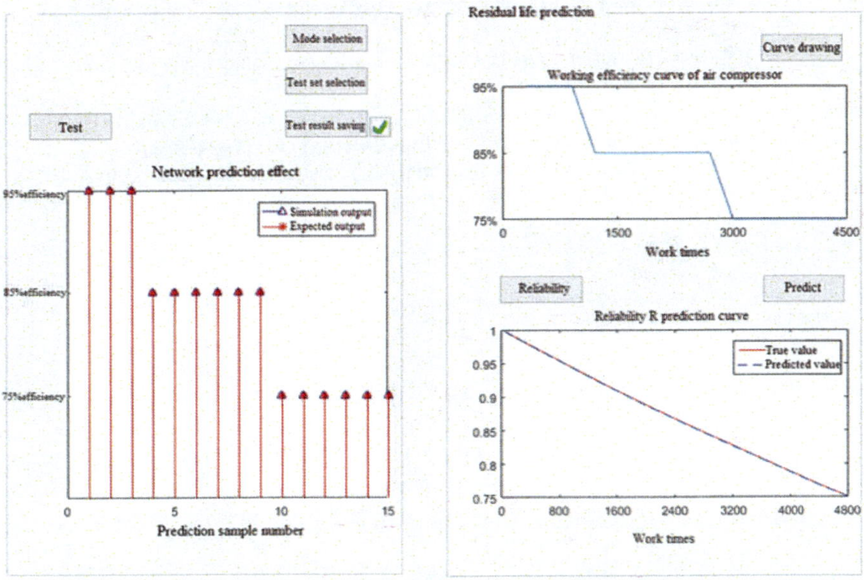

Fig. 13.10 Air compressor fault prediction result

13.5 Conclusion

By analyzing the fault modes of the air compressor, determining the feature parameters, using AMESim software to establish a simulation model of the water supply system, and obtaining sample data, a wavelet neural network was trained. By using the feature parameters obtained from the comprehensive control platform, the relationship between the number of times the air compressor is used and its remaining life was obtained, which is convenient for ground support personnel of airlines to take measures in advance for maintenance and support, improve the reliability and safety of the system, enhance passenger comfort experience, and reduce operational losses, thereby improving market competitiveness [9].

References

1. Lei, M.L.: Research on the selection of water system for new civil electric passenger aircraft. Aeronautical Eng. Progress **3**(1), 229–234 (2012)
2. Xing, C.X.: Modeling and Optimal Control of Compressed Air System. Hangzhou Dianzi University, Hangzhou (2016)
3. Su, C.G.: Research on Condition Monitoring and Fault Diagnosis System of Air Compressor. Xi 'an University of Architecture and Technology(2010)
4. Zhou, W.Q., Yin, S., Gao, R.B., et al.: Failure mode analysis of boeing 787 cabin air compressor. Aviation Maintenance Eng. **7**, 96–100 (2022)
5. Ni, L.M.: Remote Fault Diagnosis and Prediction Technology of Air Compressor. Zhejiang University of Science and Technology (2019)
6. Gao, J.: Principle and Simulation Example of Artificial Neural Network. China Machine Press, Beijing, pp. 45–52 (2005)
7. Tang, S.L., Luo, Y.F.: The development and applica-tion of artificial neural network technology. Comput. Dev. Appl. **10**, 59–61 (2009)
8. Liu, B., Yang, J.Y., Wu, Z.D., et al.: Research on wavelet neural network predictive control and its application in oil field production. Control. Eng. **29**(10), 1793–1799 (2022)
9. Ye, Q.F.: The development direction of maintainability design from the perspective of B787. Civ. Aircraft Des. Res. **3**, 51–53 (2010)

Chapter 14
The Development of Intelligent VR Systems Based on Deep Learning

Yaohui An, Xinkai Wang, Wei Wu, and Jiayu Lei

Abstract By applying deep learning technology to smart Virtual Reality (VR) devices, it can achieve continuous iterative upgrades and provide users with a better immersive design experience. Based on deep learning technology, this research builds an image difference prediction calculation model and creates prediction models for pictures using data from visual interaction screens after analysis and processing. For immersive display design screens, an image prediction computational model is first built. In the experiment, numerous screenshots of the display device are processed, superfluous pixels are removed, and multiple colour display judgements are made using the image difference prediction model. The quality level of the display effect has successfully been raised based on the data from the initial colour sample points. Additionally, the enhanced display screen allows for specular reflection to arise from real-time algorithms rather than just simple mirror mapping; further enhancing the visual effect. This has great significance for the application of VR technology in education, entertainment and other fields.

14.1 Introduction

Because of the fast advancement of technology and public demand for immersive experiences, virtual reality (VR) technology is becoming an increasingly compelling field of innovation and research. However, traditional VR experiences are still limited by the interaction and realism between the user and the virtual environment. In order to overcome these challenges and achieve a more realistic, natural and intelligent virtual reality experience, deep learning technology is gradually being introduced into the VR field [1]. Deep learning is a machine learning method based on artificial neural networks, which simulates the working mode of the human brain through a

Y. An (✉) · X. Wang · W. Wu
School of Aeronautics and Astronautics, Shenyang Aerospace University, Liaoning, China
e-mail: 3265065858@qq.com

J. Lei
School of Energy and Environment, Shenyang Aerospace University, Liaoning, China

R. Kountchev et al. (eds.), *Multidimensional Signals, Augmented Reality and Information Technologies*, Smart Innovation, Systems and Technologies 374,
https://doi.org/10.1007/978-981-99-7011-7_14

multi-level neural network structure so that computers can learn and extract features from large-scale data. The introduction of this technology brings great opportunities and challenges to virtual reality. Deep learning can be used to enhance the perception and interaction capabilities of VR environments, making them more intelligent and personalized. The application of deep learning in virtual reality can improve perception [2]. Traditional VR systems rely on pre-designed and rendered virtual environments, which limit users' perception and interaction with the real world. Deep learning, on the other hand, can understand user behaviour and the environment in real time by analyzing sensor data, such as camera and voice input [3].

14.2 Immersive Display of VR Technology

VR (Virtual Reality) technology is an innovative technology that immerses users in a virtual environment. By using a head-mounted display and perception devices, VR technology can create an immersive experience that makes users feel like they are in a completely fictional world. This kind of immersive display brings unprecedented sensory experience to users and also provides many opportunities for various industries. In the field of entertainment, the immersive display of VR technology has achieved great success in games, movies, and virtual tourism [4]. In addition, the immersive display of VR technology is also playing an important role in the field of education and training. Finally, the immersive display of VR technology has also revolutionized the display of tourism and cultural heritage. By using VR technology, people can experience places of interest and cultural heritage around the world at home or in the office. This immersive display of cultural heritage not only provides convenience for users.

14.3 Image Showing Difference Prediction Calculation

Integer wavelet coefficients and picture prediction mistakes may be used to modify the procedure, which has a high degree of flexibility and can give a huge capacity of embedding values. This work enhances the conventional picture difference prediction approach and makes full use of the connection between colour components. Such an enhancement not only lowers the algorithm's computational burden, but also greatly raises prediction calculation accuracy [5]. The display design may now better include greater immersion thanks to this development.

14.3.1 Standard Image Comparison Prediction Techniques

It is possible to use conventional image similarity algorithms to forecast nearby colour pixels. The expressions that follow can be used to represent the mean and difference of two colour pixels.

$$l = \frac{x + y}{2}, h = x - y \tag{14.1}$$

The value range is 0–255, the mean value is represented by l, the difference value is represented by h, and x and y are the grey values of two adjacent colour pixels.

$$\begin{cases} x = l + \frac{h+1}{2}, y = l - \frac{h}{2} | h \geq 0 \\ x = l + \frac{h}{2}, y = l - \frac{h-1}{2} | h < 0 \end{cases} \tag{14.2}$$

The discrepancy-extended prediction equation is then created by inserting a difference of 1 pixel in the formula that follows, h_ω is the extended predicted value and ω is the constant term [6]

$$h_\omega = 2h + \omega. \tag{14.3}$$

14.3.2 Inaccuracy in One Aspect of the Image Expectation

Prediction linear colour compression is based on the visual depiction of high correlations between pixels on a screen. The precision of the forecasting of nearby pixels is better than that of far-off pixels because this relationship, which initially exhibits a negative correlation as the distance between pixels rises, gradually deteriorates [7]. The task at hand tries to choose a pixel of any given colour on the VR display device and determine its surrounding pixels in order to calculate the forecasting error generated by this knowledge. In the experiment, the colour of the pixel is predicted by the correlation. By observing the relationship between a pixel and its neighbours, it is possible to infer the predicted value of the pixel and calculate the error from the actual value [8]. This approach can help understand the correlation between pixels and provide a reference for display design and image compression techniques. By analyzing the prediction error, conclusions about the correlation between pixels can be drawn, and the prediction algorithm can be further optimized to improve the accuracy and effect of the display, Prediction error is p_ω, ω is a constant term

$$p_\omega = 2p + \omega \tag{14.4}$$

α_ω is the predicted pixel, The value of the predicted pixel is

$$\alpha_\omega = \overline{\alpha} + p_\omega. \tag{14.5}$$

The pixels on the chosen screen likewise contain separate components, in accordance with the idea that the RGB elements included in the colour map's triangle are all independent [9]. The above, lower, left, and right neighbouring pixels to the pixel that has to be forecasted are utilized as reference points, and the average value is calculated, in order to increase the accuracy of picture prediction and benefit from the strong correlation between nearby pixels. The pixel average $\overline{S}$:

$$\overline{S} = \frac{1}{4} \times \begin{vmatrix} & 1 & \\ 1 & . & 1 \\ & 1 & \end{vmatrix}. \tag{14.6}$$

The black dot in the centre of the prediction template in the formula above stands in for the pixel to be predicted.

14.3.3 The Differential Growth of a Single-Component Picture Prediction

The mean and difference of errors are used to estimate the colour components of any two pixels on the device's screen

$$l = \frac{P_1 + P_2}{2}, h = P_1 - P_2. \tag{14.7}$$

The aforementioned equation may also be produced via integer wavelet inverse transformation and refers to the conventional image difference prediction approach:

$$\begin{cases} P_1 = l + \frac{h+1}{2}, P_2 = l - \frac{h}{2} | l \geq 0, h \geq 0 \\ P_1 = l + \frac{h}{2}, P_2 = l - \frac{h-1}{2} | l \geq 0, h < 0 \end{cases} \tag{14.8}$$

$$\begin{cases} P_1 = l + \frac{h}{2}, P_2 = l - \frac{h+1}{2} | l < 0, h \geq 0 \\ P_1 = l + \frac{h-1}{2}, P_2 = l - \frac{h}{2} | l < 0, h < 0 \end{cases} \tag{14.9}$$

Then embed the predicted pixels into

$$h_\omega = P_1 + P_2\sqrt{2h + \omega}. \tag{14.10}$$

The total and difference of the two pixels are added to prevent mistakes brought on by rounding down [10]

$$a = \frac{(P_1 + P_2)^2 + (P_1 - P_2)^2}{2},$$

$$h = 2\sqrt{(P_1 + P_2)^2 + (P_1 - P_2)^2}. \tag{14.11}$$

The inverse transform of the associated integer wavelet is then translated into two separate sets of mathematics:

$$\begin{cases} P_1 = l \pm \frac{h+1}{2}, P_2 = P_2 \pm \frac{h}{2} | \alpha \geq 0, h \geq 0 \\ P_1 = l \pm \frac{h}{2}, P_2 = P_2 \pm \frac{h-1}{2} | \alpha \geq 0, h < 0 \end{cases} \tag{14.12}$$

$$\begin{cases} P_1 = l \pm \frac{h}{2}, P_2 = P_2 \pm \frac{h+1}{2} | \alpha < 0, h \geq 0 \\ P_1 = l \pm \frac{h-1}{2}, P_2 = P_2 \pm \frac{h}{2} | \alpha < 0, h < 0 \end{cases} \tag{14.13}$$

The two equations can be operated on to further determine the minimal effective pixel function

$$\omega = LSB(h_\omega), h = \frac{h_\omega - \omega}{2} + \omega. \tag{14.14}$$

The grey value range of the anticipated pixel is also constrained in order to increase the saturation and accuracy of the device's display screen and avoid pixel overflow [11]

$$P_\omega \in [-a, -255 - \overline{a}]. \tag{14.15}$$

To get the final picture difference prediction model shown on the VR device's screen, determine the colour component of each pixel that has to be forecasted within this range. Then, take into account the inaccuracy discovered above, $P_{n\omega}$ is the image difference prediction value [12]

$$\begin{Bmatrix} |P_{n\omega}| = |P_n| - 1 | P_n > T_n \\ P_{n\omega} = P_n + 1 | \overline{a} < 128, P_n < T_n \\ P_{n\omega} = P_n - 1 | \overline{a} \geq 128, P_n < T_n \end{Bmatrix}. \tag{14.16}$$

14.4 Conclusion

First of all, multiple screenshots are sampled on the display screen of the VR device, and an image processing model is used to analyze the effective display screen pixels and redundant display screen pixels.

The effective pixel area, or the pixel block that the image difference model may use, is known in image processing as the white area. The unnecessary pixel region, shown by the orange portion, must be removed from the prediction computation. Redundant pixels include those that include worthless information and those that overlap valid pixel information. Both sorts of pixels must be avoided since they slow down and obstruct the model's processing [13].

The original display device screen pixels that were collected as well as the pixels that were received after completing the prediction are computed and taken into account. The steady and generally linear increase of the initial pixel space complexity shows that the initial pixel data retrieved for this experiment is stable [14]. Additionally, the pixel data for the colour to be forecasted exhibits exponential development, showing that the more information used in the prediction, the greater the level of spatial complexity and the more accurate the outcome of the prediction. Figure displays a thorough visualization of the data structure [15].

Colours close to the vertical axis are viewed as black no matter what their hue when saturation is zero and brightness is likewise very close to zero [16]. Colours close to the horizontal axis are viewed as transitioning from black to grey to white as brightness rises, but saturation is still at 0 and hue is almost at 0.

When the saturation and value of a colour are both 0, the brightness is normalized from 0 to 1, showing a gradient effect from black to grey to white. This display method makes the difference that people perceive most intuitively is the difference in brightness. Therefore, trapezoidal blurring of the non-colours using the blurring method again yields the effect shown in the right panel [17]. By improving the image difference prediction model, the display effect of VR devices has been significantly improved. The improved display makes colours in bright areas more vivid, and details remain clearly visible when brightness is increased. Additionally, the hierarchy of the presentation of dark regions is improved, and places without light might seem completely black. The material's texture in the image has also been much enhanced. In the virtual environment, plant stems and leaves have a more delicate and life-like feel. Each leaf's texture is randomly distributed using the algorithm [18]. The depiction of materials like fabric, metal, and plastic in clothes and props is more accurate due to the incorporation of more polygonal forms, which also makes the fur of animals and characters more realistic and soft. The consequence of a real-time algorithm, a specular reflection is now more than just a mirror image. The enhanced display screen makes it possible for VR technology to be used in the education sector in more ways, and students may become more involved in the information while studying [19]. To further improve the learning experience, users may visualize and engage with the abstract concepts in the book using VR.

14.5 Discussion

Deep learning intelligent VR technology research has made significant advancements and enhancements to the virtual reality experience. Through the application of deep learning algorithms, a more intelligent, real and personalized virtual environment display can be realized. In this study, The visual dynamic screen information has been analysed and processed, and an image forecasting framework is created based on the image difference prediction calculation generated by deep learning. Deep learning is used to enhance perceptual abilities in the world of virtual reality, making it seem more genuine and lifelike. By analyzing sensor data and user behaviour, deep learning algorithms can sense user actions, emotions, and intentions in real time [20].

In future research, the development of intelligent virtual reality (VR) systems based on deep learning will explore the following aspects:

1. Emotion recognition and self-adaptation: intelligent VR systems will aim to better understand users' emotional states and adjust the virtual reality experience according to their emotional states. Through deep learning technology, the system can analyze the user's facial expressions, voice, heart rate and other physiological indicators to further provide personalized VR content and interactions [21].
2. Autonomous decision-making and behaviour simulation: future intelligent VR systems will have more advanced decision-making and behaviour simulation capabilities. Through deep reinforcement learning and other technologies, the system can learn and simulate human behaviour patterns to provide more realistic and interactive virtual scenes.
3. Multimodal perception and interaction: Intelligent VR systems will further integrate multiple perception technologies, such as visual, auditory and haptic, to provide a more immersive and realistic experience. Deep learning can be used to process and fuse data from different perceptual modalities to achieve multimodal interaction and feedback.
4. learning and adaptive capabilities: future intelligent VR systems will have learning and adaptive capabilities and be able to improve system performance based on user feedback and usage habits. Deep learning techniques will be used to build models with enhanced learning and adaptive mechanisms, enabling the system to continuously optimize and personalize the user experience.
5. Realism and Simulation: Future research will focus on further improving the realism and simulation of intelligent VR systems. Deep learning techniques can be used to generate realistic models of virtual environments, characters and objects, thus providing a more realistic virtual reality experience.
6. Cross-domain integration and applications: Intelligent VR systems will be combined with technologies from other domains, such as robotics, computer vision, natural language processing, etc., to achieve a wider range of applications. Deep learning will be one of the key technologies to achieve these cross-domain integrations.

In conclusion, future research will be devoted to further promoting the development of intelligent VR systems through deep learning technologies, improving the intelligence, realism and user experience of the systems, as well as enabling applications and innovations in more domains.

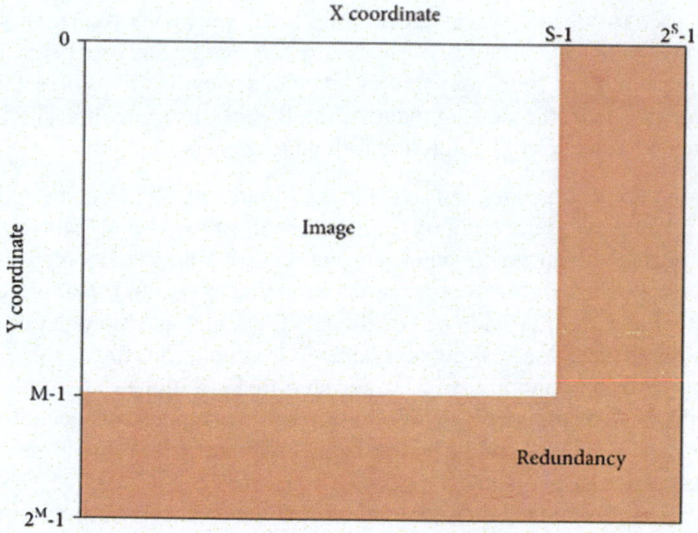

Fig. 14.1 Shows the redundant ceramic colour pixel map and the effective ceramic colour pixel map that were produced after model processing

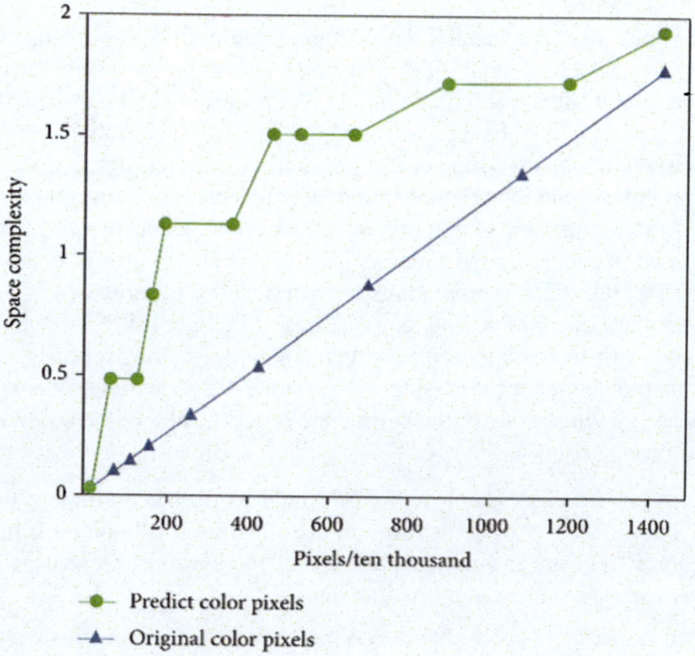

Fig. 14.2 The difference in spatial complexity between the pixels acquired after prediction and the original colour-matched pixels

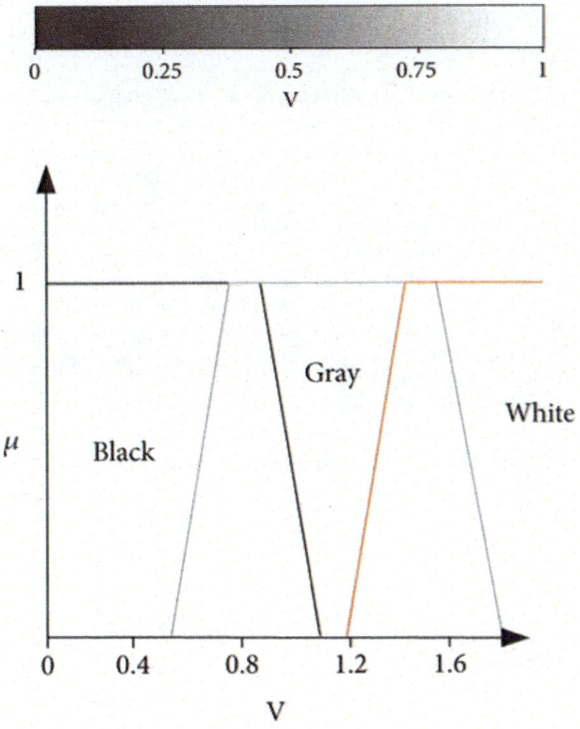

Fig. 14.3 V and V-component fuzzy set partitions as a result of the ceramic material colour

Acknowledgements Funded by the National Undergraduate Innovation and Entrepreneurship Training Program Support Project (202210143020).

References

1. Zhang, Z., He, T., Zhu, M., Sun, Z., Shi, Q., Zhu, J., Lee, C.: Deep learning-enabled triboelectric smart socks for IoT-based gait analysis and VR applications. NPJ Flexible Electron. **4**(1), 29 (2020)
2. Chen, M., Challita, U., Saad, W., Yin, C., Debbah, M.: Artificial neural networks-based machine learning for wireless networks: a tutorial. IEEE Commun. Surveys Tutorials **21**(4), 3039–3071 (2019)
3. Herasymova, A., Chumachenko, D., Padalko, H.: Development of intelligent information technology of computer processing of pedagogical tests open tasks based on machine learning approach. In: MoMLeT+ DS, pp. 121–131 (2020)
4. Chen, F., Du, Y., Qiu, T., Xu, Z., Zhou, L., Xu, J., Sun, J.: Design of an intelligent variable-flow recirculating aquaculture system based on machine learning methods. Appl. Sci. **11**(14), 6546 (2021)
5. Tai, Y., Gao, B., Li, Q., Yu, Z., Zhu, C., Chang, V.: Trustworthy and intelligent covid-19 diagnostic iomt through xr and deep-learning-based clinic data access. IEEE Internet Things J. **8**(21), 15965–15976 (2021)

6. Khanzode, K.C.A., Sarode, R.D.: Advantages and disadvantages of artificial intelligence and machine learning: a literature review. Int. J. Libr. Inf. Sci. (IJLIS) **9**(1), 3 (2020)
7. Li, X., Liu, H., Wang, W., Zheng, Y., Lv, H., Lv, Z.: Big data analysis of the internet of things in the digital twins of smart city based on deep learning. Futur. Gener. Comput. Syst. **128**, 167–177 (2022)
8. Baduge, S.K., Thilakarathna, S., Perera, J.S., Arashpour, M., Sharafi, P., Teodosio, B., Shringi, A., Mendis, P.: Artificial intelligence and smart vision for building and construction 4.0: machine and deep learning methods and applications. Autom. Construct. **141**, 104440 (2022)
9. Winkler-Schwartz, A., Bissonnette, V., Mirchi, N., Ponnudurai, N., Yilmaz, R., Ledwos, N., Siyar, S., Azarnoush, H., Karlik, B., Del Maestro, R.F.: Artificial intelligence in medical education: best practices using machine learning to assess surgical expertise in virtual reality simulation. J. Surg. Educ. **76**(6), 1681–1690 (2019)
10. Zhou, X., Liang, W., Kevin, I., Wang, K., Wang, H., Yang, L.T., Jin, Q.: Deep-learning-enhanced human activity recognition for Internet of healthcare things. IEEE Internet Things J. **7**(7), 6429–6438 (2020)
11. Murugan, S., Sampathkumar, A., Kanaga Suba Raja, S., Ramesh, S., Manikandan, R., Gupta, D.: Autonomous vehicle assisted by heads up display (HUD) with augmented reality based on machine learning techniques. In: Virtual and Augmented Reality for Automobile Industry: Innovation Vision and Applications, pp. 45–64. Springer International Publishing, Cham (2022)
12. Herasymova, A., Chumachenko, D., Padalko, H.: Development of intelligent information technology of computer processing of pedagogical tests open tasks based on machine learning approach. In: MoMLeT+ DS, pp. 121–131 (2020)
13. Syam, N., Sharma, A.: Waiting for a sales renaissance in the fourth industrial revolution: machine learning and artificial intelligence in sales research and practice. Ind. Mark. Manage. **69**, 135–146 (2018)
14. Ciolacu, M., Tehrani, A.F., Binder, L., Svasta, P.M.: Education 4.0-Artificial Intelligence assisted higher education: early recognition system with machine learning to support students' success. In: 2018 IEEE 24th International Symposium for Design and Technology in Electronic Packaging (SIITME), pp. 23–30. IEEE (2018)
15. Alam, A.: Should robots replace teachers? Mobilisation of AI and learning analytics in education. In: 2021 International Conference on Advances in Computing, Communication, and Control (ICAC3), pp. 1–12. IEEE (2021)
16. Shakarami, A., Ghobaei-Arani, M., Shahidinejad, A.: A survey on the computation offloading approaches in mobile edge computing: a machine learning-based perspective. Comput. Netw. **182**, 107496 (2020)
17. Aggarwal, K., Mijwil, M.M., Al-Mistarehi, A.H., Alomari, S., Gök, M., Alaabdin, A.M.Z., Abdulrhman, S.H.: Has the future started? The current growth of artificial intelligence, machine learning, and deep learning. Iraqi J. Comput. Sci. Math. **3**(1), 115–123 (2022)
18. Alam, A.: A digital game based learning approach for effective curriculum transaction for teaching-learning of artificial intelligence and machine learning. In: 2022 International Conference on Sustainable Computing and Data Communication Systems (ICSCDS), pp. 69–74. IEEE (2022)
19. Haidine, A., Salmam, F.Z., Aqqal, A., Dahbi, A.: Artificial intelligence and machine learning in 5G and beyond: a survey and perspectives. Moving Broadband Mobile Communications Forward: Intelligent Technologies for 5G and Beyond; IntechOpen: London, UK, p. 47 (2021)
20. Khanna, A., Sah, A., Choudhury, T.: Intelligent mobile edge computing: a deep learning based approach. In: Advances in Computing and Data Sciences: 4th International Conference, ICACDS 2020, Valletta, Malta, April 24–25, 2020, Revised Selected Papers 4, pp. 107–116. Springer Singapore (2020)
21. Rahman, M.M., Khatun, F., Uzzaman, A., Sami, S.I., Bhuiyan, M.A.A., Kiong, T.S.: A comprehensive study of artificial intelligence and machine learning approaches in confronting the coronavirus (COVID-19) pandemic. Int. J. Health Serv. **51**(4), 446–461 (2021)

Chapter 15
Research and Implementation of Multi Fusion Data Model Construction Technology for Distribution Network Digital Twins

Junfeng Qiao, Lin Peng, Aihua Zhou, Sen Pan, Pei Yang, Yanfang Mao, and Fuyun Zhu

Abstract The core of digital twin is the digital reconstruction of physical space, that is the physical space perception and processed data are applied to the virtual space to achieve real-time mapping of physical space through full integration of data. With the rapid development of digital twin technology in the power field, in order to achieve the purpose of real-time data fusion of physical space and virtual space in the digital twin of distribution network, the virtual and real data fusion processing technology for digital twin has become the focus of research in the power field. In this paper, firstly, the characteristics and ontology technology of power digital twin virtual and real data are studied, and an ontology-based virtual and real data fusion architecture is proposed. Then, a semantic feature extraction method is proposed for the structured and unstructured virtual and real data of power digital twin, and the extracted semantic features are formatted, described, and stored. Then the overall architecture of the distribution network digital twin virtual real data fusion system is designed, the technical route of the system is analyzed and discussed, and the virtual real data fusion processing is completed using the similarity algorithm and fusion rules based on the elements between ontology. Finally, the functional modules of the distribution network digital twin virtual real data fusion system are divided and introduced, and the digital twin virtual real data fusion of the fan hub casting process is verified as an example.

J. Qiao (✉) · L. Peng · A. Zhou · S. Pan · P. Yang
State Grid Laboratory of Power Cyber-Security Protection and Monitoring Technology, State Grid Smart Grid Research Institute Co.,LTD, Nanjing 210003, Jiangsu, China
e-mail: 1318558905@qq.com

Y. Mao · F. Zhu
Nantong Power Supply Company, State Grid Jiangsu Electric Power Supply Co., LTD, Nanjing 210014, Jiangsu, China

R. Kountchev et al. (eds.), *Multidimensional Signals, Augmented Reality and Information Technologies*, Smart Innovation, Systems and Technologies 374,
https://doi.org/10.1007/978-981-99-7011-7_15

15.1 Introduction

One of the goals of the development strategies formulated by German Industry 4.0 and American Industrial Internet is to use a new generation of information technology to establish connectivity and intelligent operations between the physical world and the information world in the industrial digital twin, thereby realizing industrial intelligence [1]. Subsequently, under the strong background of "network power" and "strategic power", China has successively introduced the industrial development strategies of "China's Industrial Manufacturing 2025" and "Internet Plus". At the same time, in the report of the 19th National Congress of the Communist Party of China, it was also clearly proposed that promoting the deep integration of the new generation of information technology (industrial Internet, big data, artificial intelligence technology) and industry is the key technology for the development of industrial intelligence [2]. Therefore, the core bottleneck facing both at home and abroad is how to realize the integration and interaction of physical space and information space in the industrial production process. The development of digital twin technology provides an effective solution to this problem.

At present, many internationally famous enterprises have begun to explore the specific application of data fusion technology in industrial product research and development, manufacturing, production, and service [3]. For example, applying the fused data to the product development stage will improve the accuracy of product design; In the manufacturing stage, it will provide optimized production decisions for products [4]. In the product maintenance stage, it will provide a favorable decision-making basis for product maintenance and fault prediction, and effectively improve the reliability and availability of the product. How to realize the efficient application of data in the whole life cycle of power equipment products through data fusion technology and provide real-time and intelligent decision optimization for power digital twins has become an urgent problem to be solved [5]. At present, digital twin technology has the following problems in the use of data:

(1) A large number of real-time data will be generated in power production. Although traditional data storage methods can store and retrieve these data, they cannot meet the requirements of real-time query and rapid response to data. For power equipment with high real-time, such data has a large amount of delay.
(2) The stream data in the power digital twin is highly complex, and its main characteristics are data instability, high coupling, and strong correlation. Due to the influence of the acquisition environment, technology, equipment, and other factors, the collected data is often of low quality, incomplete, noisy, and redundant. If the data is not effectively processed, it will have a great impact on production management, optimization, diagnosis or decision-making.
(3) Distribution network data sources include not only historical data of equipment operation, but also multi-source heterogeneous data such as real-time perception data, environmental data, sensor information, virtual model information, etc. In

order to achieve data fusion application, multi-source heterogeneous data needs to be uniformly represented.

15.2 Related Work

Many technology companies in the United States have provided a unified digital twin operation interface to realize the digitization of the whole process from product design to maintenance of electrical equipment in the physical space [6]. In addition, many institutes have established a 3D experience platform based on digital twins for user interaction needs of complex products and continuously improved the product design model using user feedback information to optimize the product entity in the physical space [7]. In China, more and more attention has been paid to the development of electric power digital twins. In terms of theory, universities in China have studied and summarized the interaction and integration of physical space and virtual space in digital twins, laying a solid foundation for the subsequent use of twin technology. In terms of application, the method of twin data interaction and integration in digital twin workshop has been realized on the basis of theory in China. In 2017, digital twin technology was listed as one of the top ten scientific and technological developments in intelligent manufacturing in the world at the World Intelligent Manufacturing Conference of the Intelligent Manufacturing Academic Consortium of China Association for Science and Technology [8].

As the collected power data is characterized by large quantity, real-time, multi-dimensional, and multi granularity, it is necessary to store the data up to 1000 megabytes or one trillion bytes level to ensure that the query and retrieval results of massive data can quickly respond to users [9]. Because the traditional data storage mode can no longer meet the requirements for real-time storage and efficient query access of massive data, the memory-based real-time database storage mode is widely used in the power production process. At present, researchers at home and abroad have conducted in-depth research on real-time databases, and many products applying real-time databases have emerged [10].

15.3 Research on the Construction Technology of Multiple Fusion Data Model of Distribution Network Digital Twins

15.3.1 Real-Time Data Acquisition and Storage Access Mechanism for Power Production

As the first access to real-time data, industrial data acquisition is the basis for realizing the physical world into the digital world. At the same time, in order to realize the complete use of industrial data, it is necessary to store the collected data in real time. At present, in the process of industrial data acquisition, there are problems such as inconsistent data frequencies, various data communication protocols, and poor compatibility. Although traditional databases can meet the storage requirements for real-time data, they cannot quickly respond to real-time data reading and continuous query access. These problems make the digital twin system unable to obtain a large number of complete and real-time data. To solve the above problems, this section will study the industrial real-time data acquisition and storage access mechanism.

As long as the perception data is obtained within the expected time range, it belongs to industrial real-time data. It meets the timeliness, real time, and integrity, and the timeliness expectation also provides a certain basic theory for the collection time of industrial real-time data.

(1) Real-time data appears in an appropriate time range: because it meets the expected time range of time effectiveness, it is collected within an effective time range. If it exceeds this time range, the data will lose its real-time characteristics, and it is worthless to users.
(2) Real-time data is sequential: the data in industrial real-time data is achieved in chronological order, and the order in which each real-time data arrives is the order in which data points are obtained. Assume that there are two data points, X_i and X_j, and the corresponding time marks of the two data points are T_1 and T_2. It is specified that $T_1 < T_2$, and the collected data points X_i must be reached before X_j.
(3) Integrity of real-time data: integrity is the most basic meaning of real-time data. Only by providing complete and valuable information can we provide data users with the basis for decision optimization and reference.

As is shown in Fig. 15.1, the data acquisition layer mainly uses ubiquitous sensing technology to access industrial data of various industrial production equipment or intelligent products. However, due to the existence of private industrial protocols for various industrial equipment, the protocol standards are not unified and incompatible with each other, which cannot meet the data collection requirements of the whole production process.

As a bridge between the industrial data perception layer and the processing layer, the data processing layer mainly realizes the real-time storage and continuous query

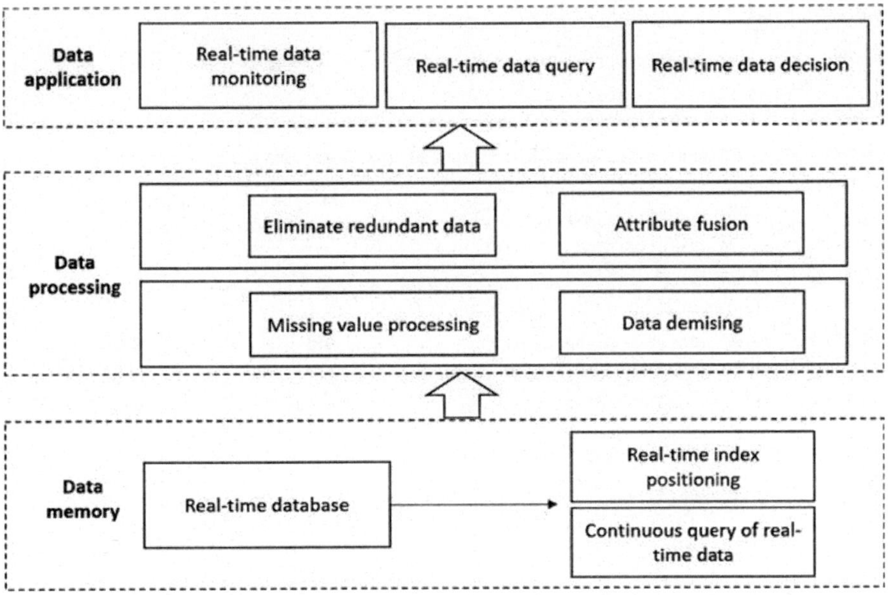

Fig. 15.1 Electric real-time data acquisition and processing architecture

and access of data. Real-time storage is mainly to meet the needs of rapid response to industrial stream data, and continuous query access is mainly to meet the needs of continuous query of streaming data.

15.3.2 Data Fusion Technology for Electric Distribution Network

Because of the heterogeneity of perception data and virtual data, this paper first analyzes the characteristics of virtual and real data in the power digital twins, uses ontology technology to uniformly describe the virtual and real data with multi heterogeneity, and elaborates the definition, description language, and construction method of ontology in detail.

The sources of power digital twin virtual and real data are often multiple different data sources. Because the data sources are independent of each other and the information standards are not unified, the data structures generated are different and the storage methods are different, making the industrial digital twin virtual and real data have the characteristics of multi-source heterogeneous. At the same time, due to the different sources of power digital twin virtual and real data, different production processes in the power field have made different interpretations of the same object information. For example, the same data is stored in different tables of different databases, which makes the same data described with different database names,

Table 15.1 Basic elements of electric twin data

Basic elements	Description
Concept	Ontology class is also called concept, which represents a collection of concepts
Relation	The relationship in ontology represents the relationship between concepts
Function	Constraint function is a formal expression of the relationship between concepts, and the relationship between concepts can be inferred from the mapping relationship existing in the function
Instance	Instances in ontology are also called individuals, which are materialized objects corresponding to ontology classes
Attribute	Attributes refer to the properties and characteristics of the concepts in the field, or the instances related to them
Axion	As the reference of ontology reasoning, it is a recognized knowledge rule

table names, and data types. This description method will make people have different understandings of the same object, which makes the virtual and real data of power digital twins have the characteristics of heterogeneous semantics.

From the above analysis, it can be seen that the multi-source heterogeneity and semantic heterogeneity of virtual and real data in the power digital twins have brought obstacles to data fusion. Compared with the elimination of semantic heterogeneity, multi-source heterogeneity is easier to implement, and only the mapping relationship between data and data sources needs to be implemented in specific operations, while semantic heterogeneity needs to consider the source of data and the structural information of data to achieve the relationship mapping between fields or records.

As shown in Table 15.1, according to the above definition of ontology, the virtual and real data in the power field are defined as instances in the ontology, and the classification of data is defined as concepts or ontology classes. By establishing an instance ontology, the virtual and real data of the electric digital twins with multi-source heterogeneity and semantic heterogeneity can be uniformly and normatively represented.

15.4 Technical Realization of Multi Fusion Data Model Construction of Distribution Network Digital Twin

15.4.1 Overall Architecture Design of Multi Fusion Data Model

The electric power digital twin virtual real data fusion system should have all the basic functions of the general power data platform for data lifecycle management. From the perspective of system functions, it not only needs to support the key technologies such as power data acquisition, storage, processing, fusion, and visualization, but also needs to support power applications in the digital twin environment with high complexity, multi-mode, and strong correlation characteristics.

The system uses B/S software architecture. This paper divides the overall architecture of the power digital twin virtual real data fusion system into five layers, namely, data acquisition layer, data storage layer, data processing layer, data fusion layer, and data application layer. The overall architecture of the system is shown in the figure below.

As is shown in Fig. 15.2, the data acquisition layer mainly uses ubiquitous sensing technology to realize data acquisition of industrial equipment. Due to the variety of modern power equipment protocols and different standards, there are high delays and incompatibilities in data acquisition of various power equipment. Therefore, this paper will unify the protocols of power equipment to achieve the data acquisition drive under the same protocol.

The data storage layer, due to the large amount of power data, the traditional relational database can only query the power data according to the strip when storing and managing the power data, and the data query speed and reading speed are slow, which cannot meet the high response requirements of users to real-time data. Therefore, this paper will use the memory database to store the real-time data of electric power, and its internal index structure is a tree to organize and manage the data, which can not only effectively avoid the risk of data loss, but also quickly realize the index positioning of real-time data, and improve the efficiency of data reading. At the same time, the sliding window mechanism is used to solve the problem of continuous query and access of real-time data, realizing real-time query and real-time response to data.

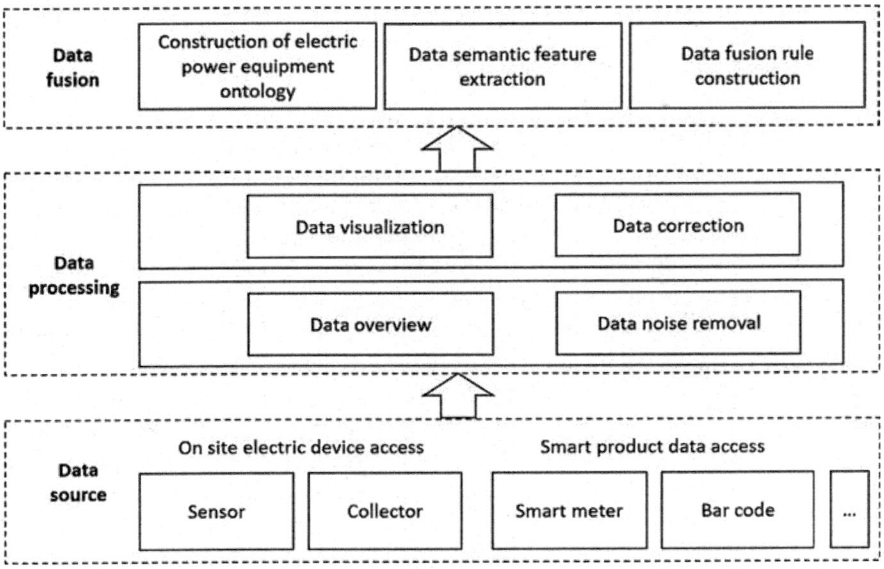

Fig. 15.2 Architecture design of multi fusion data model

Data fusion aims to process and combine virtual and real data from different sources to obtain more effective data information and use these data information to provide optimal auxiliary decision-making for power digital twins. In order to achieve the fusion of heterogeneous data, it is first necessary to use ontology technology to uniformly and normatively represent the virtual and real data with multiple heterogeneities, and then, according to the characteristics of the elements between ontology after construction, ontology mapping is implemented using the method based on similarity calculation to integrate different data representing different perspectives into a unified panorama. Finally, the data describing the same object are fused according to the fusion rules, which provide the most effective data for the power digital twins. The purpose of data application is to use the fused data to provide more accurate prediction for fault prediction, health management, status detection management, etc., of equipment in the industrial digital twins.

15.4.2 The Implementation of Multivariate Fusion Data Model for Distribution Network Digital Twin

The encapsulation and operation of the algorithm are based on the imported industrial data, Python algorithm, and MATLAB algorithm. First, configure the parameters of each algorithm on the front page of the system, and verify the legitimacy and standardization of the input parameters on the back end. Then the association mechanism between algorithms is constructed to arrange the algorithm services reasonably, so as to ensure the correctness of the associated algorithm for data processing. Then, the Python language algorithm and MATLAB language algorithm are encapsulated and submitted in the back-end logic, and the process data obtained after the algorithm processing is visualized.

As is shown in Fig. 15.3, firstly, input the algorithm files behind the scenes into the system build, and then allocate tasks to the system's computing nodes through the configuration file. After a series of transformations, the algorithm file that undergoes format conversion first calls the function interface. Finally, the algorithm is used to write the data stream into the returned result and return it to the function result.

The data processing module of the system mainly includes three sub-modules: system data management, data process and redundancy processing, and data visualization. System data management is to manage the data of the whole system, which is divided into three parts. The first is to enter the original data to be processed in the system through the acquisition drive mode, which is called the existing data. Second, the data accessed by the system through the data source add button is called new data. Third, the intermediate data or final results obtained after data processing are called result data. The system data management includes operations such as deleting, previewing, editing, and retrieving these data. The data processing page includes three areas: data and algorithm selection on the left, processing flow construction of the middle canvas, and parameter setting on the right. The left area contains the

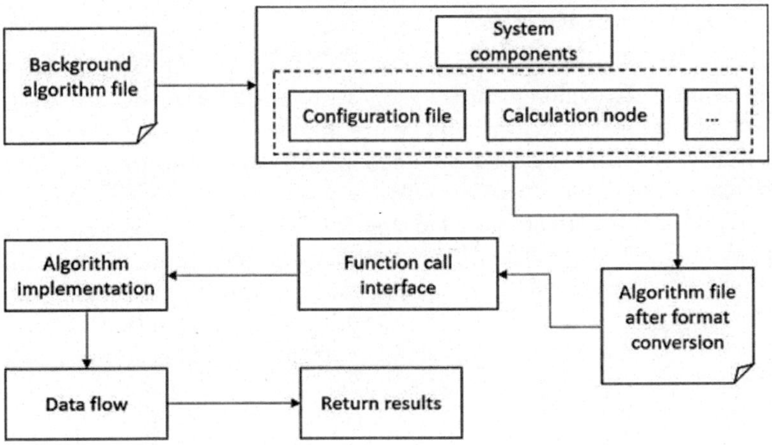

Fig. 15.3 Tree search algorithm connectivity analysis process

integration of original data in the system, the algorithm encapsulation of the algorithm library, and the export of target data. In the middle area, the data processing flow is constructed by importing initial data, selecting missing value processing, demising, and redundancy processing algorithms; the area on the right includes algorithm parameter setting, attribute description, and algorithm introduction to facilitate understanding the algorithm and selecting appropriate algorithm for data processing. Finally, define the output data source to complete the data processing.

The data fusion module of the system mainly includes ontology construction, data semantic relation extraction, ontology mapping, and data fusion. The result shows four modules. Data semantic relation extraction module: This module is divided into four areas: database configuration, data selection, data overview, and semantic file list. The database configuration is mainly to connect to the corresponding database by entering the configuration information of the database and filling in the semantic file name column. The data selection area loads the tables in the database. The data overview area previews the data in the tables; Click Extract to extract the semantic relationship and display the list of the last extracted files.

Ontology mapping module: mainly select the local ontology that needs to establish a mapping relationship, select the relevant data in the ontology, filter out the relevant fields, and select the fusion rules to complete the data fusion. Fusion Result Display: mainly manages the fused data results, including the result data name, fusion rules, creation time, etc. Click View to display the result data.

15.5 Application Results

In order to meet the requirements of rapid response, continuous query, and high quality of massive real-time data in the power digital twins, this paper first proposes a memory-based real-time database storage mechanism to effectively organize and manage massive real-time data. Secondly, the data fusion method based on digital twins mentioned in the article is used to complete the fast index location and continuous query access of data, and then Lagrange interpolation, non-local mean demising, and Kalman filtering algorithms are used to process missing values, demising, and redundancy of data. It lays the foundation for the efficient fusion of the subsequent virtual and real data. Aiming at the problem of multi-source heterogeneity and semantic heterogeneity of industrial virtual and real data, this paper proposes an ontology construction method, designs a virtual and real data fusion architecture based on this method, and realizes the semantic feature extraction of power digital twin structured and unstructured virtual and real data. At the same time, the similarity calculation method is introduced to realize the ontology mapping and the virtual and real data fusion is completed according to the fusion rules. Finally, the digital twin virtual real data fusion of the fan hub casting process is taken as an example to verify.

Through the research and implementation of the power digital twin virtual real data fusion system, we have a deep research and understanding of the real-time data storage, continuous query access, real-time processing, and heterogeneous data fusion methods of the power digital twin. However, due to time, personal knowledge depth, learning ability, and other factors, the power digital twin virtual real data fusion system studied in this paper has many deficiencies. The data security needs to be improved in the future. What the system accesses is the data in the actual power production process, which will be directly associated with the power equipment. The data may have security vulnerabilities, so we need to focus on the security of power data in the next step of research. In the subsequent system version iteration process, we need to optimize the overall system architecture. Realize power data isolation and security protection.

Acknowledgements This work was supported by State Grid Corporation of China's Science and Technology Project (5108-202218280A-2-396-XG) which is 'Research on rapid construction and linkage analysis technology of distribution network facilities digital twins integrated with electrical topology'.

References

1. Gong, Q., Wang, M.G., Lan, W.Y., et al.: Research on digital scheme of storage power station based on digital twin. Electromech. Technol. Hydropower Station **45**(11), 3 (2022)
2. Zhang, C., Tang, L., Huang, X., et al.: Research and application of smart substation based on digital twin technology. Power Equipment Manage. (19), 3 (2022)

3. Wang, R.G., Ding, J., Wang, Y.S., Liu, B., Zhang, J.M., Jia, B., Li, J.T.: Research on data reliability of distribution switch cabinet temperature sensor based on digital twin. Guizhou Electric Power Technol. **024**(006), 27–35 (2021)
4. Huang, Y., Mao, L.S., Zhang, X.F., Sheng, S.S., Huang, S.M., Zhang, G.J.: Salary analysis and process design of online immersive teaching system based on digital twin platform. J. Dist. Educ. **039**(001), 51–62 (2021)
5. Wan, L.Y.: College maker space with digital twin: implication, framework and function—evolution from physical space, mixed space to mapping space. J. Dist. Educ. **38**(3), 10 (2020)
6. Gao, Y., He, X., Ai, Q.: Multi agent coordinated optimal control strategy of smart micro grid based on digital twin drive. Power Grid Technol. **45**(7), 9 (2021)
7. Li, H.F., Wang, W.: Digital twin smart learning space: connotation, model and strategy. Modern Dist. Educ. Res. (2021)
8. Yang, G., Zhou, H.Y., Wang, B.C.: Intelligent man-machine cooperation driven by digital twins: theory, technology and application. J. Mech. Eng. **58**(18), 279–291 (2022)
9. Li, L.H., Mao, C.L., Lei, B.B.: Digital twin driven multi station linkage optimization of IoT mixed single packaging line. Packaging Eng. **42**(1), 163–170 (2021)
10. Zhao, Y., Fu, X.L., Liao, Q.M., et al.: Intelligent pulse control based on digital twins. Aviation Manuf. Technol. **63** (1/2) (2020)

Chapter 16
Research and Implementation of Rules Extraction Technology for Digital Twin Objects in Distribution Network Based on Semantic Understanding

Junfeng Qiao, Lin Peng, Aihua Zhou, Sen Pan, Pei Yang, Zhujian Ou, and Jianguang Yao

Abstract The digital twin of electric topology grid refers to the digital twin of topology grid, mapping the real world topology grid into a virtual digital space, and then fully reflecting the dynamic changes of entities and their relationships in the whole life cycle time scale through full coverage of high-density dynamic data, so as to achieve intelligent analysis, dynamic decision-making and mutual inductance cooperation based on the full model data of digital twin. The digital topology grid can affect the twin interaction of real reality and digital reality by constructing the closed loop data flow of the electric topology grid. The digitization electric network constructed by it relies on the efficient real-time data flow to realize the situation awareness, virtual deduction, and operation management regulation of the electric grid. Digital twin grid is an inevitable stage for the transformation of fresh forms of electric energy, for example, the smart electric grid, energy Internet of Things, active topology grid, etc., in recent years, and is also an effective test method for new digital infrastructure tasks of electric power from 2020. From the perspective of topology grid digital twin object rules, this paper uses a semantic understanding method to extract topology grid digital object rules, to meet the needs of topology grid digital technology, and promote the business appliance of topology grid digital technology.

J. Qiao (✉) · L. Peng · A. Zhou · S. Pan · P. Yang
State Grid Laboratory of Power Cyber-Security Protection & Monitoring Technology, State Grid Smart Grid Research Institute Co., LTD., Nanjing 210003, Jiangsu, China
e-mail: 1318558905@qq.com

Z. Ou · J. Yao
Nantong Power Supply Company, State Grid Jiangsu Electric Power Supply Co., LTD, Nanjing 210014, Jiangsu, China

R. Kountchev et al. (eds.), *Multidimensional Signals, Augmented Reality and Information Technologies*, Smart Innovation, Systems and Technologies 374,
https://doi.org/10.1007/978-981-99-7011-7_16

16.1 Introduction

Digital Twin is an advanced simulation and analysis technology that integrates digital information technology, interactive technology, mathematical logic analysis technology, and real-time simulation analysis, which helps solve the electric bottleneck problems tucked by the current power system development [1]. The key to digital twin lies in the deep coupling, joint simulation, and panoramic presentation of information flow and energy flow, to improve the transformation of electric grid from control based on simple signals to intelligent control based on information physics. However, the fusion and interaction of information flow and energy flow will inevitably lead to the transmission of the consequences of network attacks suffered by the information system to the electric grid, which is likely to cause the chain failure reaction of the information physical system and cause the expansion and severity of the failure [2]. With the launch of the ten major tasks of the new digital infrastructure of electric power in 2020, digital twins will have broad application prospects in the electric grid, and will definitively promote the emergence of digital twin dispatching system, digital twin generation system, digital twin substation, digital twin topology grid, and other new forms [3]. Among them, the topology grid is the best practice scenario that can really be implemented by the digital twins in terms of coverage, user experience closeness, transformation investment, and effectiveness. However, due to the large scale of the topology grid, complex network structure, a lot of terminals, and various communication modes, its information system and communication network are facing great security challenges [4].

Based on digital twin technology and semantic understanding method, this paper designs a digital twin-based ubiquitous electric grid network of equipment model architecture, and studies and describes the digital twin ubiquitous power internet of things model composition (including physical object, virtual object, ubiquitous electric power internet of things twin data, service system), operation mechanism, etc., to achieve the optimization of fast calculation, equipment chain management technology of electric grid internet of things system [5].

16.2 Related Work

Before the concept of digital twins was put forward, there were some similar definitions, such as digital model, digital shadow, etc. In order to avoid confusion among readers, this paper gives clear differences between digital models, digital shadows, and digital twins [6]. Digital models are digital representatives of real reality, while whether the data collection is real-time enough is uncertain, so they belong to static models. There is an inevitable correlation between the virtual model and real model [7]. The two affect each other, and the virtual digital model can provide guidance for the correct status of the actual power equipment and facilities. However, in real data description entities, it is unknown whether the data description is correct. Digital twins are not only the description of physical entities, but also two-way data transmission between them [8]. From one perspective, the situation of real reality is transferred to virtual twins to achieve actual-time rectification of virtual twins. On the other hand, the imitation and optimization results completed by Digital Twin in the virtual model can also reflect real reality to correct real results. So, different data sets interaction with real reality is the most important factor of virtual twins and an important measure to ensure the synchronization of virtual twins and real reality in the state.

In recent years, virtual twin has been proposed in the top five important technology which respected by company of Gartner for many years (2019, 2020, and 2021). Meanwhile, American virtual market research institutions emphasize that the scale of the virtual twin market will reach 65 billion dollars by 2030 [9]. Up to now, there have been more than 100 academic articles on digital twins, most of which were published after 2016, and the fields involved have gradually expanded from the initial aerospace field to manufacturing, navigation, automobile, petroleum, and other fields [10]. Virtual technology is a useful technology to realize the fast decomposition and analysis of the real reality and the virtual model; a method that combines big data governance, artificial intelligence technology, and digital construction technology. With virtual digital technology, electric power equipment is abstracted into a virtual three-dimensional model, and then through the analysis of the three-dimensional model, the problems in the power operation and business process are found, which will react to the improvement and improvement of the power business.

16.3 Key Techniques of Rule Extraction for Digital Twin Objects in Topology Grid

In this article, the basic architecture of the virtual model in an electric topology grid is divided into four modules: topology grid physical object, information object, and service system and twin data. Physical objects are entities, processes, and systems that exist objectively in complex electric power network topology. The information object is a multi-dimensional digital simulation of physical objects. It mainly simulates, optimizes, and evaluates the various parts of the complex electric power network topology, and conducts fast electric power operation index prediction, regulation, and evaluation of its operation process. The service system is the general name of all kinds of data-driven business functions for different power customers and relevant government departments. It is mainly responsible for providing support and services for the intelligent control of complex electric power network topology driven by twin data. The data contained by virtual objects, service systems, and physical objects, as well as their multi-dimensional fusion data sets, constitute the twin data system of complex electric power network topology, which is the driving force for electric grid operation and interaction.

16.3.1 Digital Twin Physical Object and Information Object of Topology Grid

Physical objects can acquire real-time and accurate data and evaluate the operation method of the electric grid. Compared with the current power system, the complex electric power network topology which was built by the digital twin method, in addition to the existing functions, also has the ability to collect and integrate multi-dimensional real-time data for its physical objects, as well as the ability to integrate all elements of the complex electric power network topology in the "customer grid computer environment". All parts of the complex electric power network topology can not only design their own action mechanism according to operation data, state sensing data, and interference data, but also carry out collaborative control and optimization of their respective functions under the goal of global optimization.

For semantic understanding, it is important to first establish a plan semantic understanding model, establish a plan entity analysis model based on regular expressions, and establish a plan entity standardization model based on a text similarity algorithm. By integrating the above models, the semantic understanding, semantic analysis, semantic standardization, and other functions of the power grid fault handling plan are realized, including notification, startup and shutdown, unit output adjustment, power flow control, and control voltage. The entity library of fault handling plans such as control frequency strongly supports the application of fault handling plans in regulating business scenarios.

The information object in the article mainly refers to the collection of digital models. We build these models from three aspects: role, behavior, and indicators. The complex electric power network topology based on digital twins collects and processes the real-time data and operation status of real entities during operation, and conducts real-time and dynamic regulation and improvement of its operation process through digital simulation under the condition that the physical prototype is highly consistent. At the same time, the digital complex electric power network topology model and related information can interact with real reality in real time to achieve the integration of digital complex electric power network topology and physical objects.

As is shown in Fig. 16.1, the digital twin information object consists of the grid layer and platform layer of complex electric power network topology. The network layer uses edge computing to reprocess the terminal sensing data, distribution terminal data, regional user load data, power Internet of Things data, etc., and upload them to the remote layer through 5G and other intelligent sensing technologies of wireless networks and advanced technologies of wired networks.

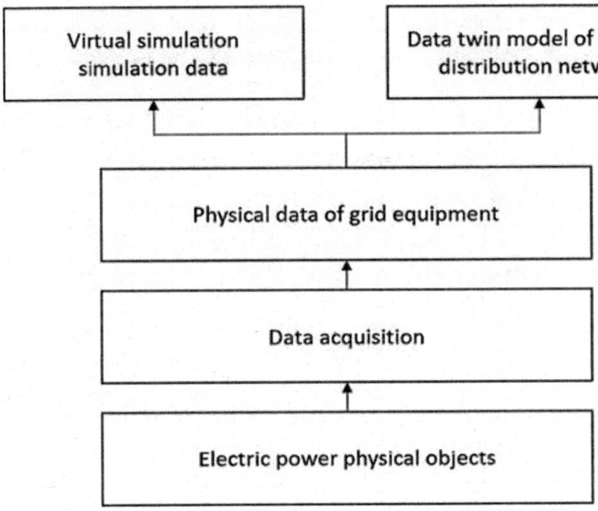

Fig. 16.1 Data extraction of digital twin model in topology grid

16.3.2 Electric Topology Grid Digital Twin Service System and Twin Data

The topology grid digital twin service system refers to the collection of all functions that drive the complex electric power network topology system to fully operate by using the data obtained from information objects and the model built, such as equipment operation situation awareness, load density analysis and power flow of the topology grid system, active perception of low-voltage faults in the operation and distribution of the complex electric power network topology and early prediction of electric grid risks.

The complex electric power network topology twin data mainly consists of four parts: data perceived by physical objects, data simulated by information objects, data processed by service systems, and data generated by the integration of the three. The twin data of physical objects are composed of complex equipment files related to power, main network of the power system, and equipment data of the distribution system. The twin data of the information object includes the data received in real time during the operation of the complex electric power network topology and the data perceived and constructed during the operation. The data related to the service system includes the data from top management to down operation control. The continuous data fusion of complex electric power network topology twin data and the iteration and increase of its data are the driving forces to complete the operation of physical objects, service systems, information objects, and the mutual fusion of the two.

The natural language semantic understanding algorithm is generally applied in the field of power grid fault identification due to its excellent feature extraction capability. The semantic understanding of an equipment fault handling plan is essentially a text classification task. The text data is treated as a dimensional time series and processed according to one-dimensional images to achieve text semantic understanding modeling.

16.4 Implementation of Digital Twin Technology for Application Requirements of Topology Grid Operation Scenarios

The virtual model of the topology grid mainly collects the basic archive data and status operation data of the physical measurement and control equipment, protection measurement and control equipment, temperature monitoring equipment, operating environment monitoring equipment, and other equipment in the topology grid, and then perceives the basic information data of physical objects through high-precision micro intelligent sensing technology, distributed advanced sensing technology, and land navigation technology. Then in the network layer, edge computing is carried out for distribution terminals, load management terminals, and physical gateways through terminal awareness. Wired network transmission and wireless network transmission based on stereo sensing technology are realized through peer Ethernet and power carriers. The digital twin technology of the topology grid simulates the power grid through real-time control, multi-network coordination, and full-service monitoring in the digital twin processing system after collecting the power grid information and other relevant real-time state information. Finally, on the basis of multi-dimensional data processing and fusion, the knowledge base and experience base based on simulation data are constructed.

16.4.1 Power Consumption Prediction of Topology Grid Based on Digital Twin Technology

The power consumption of users is related to economy, humidity, temperature, season, and human factors, and the effect on electricity customers is different in different environments. In the article, the most important influencing factor is to describe the differences between users' consumption and distinctive effects, and the contraction of each affective factor is statistically sorted to remove the factors with low coupling and is very important for reducing power loss in transmission lines. In the statistical calculation of the relevance coefficient, parameter a is the electric power consumption situation, parameter b represents the electricity consumption index. The similarity between the two data sets depends on the curve fitting degree of the two data sets. Parameter C is an ordered even set, including a variable and b variable, and the parameter C|E represents the proportion of C in parameter E.

$$\mathrm{FID(C)} = \max_{ab<B(|C|)} \frac{I*(C,a,b)}{\log_2 \min\{a,b\}} \tag{16.1}$$

As is shown in formula (16.1), I*$(C, a, b) = \max_E(C|E)$ is the FID under area E. F is a monotone increasing function and satisfies F(i) = o(i). For data set C_k (k = 1, 2, …, n), FID is used to evaluate the correlation degree between various influencing factors and the electric operation index. The solution process of this data set is divided into several steps, ranking each effective factor, and influencing factors of data value. Due to different input parameters, extract several indicators with typical power operation characteristics into this data set. The data set composed of these indicators can fully display and describe the consumption of electricity customers.

For the data set C = $\{a_i, b_i\}$ containing l data set and j role electricity characteristics, where |C|=l, the integration model of the X training tree uses k sub-functions to get the following calculation result:

$$\mathrm{b} = \varphi(a_i) = \sum_{k=1}^{n} f_k(a_i),\ f_k \in \mathrm{M} \tag{16.2}$$

In formula (16.2), M = $\{f(a) = wq(a)\}$. Where M represents the node position at the beginning of traversal, q represents the data pattern of this model, and parameter Y represents the order of the corresponding terminal of the structure. Each Mk represents the connectivity path between any two nodes. Use w_i to calculate the value range of each terminal node. Given a sample, use the calculation configuration of the data structure to be divided into multiple calculation steps. At the same time, the sum of the scores of the related terminal nodes is added as the last calculating output.

$$\begin{cases} N(\varphi) = \sum_{i=1}^{k} n\left(\hat{b}_i, b_i\right) + \sum_k \Omega M_k \\ \Omega(M) = \gamma Y + \frac{1}{2}\lambda||a||^2 \end{cases} \tag{16.3}$$

As is shown in the formula (16.3), N represents the key factor for calculating line loss, which is the difference between the predicted value of electricity consumption and the target value, and Ω is a restricted interval to determine the time interval and performance of the calculation process.

16.4.2 Implementation of Digital Twin Technology for Power Topology Grid

The following describes the operation process of the twin electric topology grid from three aspects: key equipment, overall regional power consumption, and middle-voltage electric supply scope of influence. The operation process of the analysis is shown in Fig. 16.2, then the specific process is shown below.

As is shown in Fig. 16.2, high-performance data acquisition equipment is used to collect real-time massive data during electric operation and maintenance, equipment failure, and use in the electric grid topology. Based on the gathered real-time data of power grid equipment and the historical data of the service cloud platform

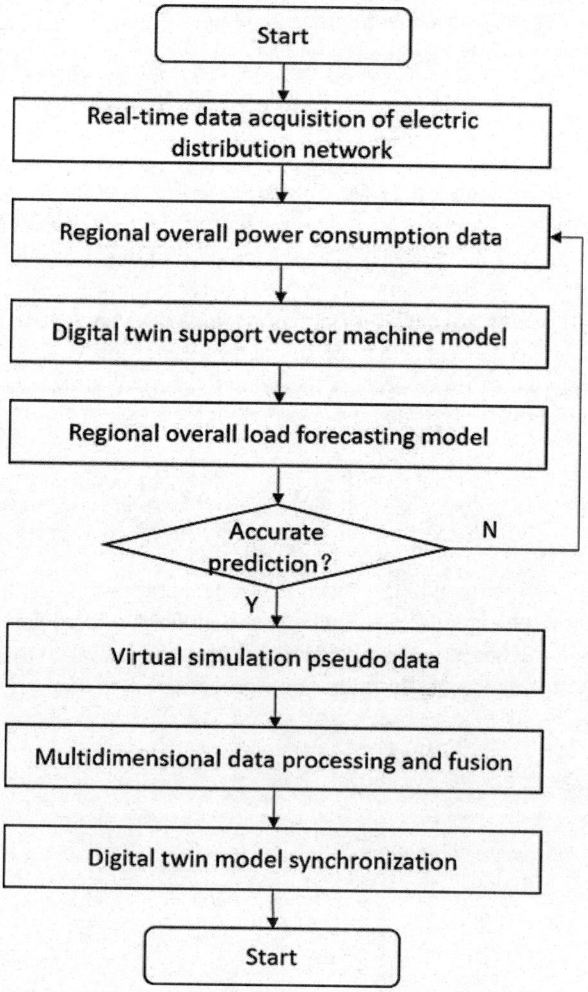

Fig. 16.2 Digital twin realization of topology grid for power application

of the electric device, the real-time data of the electric grid device is collated, the quality comparison algorithm is used to compare with the historical data of the service platform, and the specification data of the service cloud platform of power grid equipment is used to judge whether the power grid equipment in operation meets the quality requirements. If it does not meet the quality requirements, it will convey to the probable electric device with virtual data and find the three-dimensional maintenance instructions in the complex electric grid graph topology experience database. The three-dimensional maintenance instructions are associated with the relevant maintenance operation instructions through the actual equipment maintenance services, and a digital twin model based on high fidelity is built. This model forms the three-dimensional maintenance instructions by simulating the maintenance process of electric multi-type devices and improves the accuracy of maintenance.

The data of the power grid device after maintenance shall be collected and iterated interactively with the matching twin model. When all requirements of power supply quality control are completed, it shall be iterated interactively with the simulation data of shafting equipment consumption, value depreciation and failure in daily operation, etc. The virtual model can detect these data, as well as find out the cause of the accident through multi-dimensional cause analysis, analyze, and find out potential power grid operation faults. Compare and interact with the current expected state defined by the digital twin model of ubiquitous power Internet of Things, evaluate the current operating state of ubiquitous power Internet of Things, and achieve equipment whole-process management, equipment fault analysis, equipment network access status detection technologies of ubiquitous electric grid frame topology. According to the acquired user profile information, process the real-time data and historical data of power consumption, make statistics on all relevant influencing factors, train the corresponding power consumption prediction model for power users 1, 2,..., and n using the twin support vector machine algorithm, and fuse all the prediction results to obtain the overall power consumption prediction model for each region. When there is a load problem in the whole area, it interacts with the matching power twin model, finds a three-dimensional maintenance guide in the ubiquitous electric grid frame topology experience database, and simulates the electric power flow calculation process to improve the prediction and analysis of the matching degree of power customer's power load and power supply.

If there is no topology error and insufficient power supply in the whole area, the simulation data of shafting fatigue wear, torsional vibration, thermal stress, and other simulation data generated by the digital twin model will be iterated interactively to complete the assessment of power grid recover-ability and health status of the electric grid. Compare and interact with the current expected state defined by the digital twin model of complex electric grid frame topology Internet of Things, evaluate the current operating state of complex electric Internet of Things, and achieve iterative optimization of fault diagnosis, prediction, and health management technologies of ubiquitous power Internet of Things system. Based on real-time measurement data in public transformer areas and special transformer areas, weather, environment, and other data of other relative data, as well as the historical data of low-voltage power supply service cloud platform, access the fault recording data by connecting the digital twin knowledge base and experience base. When problems occur in the low-voltage power supply area, it interacts with the matching low-voltage power supply system twin model, finds three-dimensional maintenance instructions in the ubiquitous power Internet of Things experience database, and simulates the fault maintenance process on the digital twin model. When there is no problem in the low-voltage power supply area, it interacts with the real data generated by the digital twin model and the fusion data obtained through multi-dimensional data processing and fusion to realize active awareness of low-voltage faults in the operation and distribution deployment process. Compare and interact with the current expected state defined by the digital twin model of the topology grid, evaluate the current operation state of the complex electric Internet of Things, and realize the equipment networking posture, sudden probability of equipment operation failure, and equipment operation situation awareness of the complex electric Internet of Things system.

16.5 Application Results

A digital twin model is built for generating units in power plants to analyze and predict their working life and operating conditions. The intelligent management system of the generator set based on a digital twin monitors the bearing vibration and shaft speed of the entity in real time by current power sensor, angular speed sensor, etc. Based on the specification data, stored historical data, and real-time collected data, the system describes the components of the generator set from three dimensions of physics, information, and behavior on the cloud platform, and integrates the high-fidelity digital twin model of the generator set. The entity of the generator set and its digital twin information model operate together. The twin data generated include the data sensed during operation, as well as the simulation data of shafting fatigue wear, torsional vibration, thermal stress, etc., generated by the digital function.

The virtual technology of the generator set includes condition monitoring service, equipment operation service, optimized operation service, and maintenance guidance service. The condition monitoring model is mainly used by the operation and maintenance personnel to check the internal operation status of the generator set through the digital twin information model built, compare and integrate the expected operation status of the twin digital model with the actual operation status of the generator set, and control the overall working status of the generator set.

The semantic understanding-based digital twin rule extraction method proposed in this article can make the construction of digital twins in power equipment more efficient and easy, reduce manual intervention, and reduce the errors of digital twin models, providing technical support for future automated digital twin model construction. The fault diagnosis service comprehensively analyzes the digital twin model of the generator set and the relevant data of the generator set entity to accurately warn the generator set of overspeed, broken shaft, and other accidents. The optimization operation service mainly uses the dynamic interaction between the digital twin model and the physical entity of the generator set to complete the iterative optimization of the bearing opening and operation parameters. The maintenance guidance service mainly simulates the maintenance process of the generator set based on the digital twin model by querying the maintenance operation experience database, forming a three-dimensional operation experience database, which provides an effective way for the reliable operation and maintenance of the generator set.

Acknowledgements This work was supported by State Grid Corporation of China's Science and Technology Project (5108-202218280A-2-396-XG) which is 'Research on rapid construction and linkage analysis technology of distribution network facilities digital twins integrated with electrical topology'.

References

1. Zhang, K., Lu, G.Y., Wu, L., et al.: Validation and analysis of image information for sentence semantic understanding and representation. J. Comput. Sci. **44**(3), 15 (2021)
2. Huo, T.T.: Design and application of virtual simulation technology for power system relay protection. J. Heilongjiang Inst. Technol. Compreh. Edn. (007), 022 (2022)
3. Sun, P.Y.: Research on Network Attack Reconstruction and Simulation Technology of Power Information Physical System. Zhejiang University (2020)
4. Cao, Y., Xu, A.D., Tao, W.W., et al.: Performance Analysis and Simulation of Power Internet of Things LPWAN. China Southern Power Grid Technol. (008), 014 (2020)
5. Zhao, H., Miao, K., Li, Z., et al.: Research on edge sensing adaptive data processing technology of power Internet of things. Single Chip Microcomput. Embedded Syst. Appl. **21**(011), 34–37, 41 (2021)
6. Cheng, Y. M., Jiang, M., Ma, Y. L.: Design of real-time monitoring of power customer flow data based on metadata. Inf. Techno. **44**(4), 5 (2020)
7. Chen, J.X., Liang, L., Fu, J.F., et al.: Research on smart grid data dispatching and fast distribution method based on big data. Electric Measur. Inst. **57**(6), 6 (2020)
8. Zhao, G.S., Liu, D.W., Chen, S.Y., et al.: Power system transient stability boundary analysis from the perspective of data. Power Gener. Technol. **41**(2), 167–174 (2020)
9. Zhao, Q., Wang, X.Y., Qiao, J.: Key technologies for data-driven modeling and simulation of energy Internet. Power Inf. Commun. Technol. **18**(1), 7 (2020)
10. Zeng, J.R., Li, P., Gao, L., et al.: Detection of false data injection attacks in smart grid based on TNPE. Sci. Technol. Work Safety China (2021)

Chapter 17
Analysis of Urban Landscape Plants' Configuration Based on Virtual Reality

Huishan Wang and Yanmin Liu

Abstract In order to effectively improve the quality of urban landscaping projects, it is necessary to pay full attention to the allocation of garden vegetation and adhere to the principle of adapting to local conditions to carry out reasonable collocation and effective integration of garden plants, so as to promote them to form an organic whole. Nowadays, VR (virtual reality), as a comprehensive science and technology in the computer field, has been involved in many research and application fields and is recognized as an important development discipline in the twenty-first century and one of the important technologies that affect people's lives. In this paper, the configuration analysis of urban landscaping plants based on VR is carried out. In this paper, firstly, the garden scene is modeled by 3DSMAX software, and the configuration, construction and simplification methods of plant models are emphatically studied. Then, the progressive mesh simplification algorithm is used as the generation method of continuous LOD (Levels of Detail) model, and the complexity of the scene is controlled by providing different LOD descriptions for the scene, which improves the rendering efficiency of the scene and the realism of the landscaping plant configuration. This not only achieves the high efficiency of display but also avoids redundant data. This method is particularly effective in the construction of large-scale scenes.

17.1 Introduction

For a city, if it lacks unique shade and flowers, it can't achieve the purpose of reducing noise and fog and dust pollution, and urban residents can't enjoy the delicate mountains and blue sky. As a living organism, plants are the most unique vitality elements in the landscape system. With the rapid development of the city, the change of urban ecological space has more and more obvious influence on plant landscape [1, 2].

H. Wang (✉) · Y. Liu
School of Building Science and Civil Engineering, Xiamen Institute of Technology, Xiamen 361021, China
e-mail: zxbtianyi@vip.qq.com

R. Kountchev et al. (eds.), *Multidimensional Signals, Augmented Reality and Information Technologies*, Smart Innovation, Systems and Technologies 374,
https://doi.org/10.1007/978-981-99-7011-7_17

However, due to different concepts and insufficient management in construction, maintenance and other related links, resources are wasted to varying degrees. In recent years, with the gradual enhancement of people's awareness of environmental protection, more attention has been paid to the economic and ecological benefits brought by plant landscape in landscape architecture design, and plant configuration and planning have played an important role in ecological balance, greening and appreciation of gardens.

The arrangement of plants in urban landscaping projects includes two aspects: on the one hand, the arrangement of various plants, considering the choice of plant species, the combination of trees, the composition, color, season and garden artistic conception of the plane; on the other hand, it is the configuration between garden plants and other garden elements [3]. Related research takes the road green belt as the research object and makes full use of the location and width of the green belt to establish a model to evaluate the landscape effect of the green belt [4]. Another study pointed out that the design principle of urban roads and rural roads can follow the design principle of highway appearance, and rural roads cannot destroy nature, mainly by integrating roads into the background [5]. Some scholars have investigated the public's preference and cognition of roadside landscape of urban expressway by showing the subjects pictures depicting the road landscape [6].

In order to effectively improve the quality of urban landscaping projects, we must pay full attention to the allocation of garden vegetation, and adhere to the principle of adjusting measures to local conditions to carry out reasonable collocation and effective integration of garden plants, so as to promote them to form an organic whole [7, 8]. Nowadays, VR (virtual reality), as a comprehensive science and technology in the computer field, has been involved in many research and application fields, and is recognized as an important development discipline in the twenty-first century and one of the important technologies that affect people's lives. At present, VR technology has been widely used in commerce, aviation, military, manufacturing, engineering construction, medical care, geoscience visualization, education, entertainment and other aspects. The simulation system based on VR technology is to create an information environment with immersion, human–computer interaction ability and help participants to inspire ideas, so as to achieve the ultimate goal of participants to acquire knowledge and form concepts in the virtual environment.

17.2 Principles of Plant Disposition in Urban Landscaping

The principle of sustainability. In the process of urban landscaping plant allocation, we should make a comprehensive analysis with the natural environment as the core, and based on the relevant theories of ecology, scientifically and rationally allocate urban landscaping plants under the condition of deeply grasping the biological and ecological characteristics of various plant species, which can adapt to the local environment and form an all-round and stable complex ecosystem with other organisms.

The principle of plant diversity. Due to the lack of reasonable planning in cities, the species of plants are single and the structural layout is unreasonable. In this regard, it is suggested that the principle of diversity of greening plants should be adhered to in the selection and configuration of urban landscaping plants, so as to play the role of environmental protection and improve the aesthetics of the city [9]. Pay attention to the seasonality of urban landscaping plants and choose diversified plants according to the seasonal and climatic factors of your own city. You can't just choose plants according to the winter standards, try to reflect the differences in colors and highlight the landscaping effect of the garden.

Follow the principle of beauty. A high degree of unity of science and artistry is a must for high-level plant landscape design. Only when plants can fully adapt to the environment and make use of it, but also when they are reasonably collocated, can the beauty of modeling and artistic conception of plants and groups be reflected.

17.3 Configuration Method of Urban Landscaping Plants

17.3.1 Garden Scene Modeling

There are many kinds of entities in gardens, covering a wide range of complex and changeable contents, ranging from scenic spots to the design of courtyards. According to different functions, there are parks, forest parks and botanical gardens. Garden scene is composed of garden model, and the quality of garden model directly affects the effect and fidelity of garden scene operation. For us to make garden scenes, 3DSMAX should be the first choice, and 3DSMAX is adopted for modeling because the model established with it has a strong three-dimensional simulation effect. Because they are invisible when browsing the scene, removing them will not affect the visual effect of the model, but eliminating these redundant polygons can greatly reduce the complexity of the whole scene.

In the allocation of urban landscaping plants, it is mainly necessary to follow the principles of sustainability, culture and ecological priority, so as to show the urban characteristics. There are many forms of urban landscape construction. Different locations, occasions and regions need to realize different planting methods and plant composition to meet different functional requirements. Static characters cannot move in the scene, but they can participate in the collision [10]. No matter through dynamic simulation, script setting or mouse dragging, it can not be deleted in the running state, because the collision volume of static characters will be optimized in a special format. Based on this, the rational allocation of plant science in urban gardens, which can reflect humanistic connotation and represent people's spiritual character and a certain historical period, will have an important impact on the formation of landscape with regional characteristics in various cities in the future.

Beautiful and reasonable landscape of road plant community can make urban green space exert the greatest ecological benefits, and the diversity of road greening plants is an important factor to maintain ecological balance and enrich urban landscape [11]. A good plant configuration model can not only reflect the regional culture of the city but also enhance the driver's sense of security.

This paper analyzes the diversity of plant communities from the following indicators:

$$S = \text{Species appearing in sample plots} \tag{17.1}$$

Gleason exponent formula:

$$d = \frac{S}{InA} \tag{17.2}$$

Margalef index formula:

$$d = \frac{S-1}{InN} \tag{17.3}$$

In the above formula: S is the number of species in the sample plot, A is the area of the sample plot, and N is the total number of individuals in the sample plot. The three indexes are all diversity indexes, but the calculation methods are different. The higher the diversity index, the richer and more diverse the plant species in the community are. Using three calculation methods can strengthen the argument and improve the accuracy of the conclusion.

When designing landscape architecture, we should comprehensively consider the artistic connotation and cultural communication function, so that the layout of landscape architecture can meet the aesthetic requirements of the public, and the layout of landscape architecture determines the main style and cultural tendency of the garden to a certain extent. In order to enhance the ornamental value of landscape architecture, it is necessary to rationally allocate the plant species to avoid that most plants in the garden will fall leaves and wither in the cold winter, which will make the garden lose its ornamental value.

Taking plants as design materials to create landscape is a unique design method in landscape design. Trees, shrubs and grasses are common plant elements in the landscape. These green plants construct space, place, visual focus and visual field, and also provide rich colors and materials for the landscape. Considering the relationship between the environment needed by plants and other plants, choosing plants properly and combining trees, flowers and grass can not only meet the landscape elements but also create artistic effects and create a changeable garden landscape. Therefore, we must follow certain principles and pay attention to artistry in the allocation of plant models.

In order to realize the garden scene with plants as the main body, we should carefully adjust and compare the parameters when generating three-dimensional plants, so as to reduce the number of faces or points. The parameter values of plants should be carefully weighed, and the optimal or simplified values should prevail. In the comparison of Fig. 17.1, the number of simplified plant faces is still very high, so the usage of 3D models in the scene should be minimized.

On the basis of meeting the visual requirements, the most common method to render plants is to use a very simple mapping model, such as plane mapping model and cross patch mapping method. Make use of the optimization and collapse technology of 3DSMAX, and make the model have as few points and faces as possible without affecting the appearance.

The woodland model can also be automatically generated according to vector data and templates. Figure 17.2 shows a simple template of the woodland model, which defines the woodland as a box and expresses the landscape of the woodland with texture mapping. Complex templates need to define various parameters, such as the surface texture, the model of individual trees, the types of trees, the density and distribution characteristics of trees, the range of random heights of trees and so on.

Fig. 17.1 Simplification of plant model

Fig. 17.2 Woodland model

17.3.2 Landscape Plant Model Generation

In the process of urban green space planning and design, we should pay attention to plant landscaping, improve the rate of green space, enhance plant diversity and enhance the ecological benefits of urban green space. The multi-layer plant community model composed of tree layer, shrub layer, hedge layer, herb layer and interlayer plants is an important form to realize the landscape and ecological function of garden green space. Influenced by China's ancient poetry culture for a long time, plants are usually endowed with some special images, which are reflected in ancient and modern landscape gardens in China. If the layering of flowers and trees is just right, it will greatly improve the effect of urban landscaping, enhance the beauty of garden landscape and give tourists a refreshing feeling. For example, when arranging poplars, maples and pines with different heights, they should be arranged in the order from low to high, so as to show the multi-layered feeling of tree species and enhance the ornamental value of plants with different colors.

The landscape simulation system based on VR technology is a real-time graphics generation system. The so-called real-time graphics generation means that when the user's viewpoint changes, the system should generate corresponding graphics immediately. The model in the three-dimensional scene is composed of a series of triangular patches, and the objects (boxes, spheres, polyhedrons and particles) seen in the scene are represented by a series of colored triangular patches in the computer. Therefore, in the plant configuration of street landscape, we should not only be rich in layers, high and low but also pay attention to the collocation and coordination between plant colors. Properly control the proportion and scale of green belt and pedestrian area, and design more reasonable street space scale to increase the comfort of walking space without affecting the landscape quality of other factors.

As far as the current level of computer graphics is concerned, as long as there is enough computing time, accurate computer images like photos can be generated. To improve the speed of graphics display, a very effective method proved by practice is to reduce the complexity of the scene, that is, to reduce the number of polygons that the graphics system needs to deal with. When there are a large number of visible faces in the picture and simplification of the scene needs to be considered, if the projection of many visible faces on the screen is less than one pixel, these visible faces can be merged without losing the visual effect of the picture [12]. LOD (Levels of Detail) technology reduces the geometric complexity of the scene by simplifying the surface details of the scene step by step without affecting the visual effect of the picture, thus improving the efficiency of the rendering algorithm.

Trees are one of the most common natural landscapes in nature, with a wide variety, diverse growth, strong structural characteristics and complex geometric structure. The tree model generated based on fractal algorithm has realistic visual effect, but the amount of data is too large for campus roaming environment, so it is difficult to visualize in real time. This section combines the advantages of the above two methods, combines the fractal model of the tree with its image replacement model

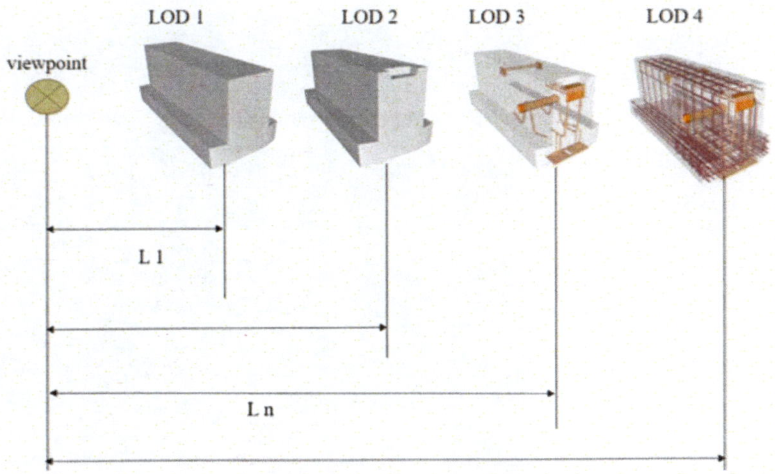

Fig. 17.3 LOD hierarchical detail model

and initially solves the visualization problem of the tree model in the arrangement of urban landscaping plants, so as to achieve realistic and high-speed visual effects.

Figure 17.3 shows the LOD hierarchical detail model. In this paper, Java3D is used as a tool to develop a 3D VR scene. Programmers can divide the display information into more display levels in order to reduce the image jump, without worrying about the increase of data. Because each additional display level is nothing more than setting a few more identification bits on the basis of the whole data.

Many plants in nature obviously grow in fractal form, and branches repeatedly split to produce smaller branches. For trees, their branches are typical fractal structures and can be connected by cylinders or cones. Self-similarity is the most important feature of fractal, and the branch structure of trees in nature has a self-similar fractal structure in statistical sense, so a self-similar structure can be extracted from the overall structure of trees as a generator to construct a tree model. It is generally believed that the growth of trees in nature is influenced by many independent random factors, each of which is very small, and the overall influence can be regarded as normal distribution. In this way, we can introduce normal distribution into the algorithm. In the process of model generation, we can control the model shape by using different parameters such as branch angle, branch length and the number of branches on the trunk to generate more realistic trees.

Trees in nature are all composed of trunks, branches and leaves, so branches and leaves have become the most basic units of trees. Therefore, in the process of 3D tree modeling, it is mainly divided into two parts. The first step is to establish tree branches by fractal method based on L system, and the second step is to add leaves to form a complete 3D tree.

In order to increase the realism of the 3D model tree, the frustum and branches are drawn with the frustum as the basic unit, and different generators are selected according to different tree species. When the binary generator is the main feature

Fig. 17.4 Spanning tree with random effect added

of the tree to be modeled, the probability is set close to 1, and only a few ternary generators are used at this time; On the other hand, the three-way generator is mainly used. As shown in Fig. 17.4, the trees generated after adding random variable control.

At a long distance, whether sparse trees or patches of trees, low-resolution tree images are used to express them, which is LOD1 at this time; At close range, when the outline branches of the tree are obvious, you need to use high-resolution images to express it, which is LOD2 at this time; When the leaves are obvious at a close distance, it is necessary to use fractal models to express them. Because there are generally few trees entering the immediate field of vision, that is, there are few fractal models to be drawn, it will not cause a sudden increase in data volume, and this time it is LOD3. As shown in Fig. 17.5, the effect comparison of tree models at three LOD levels.

The three-dimensional object described by the model is placed between two clipping planes. The front clipping plane can be regarded as a computer screen, and all three-dimensional models need to be projected on this plane. The selection of LOD level is determined by the size of pixels occupied by the models after being projected on this plane.

d is the projection of the object on the two-dimensional plane, which can be expressed by the number of pixels, D is the distance from the observation point to the front cutting surface, F is the spatial position of the model, l is its real height, and b is the inclination angle of the model caused by the change of the observation point. The size of d can be calculated as follows:

$$x = \cos b \times l/2 \tag{17.4}$$

$$y = \sin b \times l/2 \tag{17.5}$$

Fig. 17.5 Comparison of tree model effects

$$d = 4DFl\cos b(4 \times F \times 2 - l \times 2\sin 2b) \tag{17.6}$$

This can control the choice of LOD level. Corresponding to three LOD levels, three d values need to be determined: $d_1 < d_2 < d_3$. The selection of LOD is controlled by the following conditional expressions:

if d< =d1 then use LOD3

else if d1 < d < =d2 then use LOD2

else if d2 < d < =d3 then use LOD1.

Determine the tree objects located in the visible area of the viewpoint, calculate the size of the projected image cable for each tree according to its spatial position and the viewpoint position, and then determine the LOD level drawn according to the above judgment conditions. Read the corresponding LOD level data to draw the model, and control the frame rate when switching LOD levels.

17.4 Conclusion

With the rapid development of the city, the change of urban ecological space has more and more obvious influence on plant landscape. However, due to different concepts and insufficient management in construction, maintenance and other related links, resources are wasted to varying degrees. VR technology integrates digital image processing, computer graphics, multimedia technology, sensor technology and other branches of information technology, which is a higher level integration, infiltration and comprehensive application of these application technologies, and is the inevitable development direction of information display in the future. In this paper, the VR technology based on 3DSMAX is applied to the garden field, and the real-time control function of plant orientation is realized. Using keyboard to control the orientation movement of plants and make them reach the appropriate position can help landscape designers to adjust and optimize the planting design of plants conveniently.

References

1. Ao, R., Zhang, H.H.: Study on the allocation of water-saving plants for landscaping in Baotou City. Forestry Surv. Planning **44**(5), 4 (2019)
2. Lei, J.R., Xin, X., Song X.Q., He, R.X.: Landscape evaluation and structure analysis of park greenbelt plant community in Haikou City based on ahp. J. Northwest Forestry Univ. **31**(3), 7 (2016)
3. Zhang, L., Yi, Z., Zhang, C.D., Li, X.: Suitability assessment and integrated application of urban green space ecological technology. China Garden **32**(8), 5 (2016)
4. Zhao, H.: Plant configuration in urban landscape design. Urban Issues **6**, 251–251 (2015)
5. Luo, C.Y., Yue, M.: Study on species selection and landscape application of colored-leaf plants in Lintong district of Xi 'an. Acta Agron. **8**(6), 6 (2018)
6. Liang, Q.L., Zhang, N., Lu, P.: Investigation and study on landscaping plant resources in Yining City. Forestry Invest. Plann. **40**(3), 5 (2015)
7. Luo, Q., Li, C., Weng, S.F.: Analysis on the configuration and application of Erythrina in tropical gardens. Tropical Agric. Sci. **36**(3), 85–89 (2016)
8. Li, X.N. Spatial coordination method of landscape stone based on virtual reality. Modern Electron. Technol. **42**(24), 4 (2019)
9. Zhao, Z., Lei, S., Wen, Y. L.: Study on the Landscape Layout of Urban Communities in Beijing Based on Citizens' Preference. Forestry Economic Issues, vol. 38(1), pp. 6(2018).
10. Zhang, L., Yi, Z., Xue, J.H., Zhang, G.L., Zhong, Q.C., Fu, Y.: Study on the classification of urban difficult sites for landscaping in the transitional period. Modern Urban Res. **9**, 114–118 (2017)
11. Su, Y.: Application of virtual reality in landscape planning and design and optimization of level details. Modern Electron. Technol. **42**(9), 4 (2019)
12. Deng, Y., Cheng, J., Anumba, C.: Mapping between bim and 3d gis in different levels of detail using schema mediation and instance comparison. Autom. Constr. **67**(7), 1–21 (2016)

Chapter 18
3D Animation Simulation Based on Computer Virtual Simulation Technology

Honglin Huang

Abstract 3D animation involves computer VST (virtual simulation technology). As one of the important technologies of 3D animation, this technology can not only enhance the sensory effect of 3D animation but also has great significance for the successful construction of virtual environment. In this paper, character modeling and character animation generation are deeply studied. Based on the skeletal animation algorithm, the source files of 3D human skeletal animation are created in the modeling software, which mainly includes creating meshes, adding bones, skinning and editing envelopes. Then, the real motion data are obtained by motion capture technology, and the data are optimized and redirected to drive the 3D character motion. The research results show that the effect of skeleton animation is still good, and some basic actions have been completed. The improvement of this place is to add artificial intelligence elements, which makes the animation more natural and real. The research on action simulation technology of virtual characters based on skeletal animation algorithm has important engineering application value.

18.1 Introduction

With the emergence of VST (virtual simulation technology), the 3D animation in China is becoming more and more perfect. At present, many industries will start to apply VST. For example, when enterprises publicize corporate culture, they will generally use 3D animation to present it. In the process of making 3D animation, VST plays a very important role [1]. The most important thing in VST is the interaction between people and environment. The virtual environment is designed by designers, so designers can adjust or change a scene or a character in the virtual environment at will according to actual needs [2, 3]. Because the 3D animation produced by

H. Huang (✉)
Guangzhou Institute of Foreign Economics and Technology, Guangzhou, China
e-mail: 574788437@qq.com

R. Kountchev et al. (eds.), *Multidimensional Signals, Augmented Reality and Information Technologies*, Smart Innovation, Systems and Technologies 374,
https://doi.org/10.1007/978-981-99-7011-7_18

computer simulation technology has a strong sense of reality, it has been used in military, architectural and aviation fields for simulation and experiment.

Applying computer VST to 3D animation can perfectly integrate reality and illusion, which not only reduces workload but also improves work efficiency. But also make the entire animation scene more realistic and flexible, achieving scene fusion [4]. 3D character model and animation automatic generation technology is a relatively huge research topic, which involves a lot of knowledge and complicated process, such as 3D model construction, skeleton extraction, bone road animation and so on [5, 6]. Reference [6] proposed a fast algorithm for surface construction. Its algorithm is based on discrete point set processing, and its efficiency is very high. Its disadvantage is that it requires more input data and its application scope is limited. Literature [7] has made a deep research on the animation of virtual human and developed the limb animation and some emotional animation of characters in virtual environment. Literature [8] proposed to express the characteristics of human silhouette image by using shape context, and then to obtain the motion information of human joint points by using correlation vector machine and function regression.

3D animation involves computer VST. As one of the important technologies of 3D animation, this technology can not only enhance the sensory effect of 3D animation but also has great significance for the successful construction of virtual environment. Therefore, it is very important to study the application of computer VST in 3D animation production, and we need to attach great importance to it. In this paper, character modeling and character animation generation are deeply studied. Character model and animation are of great scientific research value in reverse engineering and virtual reality, and they are important users of computer graphics and computer hardware technology.

18.2 Research Method

18.2.1 Virtual Character Model Construction

In the production of 3D animation, the real-time display technology of 3D graphics is needed. With the aid of this technology, the production efficiency of 3D animation can be improved several times. Not only shortens the production time but also reduces the workload; interactive technology, also a kind of VST, mainly changes the virtual environment or things under the control of human beings and realizes the mutual blending between people and the virtual world. In order to keep the logical relationship between interactivity and real-time 3D performance stable, and ensure that all elements in the environment can respond quickly to control behavior, designers should combine real-time 3D visual database modeling and optimization tool platform to build models.

When analyzing, processing and processing human motion data, the first problem to be solved is how to extract motion information from the captured data and establish a mathematical model. If human motion is regarded as multi-particle motion, it is necessary to fit the trajectory of particles and study the motion characteristics of different parts of human body in different motions. If the interaction of multiple particles is considered, it is necessary to study and select a reference point, and then analyze the relative motion of particles relative to the reference point. In the process of human motion analysis, researchers pay more attention to human motion posture. Usually, the joint angle is used to represent the human posture during the movement.

The construction of virtual character model is mainly divided into the construction of character model, the construction of human skeleton, the calculation of skin binding weight and the rendering of map. The skeleton model is a skeleton structure built according to the physiological structure of human body. Generally speaking, some key joints can be selected [9, 10]. In this paper, the animation module, dynamic tools and rendering tools in Maya software are mainly used in the process of character modeling. The human skeleton structure is very complex, with more than 200 degrees of freedom. This paper designs a set of virtual character skeleton structure using Maya software. Figure 18.1 is the skeleton model designed in this paper, which has 30 joint points.

Space–time curve and joint angle directly model the original acquisition signal. The multi-rigid-body inverse kinematics method is usually used to convert the particle trajectory to the joint angle of human motion [11]. The 3D coordinates of human joint points can be expressed as the cascade transformation of vectors in multiple local coordinate systems, and the second matrix form is:

$$[x, y, z]^T = T^0 T^1, \cdots, T^n [x\prime, y\prime, z\prime]^T \quad (18.1)$$

Fig. 18.1 Character skeleton model

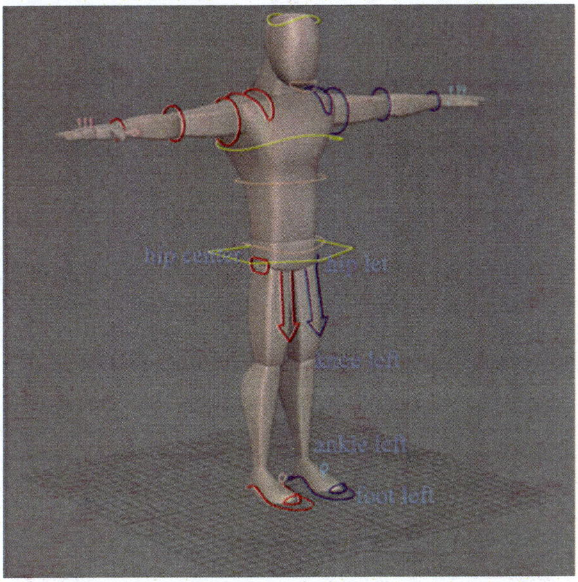

where $[x, y, z]^T$ is the coordinate of the joint point in the world coordinate system, $[x\prime, y\prime, z\prime]^T$ is the coordinate of the joint point in the n th local coordinate system, and $[x\prime, y\prime, z\prime]^T$ can calculate the rotation angle and limb length in the n th local coordinate system.

In the field of skeletal animation, we abstract people as a combination of skin and skeleton. The principle of linear mixed skin algorithm is to calculate the positions of skin vertices under the influence of a single related joint, then make a simple weighted connection, and finally get the position information of skin vertices. As shown in formula (18.2):

$$T\prime = \sum_{i=1}^{n} w_i H_i T \tag{18.2}$$

where T is the position of the current vertex (skin mesh vertex) in the world coordinate system, $T\prime$ is the position of the transformed vertex in the world coordinate system, H_i represents the transformation matrix of all joints associated with the current vertex to the world coordinate system in the initial posture, the number of previous nodes is n, and w_i is the influence weight factor of the i th joint on the current vertex.

The skeleton of a 3D character has a hierarchical structure. The geometric movements of 3D figures in space mainly include translation and rotation [12]. If a translation component is represented by a matrix, it is used to describe the translation of joint point i from the current local coordinate system to the coordinate system where its parent node is located at moment j. Matrix R_i^j is used to represent a rotation component, which is used to describe the rotation amount of the i-th joint point around its parent node at time j. Where $1 \le i \le N$.

$$P_{root}^t(x, y, z) = M_{root}^t R_{root}^t P_{root}^{t-1}(x, y, z) \tag{18.3}$$

$$P_i^t(x, y, z) = M_{root}^t R_{root}^t M_1^t R_1^t, \cdots, M_1^t R_1^t P_0^t(x, y, z) \tag{18.4}$$

where $i = 1, 2, \cdots, N$ in the above formula. When $i = root$ indicates that the node is the root node, $i = 0$ indicates that the joint point is in the initial position.

If we know the translation and rotation components of the root node in the global coordinate system, we can calculate the spatial coordinate position of the root node. Several animation frames are combined to form a motion sequence, and some animation sequences are connected to form a 3D character's skeletal animation.

18.2.2 Automatic Generation of Character Animation

There are many kinds of computer VST, such as virtual reality simulation technology, multimedia interactive design technology and graphic design technology, and each technology plays a very important role in computer VST. Therefore, only by combining all the technologies can we make the best 3D animation. 3D animation is to use it to simulate the upcoming scene in the future, and find out the possibility of future occurrence, and how to take corresponding measures to deal with it. VST is a form of 3D animation, which has a good three-dimensional sense and aesthetic value, and can display the spatial structure of buildings completely and vividly.

Computer VST has its role in many aspects. When making 3D animation, the 3D model should be established first. The 3D model includes the production of virtual scenes, 3D animated characters and action expressions. In 3D modeling, the related models should be established through the cooperation of VST and interactive technology, mainly through real simulation technology, and then some 3D animated characters should be introduced to reflect the contents that need to be displayed. Therefore, in order to quickly complete the application of VST in 3D animation, we must speed up the modeling to improve work efficiency.

Geometric model expresses the shape and appearance of computer characters, and skeleton model can realize the animation of computer characters. The movement of each character in the 3D scene refers to its displacement in the world coordinate system, while the rotation of each character refers to its transformation relative to its own coordinates. The displacement of the object can be expressed by coordinate transformation, and the rotation of the object can be expressed by skeletal animation. The 3D scene can't be a static world, it is a moving world, and many objects in it have their relative world and their own movements. The relative world movement needs coordinate transformation, while the relative self-movement needs skeletal animation technology.

With the rapid development of character animation, the generation of character model and animation also puts forward new requirements, and it is urgent to save manpower quickly. The wide application of 3D scanner in reverse engineering provides the possibility for the automatic generation of 3D character models. In this paper, 3D modeling of people is the technology used. There are generally two methods to triangulate 3D scattered point cloud data: one is to triangulate 3D scattered point cloud data directly, and the other is to project 3D scattered point cloud data onto a plane first, then to triangulate the point cloud in the plane area, and finally to map the results back to 3D space. The core Delaunay triangulation algorithm is triangulation growth method, and it is improved on this basis. It is extended by using human physiological characteristics to make the generated triangulation mesh more reasonable and efficient, and finally it is mapped back to 3D space.

The difference of all kinds of improved algorithms of triangulation growth algorithm is nothing more than the implementation of the strategy of searching for the "first point". Most scholars improve the search strategy of the third point and the data structure of the front edge list and triangle list to improve the efficiency of the algorithm. The triangulation growth algorithm is relatively balanced in both time complexity and space complexity, and it also has the advantages of simple thinking and easy programming. Compared with the point-by-point insertion algorithm, the triangle growth method does not need to spend a lot of time locating triangles, and compared with the divide-and-conquer algorithm, the triangle growth method

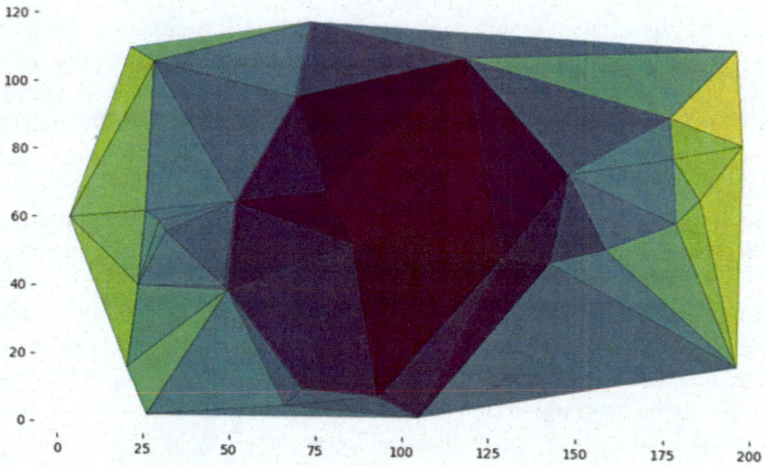

Fig. 18.2 Delaunay triangulation result diagram

does not need to spend time and effort on large-scale subset division and merging operations.

In this paper, Delaunay triangulation of scattered plane data points is carried out by using triangulation growth algorithm, which is programmed with Microsoft Visual Studio development tool under Windows 10 operating system, developed and realized by using C++ language and OpenGL graphics library function. The experimental process is to randomly generate 100 points in the 2D plane and triangulate them. The screenshot of the experimental effect is shown in Fig. 18.2.

There are many ways to find the smallest circumscribed rectangle of 3D figure point cloud data set, among which we can adopt the following methods: rotate the whole point cloud data at an angle of 10 degrees, the smaller the angle, the more accurate it is, and the sum of the total rotation angles should not exceed 90 degrees. At the same time, we should record the information of the circumscribed rectangle in the coordinate system direction every time we rotate, such as the maximum and minimum x, y, z values of its boundary points.

The matrix of coordinate transformation from coordinate system $oxyz$ to coordinate system $o\prime x\prime y\prime z\prime$ is:

$$(x\prime, y\prime, z\prime, 1) = (x, y, z, 1) * T(-x_0, -y_0, -z_0) * R \tag{18.5}$$

Therefore, the coordinate transformation matrix of point cloud data translated into standard coordinate system is:

$$(x, y, z, 1) = (x\prime, y\prime, z\prime, 1) * R^{-1} * T^{-1} \tag{18.6}$$

where R^{-1}, T^{-1} is the inverse of the matrix R, T.

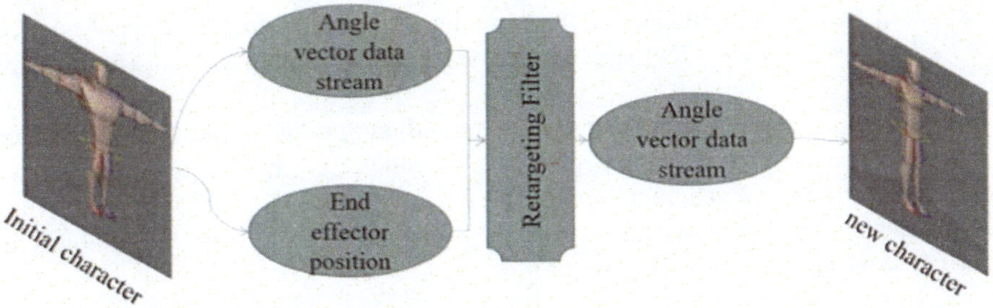

Fig. 18.3 Improved motion redirection

Because the size ratio of the virtual character may not be completely consistent with the proportion of the captured entity, or even the skeleton results are different, the action may lose its original characteristics. At this time, motion redirection can redirect the motion data to the virtual character model through a series of operations to keep the original characteristics of the action. Based on the skeletal animation algorithm, this paper completes the creation of 3D human skeletal animation source files in modeling software, which mainly includes creating meshes, adding bones, skinning and editing envelopes. Then, the real motion data is obtained by motion capture technology, and the data is optimized and redirected to drive the 3D character motion. Figure 18.3 shows an improved motion redirection.

The input is the angular vector data stream of a joint and the position of the end effector corresponding to the original model in discrete time, and the output is the angular vector data stream corresponding to the joint at discrete time, where the output data are calculated by the current joint data and the input at the previous moment. The biggest advantage of this algorithm is that the rotation and translation information of joint points is more stable compared with the data redirected by motion capture, and the error data generated during capture is reduced.

In the process of bone tracking, because of noise and hardware performance, people's movements are incoherent, and the relative position of bone joints changes greatly from frame to frame, which leads to the unreal virtual movements. Therefore, Holt exponential smoothing is proposed and applied to the smoothing process of motion data.

Holt exponential smoothing has two basic smoothing formulas and one prediction formula:

$$\begin{aligned} A_t &= \alpha P_t + (1-\alpha)(A_{t-1} + I_{t-1}) \\ I_t &= \gamma(A_t - A_{t-1}) + (1-\gamma)I_{t-1} \end{aligned} \tag{18.7}$$

$$\hat{P}_{t+T} = A_t + I_t T, T = 1, 2, \cdots, n \tag{18.8}$$

where α, γ is the smoothing parameter and P_t is the actual acquisition value of the t period; T is the number of forecast lead periods, t is the current cycle value; A_t, A_{t-1} is the estimated value of t period or $t-1$ period by using the data of t period or $t-1$

period, I_t, I_{t-1} is the estimated value of trend increment by using the data of t period or $t-1$ period, and $\hat{P}_{t+T}$ is the predicted value of $t + T$ period by using the data results of T period.

In order to ensure the real-time planting and dyeing of the system, it is necessary to reduce the number of intersection tests as much as possible, and connect the current position of 3D characters with the next position with a straight line. When 3D characters move to the intersection, they will move along the edge of obstacles until they move to the next intersection.

18.3 Result Analysis

Character animation technology mainly includes two parts of information: skin and skeleton movements. In this paper, BVH data files under different movements are selected for animation-driven test. According to the hierarchical structure of skeleton tree of human body standard posture defined in the file, the local coordinate system of each bone is obtained hierarchically, and then its transformation matrix is calculated, and then its transformation matrix is applied to its next-level bones at the same time, so as to realize the alignment of all bones.

Complete the advanced control of 3D character movement in OSG, and use the messaging mechanism to update and recall the animation at the frame rate of 30 frames per second. By changing the translation and rotation components, the position and posture of 3D characters in 3D space can be transformed to form a continuous animated picture. Figure 18.4 shows the different actions of 3D characters.

In this paper, the movement of 3D characters is controlled at the turning point of the path, and the rotation angle is updated at the same time as the displacement of 3D characters is updated, so as to form a continuous and smooth movement effect. Figure 18.5 is a screenshot of the skeleton animation effect of a character model driven by a 3D data file.

Fig. 18.4. 3D character animation renderings

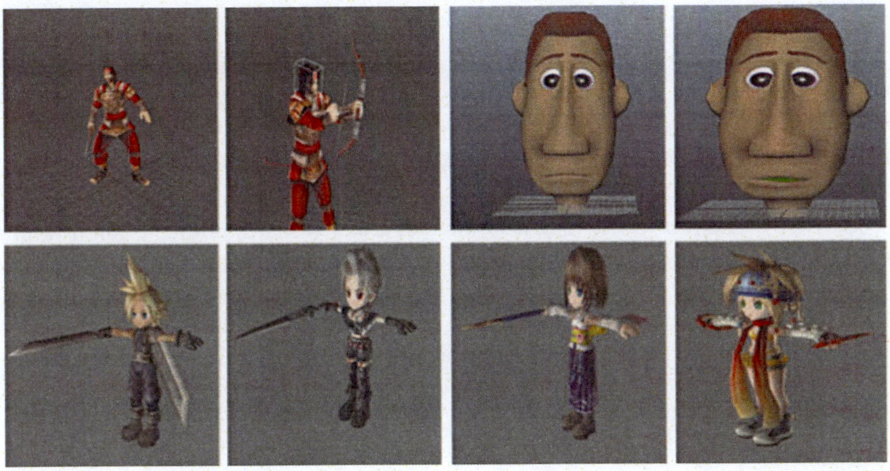

Fig. 18.5 Skeleton animation generation

It can be seen from the figure that the effect of skeleton animation is still good, and some basic actions have been completed. The improvement of this place in the future is to add artificial intelligence elements, which makes the animation more natural and real.

In the design and production of 3D animation, designers integrate network technology, computer technology and system interaction design into a whole, and then use VST to simulate it into a highly real world, so that people can freely enter it and deepen the exploration and research of the future world. In particular, the application of computer VST in 3D animation production has great practical significance. The application of computer VST in 3D animation production is conducive to people's better understanding of animation works and interpretation of real scenes. At the same time, 3D animation production plays an important role in restoring ancient times and looking forward to the future.

The virtual human action simulation based on skeletal animation algorithm may completely replace the traditional human action simulation method, and this new interactive mode is more convenient to operate. The most important thing is that the motion data of human joints can be obtained quickly and conveniently at a lower cost. Therefore, the research of virtual character action simulation technology based on bone animation algorithm has important engineering value, and will be more widely used in medicine, intelligent robots, games and 3D reconstruction in the future.

18.4 Conclusion

Applying computer VST to 3D animation can perfectly integrate reality and illusion, which not only reduces workload but also improves work efficiency. But also can make the whole animation scene more realistic and flexible, and realize the blending of scenes. In this paper, character modeling and character animation generation are deeply studied. Character model and animation are of great scientific research value in reverse engineering and virtual reality, and they are important users of computer graphics and computer hardware technology. The virtual human action simulation based on skeletal animation algorithm may completely replace the traditional human action simulation method, and this new interactive mode is more convenient to operate. The most important thing is that the motion data of human joints can be obtained quickly and conveniently at a lower cost. Therefore, the research of virtual character action simulation technology based on bone animation algorithm has important engineering use value.

Funding 2021 Guangdong Higher Vocational Computer Education Instruction Committee Education and Teaching Reform Research and Practice Project "Innovative Practice Research on the Hybrid Teaching Mode Based on" Internet plus Dichotomy "in Animation Teaching" (Project Number: JSJJZW2021053).

References

1. Mauriello, D.: Organic modeling for animation. JOM **66**(8), 1390–1398 (2016)
2. Yang, J., Yang, S.: Application of 3D reality technology combined with CAD in animation modeling design. Comput. Aided Design Appl. **18**(3), 164–175 (2020)
3. Ichim, A.E., Kadleek, P., Kavan, L., et al.: Phace: physics-based face modeling and animation. ACM Trans. Graph. **36**(4), 153 (2017)
4. Amini, R., Lisetti, C., Ruiz, G., et al.: HapFACS 3.0: FACS-based facial expression generator for 3D speaking virtual characters. IEEE Trans. Affect. Comput. **6**(4), 348–360 (2015)
5. Garrido, P., Zollhoefer, M., Wu, C., et al.: Corrective 3D reconstruction of lips from monocular video. ACM Trans. Graph. **35**(6), 219.1–219.11 (2016)
6. Yu, B.K.: 3D animation space design based on virtual reality. Modern Electron. Technol. **44**(8), 4 (2021)
7. Tan, Z.: Simulation design and implementation of three-dimensional virtual building space. Modern Electron. Technol. **41**(16), 4 (2018)
8. Xue, C.Y., Dong, H.W., Zhang, M.M., et al.: A real-time clothing simulation method based on human body fat and thin model. J. Syst. Simul. **29**(11), 9 (2017)
9. Jiang, X.B., Tan, J.W., Ren, H.X., Qiu, S.Y.: Interactive simulation system of marine windlass based on virtual reality technology. J. Chongqing Jiaotong Univ. (Nat. Sci. Edn.) **38**(006), 12–16 (2019)
10. Fu, C.Z., Automatic control system for details of 3D animation scenes based on digital technology. Modern Electron. Technol. **44**(6), 5 (2021)
11. Li, G., Wang, Z.Y.: Research on 3D landscape virtual design system under big data analysis. Modern Electron. Technol. **41**(11), 4 (2018)
12. Liu, J.X., You, Z., Huang, J.W, et al.: Research on graphic algorithm virtual simulation system based on Lushan 3D scene. Comput. Modernization (12), 0006 (2020)

Chapter 19
Lightweight Human Pose Estimation Based on Self-Attention Mechanism

Youtao Luo and Xiaoming Gao

Abstract To address the problems of a large number of parameters, high computational complexity and long detection time in many current human pose estimation network models, we adopt an hourglass structure to build a single-way lightweight network model with fewer parameters and shorter computation time. In order to ensure the accuracy of the model, we use window self-attention mechanism with fewer number of parameters, and this self-attention module that can effectively extract local and global information is redesigned on its basis to make the feature information learned by the model richer, and it merges with the inverted residual network architecture to become a separate module of WGNet (Global and Local Networks, a new structure that we construct), and finally WGNet is flexibly embedded into different stages of the model. The results of training and validation on COCO (Common Objects in Context) and MPII (Max Planck Institute for Informatics) datasets show that the model reduces the number of parameters by 25%, the computational complexity by 41%, and the inference time by nearly two times compared with HRFormer, which also uses the windowed self-attention mechanism, at the expense of only 3.5% accuracy.

19.1 Introduction

Human pose estimation is a pre-requisite task in many application scenarios such as Human Behavior Analysis [1, 2], human–computer interaction [3], medical rehabilitation assistance [4] and so on. The main approach is to predict the location of key points of the human body by building neural networks, so proposing models with high recognition accuracy is a hot research topic. The research history of this topic can be traced back to the 1970s. So far, many scholars have proposed various excellent solutions to this task. At this stage of time, the entire field of computer vision continues to develop, especially in basic research, there are already many mature

Y. Luo (✉) · X. Gao
School of Computer Science and Technology, SWUST, Mianyang, Sichuan, China
e-mail: carlousjay9011@gmail.com

R. Kountchev et al. (eds.), *Multidimensional Signals, Augmented Reality and Information Technologies*, Smart Innovation, Systems and Technologies 374,
https://doi.org/10.1007/978-981-99-7011-7_19

solutions, including image recognition, image classification, etc., and these general neural networks are applied to human body key point detection tasks, and continuously achieve higher precision. However, these algorithms [5–9] design the neural network model to be very large and redundant while pursuing accuracy, resulting in huge parameters and floating-point operations, which cause the model to consume a lot of computing resources and time in actual reasoning, almost It cannot achieve a smooth detection screen, and even freezes on GPU devices, making it even more difficult to deploy to mobile terminals, which is the current problem that needs to solved in the task of human pose estimation.

In order to reduce the number of model parameters, we can cut the depth and width of model directly, but this sacrifices quite a bit of accuracy and therefore we must design the model structure carefully. Some studies [10–12] tried to change the model based on complex models, by replacing the original convolution with a lightweight convolution structure (such as dynamic convolution network, etc.), and finally achieved certain results. In addition, with Vaswani [13] self-attention mechanism dominance on various prediction tasks, more and more researchers try to use it on computer vision tasks. Many studies introduce self-attention mechanism to the human pose estimation task making the model somewhat lighter because of its strong long-range modelling capability, simple structure and less number of parameters compared to convolutional networks. Li [14] designed vision token and keypoint token in Tokenpose, fused them together into self-attention mechanism block for computation, and combined with convolutional network, which reduce a large number of parameters, but model still suffers from the problem of large computational effort. Yuan [15] proposed a high-resolution network structure in HRFormer and introduce multi-resolution parallel design and local attention, reducing a large number of parameters and computation compared to Tokenpose, but with slow computation speed and long inference time because of its multi-branch parallel design structure.

The above-mentioned research introduces self-attention mechanism into human pose estimation and get some lightweight effect. However, it generates square-level space complexity and time complexity when in its computation process, which leads to the problems of slow computation and difficult training. These factors make it still difficult to apply the human posture estimation network to complex vision task scenarios. Therefore, we constructed a new attention structure that can efficiently extract both global and local information. And we apply the new structure to a pose estimation model with lower complexity and shorter inference times than models that also use a self-attention mechanism. The final inference validation results show that our model still has good detection results.

19.2 Related Work

There have been a number of mature studies on Network Lightweight Research. The Mobilenet series [16–18] mainly proposes deep separable convolution and inverse residual structure that can reduce a large number of parameters and increase the computational speed, which is the first choice of many lightweight models. Shufflenet [19] group the input features and then channel-mixing wash them to reduce the computational effort while ensuring that the information in each group flow.

Currently, human pose estimation is still mainly done through convolutional networks to predict each key point and many researches introduced it into pose estimation and even gradually replace convolutional networks, reducing the number of model parameters compared to convolutional networks, but there is still the problem of low computational efficiency. In addition, some studies aim at lightweight self-attention methods. A representative one is the window-attention of Swin transformer [20], in which the whole feature map is divided into multiple small windows of the same size, and each window completes the calculation of self-attention independently, thus improving the computational efficiency of the model. HRFormer [15] introduces this mechanism into multi-scale fusion model by applying it to semantic segmentation and pose estimation, reducing the number of parameters and computational effort. Methods HRFormer [15], Lite-HRNet [21] are still limited in terms of computational speed because of their multi-branch structure mainly because the operations of each branch are performed independently and cannot be parallelized. So the task of lightweight human pose estimation task still requires further research.

19.3 Lightweight Model Structure

19.3.1 Overall Network Architecture

The whole model is shown in Fig. 19.1, with a single-way structure of an hourglass and stem layer consisting mainly of depth-separable convolutions with a convolutional kernel step size of 2 and a size of 5. $s_x(x \subseteq (1, 2, 3, 4))$ (one of the stages in downsampling) is the main constituent structure of the model, consisting mainly of downsampling in steps of 2, inverted residual networks and WGNet (Global and Local Networks, a new structure that we construct, see part c in Fig. 19.1 for details) module stacked in sequence. $d_x(x \subseteq (1, 2, 3))$ (one of the stages in unsampling) recovery the size of feature, using transposed convolution, WGNet and channel attention SE(Squeeze-and-Excitation) module [22]. Each doubling of the size recovered in this phase is connected with a jump to s_x the corresponding phase, enhancing the return of the gradient during training. The WGNet consists mainly of an improved self-attention mechanism, as shown in Fig. 19.2, stacking the introduced global-attention repeatedly with window-attention, and replacing the MLP (a structure in Vision Transformer) [23] with an inverted residual network. Finally, we recover the number of channels to the number of key points by point convolution and then predict the key point position directly.

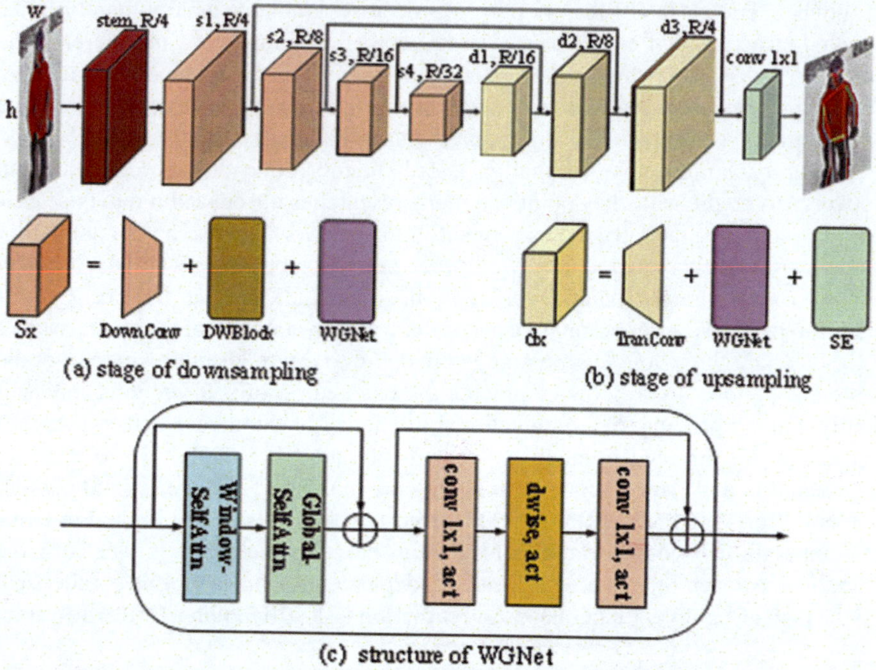

Fig. 19.1 Structure of network

Fig. 19.2 Principle of the self-attention in WGNet

19.3.2 Computational Volume Analysis of Self-Attention in WGNet

The self-attention in WGNet section is shown in Fig. 19.2. Firstly, we divide the input feature map into multiple windows of the same size, with each window calculating the self-attention separately. And then take the patch at the same position in each window and stitch together to form a new window, each of which then does the self-attention calculation.

We compare the calculation amount of self-attention in WGNet with the original self-attention (hereinafter referred to as MHSA). Assume that the initial feature height, width and channels are h, w and C. For MHSA, each patch in the feature generates query(q), key(k) and value(v) via the Q, K and V matrices, and let its length remain the same as the input feature depth C, Q, K and V are the parameters to be learned, initialize them to $W_q^{C\times C}$, $W_k^{C\times C}$, $W_v^{C\times C}$. Based on the self-attention computation process in the [13], plus the computation of generating query(q), key(k) and value(v), the MHSA computation is obtained as follows (here the number of multi-heads is assumed to be 1 and the computation of the softmax function is ignored):

$$\mathrm{Flop}_{\mathrm{MHSA}} = 3\mathrm{hwC}^2 + 2\mathrm{h}^2\mathrm{w}^2\mathrm{C} \tag{19.1}$$

(*Note* h and w are the height and width of the feature, C is channels of feature).

For self-attention in WGNet, the features are first divided into different windows, let the window size be h′ and w′, then we get $\frac{\mathrm{h}'}{\mathrm{h}} \cdot \frac{\mathrm{w}'}{\mathrm{w}}$ windows. Then extract patches from the same position in each window form a new window and calculate the amount of computation for each new window according to the calculation steps of MHSA. We let the size of each patch take 4 × 3, then multiply the number of new windows by the number of calculations per new window, we can obtain the formula for the amount of computation of self-attention in WGNet.

$$\mathrm{Flop}_{\mathrm{WGNet\_attn}} = 6\mathrm{hwC}^2 + 2hh'Cww'C + \frac{24\mathrm{h}^2\mathrm{w}^2\mathrm{C}}{h'w'} \tag{19.2}$$

(*Note* h and w are the height and width of the feature, C is channels of feature).

Assume that the input image size is 256 × 192 and the dimensions are 64 × 48 after two downsampling layers. The window size needs to be designed to be divisible by the input size. So we design it to be 4α × 3α, which gives the following table comparing self-attention in WGNet with MHSA-based calculations (Table 19.1).

When α is 1, the window size is (4, 3), which is greater than the calculation amount of HMSA. When α is 2, 4 and 8, the calculation amount is smaller than the latter. When α is 16, the method degenerates into MHSA. When α is 4, the amount of calculation is the smallest, and the value of feature dimension C is 32, which can reduce the amount of calculation by 74% compared with the ordinary attention mechanism.

Table 19.1 Comparison in FLOP of WGNET with MHSA

α	Window_size	$Flop_{win\_attn}$	$Flop_{glo\_attn}$	$Flop_{WGNet\_attn}$	$Flop_{HMSA}$
1	(4, 3)	9.2 M + 0.7N	9.2 M + 188.7N	18 M + 189.4N	
2	(8, 6)	9.2 M + 2.9N	9.2 M + 47.1N	18 M + 50.1N	
4	(16, 12)	9.2 M + 11.7N	9.2 M + 0.7N	18 M + 23.5N	9.2 M + 188.7N
8	(32, 24)	9.2 M + 47.1N	9.2 M + 2.9N	18 M + 50.1N	
16	(64, 48)	9.2 M + 188.7N	/	9.2 M + 188.7N	

(*Note* M = 1 e^3 C^2, N = 1 e^5 C, C is channels of feature)

19.4 Experimental Results and Analysis

19.4.1 Experimental Environment and Data Set

We use the PyTorch framework to build and train the network model on the i7-8700CPU platform. There are 163,961 images in the COCO training set, 149,813 images after cleaning and screening, and 5,000 images in the verification set. The MPII data set has 16 key points marked, 14,679 images in the training set, and 2729 images in the verification set. In addition, we perform various enhancements on images before training, such as random rotation, random cropping, translation transformation, scaling and other operations, which can expand the dataset and also prevent training overfitting.

19.4.2 Training Results and Analysis

We trained for 220 cycles with an initial learning rate of le^{-3} and used a cosine learning rate reduction strategy to drop to le^{-5} at cycle 170, finally obtain 71.8% of mAP in the validation data set. Finally, we compare our method with the current representative method on the COCO and MPII verification sets. The detailed data are shown in Table 19.2 and Table 19.3. As can be seen in Table 19.2, our model has 79% fewer parameters than the HRNet-w32 [9] pure convolutional network, mainly because the self-attention mechanism has fewer parameters compared to convolution. Compared with methods using the common self-attention mechanism [14, 15], our improved self-attention mechanism is less computationally intensive, 36% and 41% lower compared to TokenPose-B [14] and HRFormer-S [15], respectively, and the accuracy is 4% lower compared to HRFormer-S [15], but in terms of inference time spent on a single image, our model inference time of 24.1ms, which is nearly two times shorter than the 74.1ms of HRFormer-S [15]. Table 19.3 shows the accuracy of different models for detection at various key points on the MPII data set, our model still has a high accuracy compared to other models and our model has a lower number of parameters.

Table 19.2 Results' comparison of different models on COCO validation set

Method	Params/MB	GFlops	Time/ms	AP	AP^{50}	AP^{75}	AP^{M}	AP^{L}
HRNet-W32	28.5	7.1	36.8	74.4	90.5	81.9	70.8	81.1
TokenPose-S	6.2	11.6	28.7	73.5	89.4	80.3	69.8	80.5
TokenPose-B	13.5	5.7	30.5	74.7	89.8	81.4	71.3	81.4
HRFormer-S	7.8	6.2	70.1	75.6^{*}	90.8^{*}	82.8^{*}	71.7^{*}	82.6^{*}
Ours	$5.8_{\downarrow}$	$3.6_{\downarrow}$	$24.0_{\downarrow}$	71.8	89.6	79.5	68.7	79.6

(*Note* The mark ↓ in each column means that the parameter quantity or calculation amount index is reduced, and the mark * means that the index is the highest in this column)

Table 19.3 Results' comparison of different models on MPII validation set

Method	Params/MB	Head	Shoulder	Elbow	Wrist	Hip	Knee	Ankle	Mean
SimpleBaseline-50	34.0	96.4	95.3	89.0	88.4	88.4	84.0	79.6	88.5
SimpleBaseline-101	53.0	96.9	95.9	89.5	84.4	88.4	84.5	80.7	89.1
HRNet-W32	28.5	96.9	96.0^{*}	90.6	85.8^{*}	88.7	86.6^{*}	82.6	90.1
TokenPose-B	13.5	97.1^{*}	95.8	90.7^{*}	85.8^{*}	89.2^{*}	86.2	82.7^{*}	90.2^{*}
Ours	$5.8_{\downarrow}$	96.8	95.6	90.4	85.4	88.6	85.6	80.6	89.2

(*Note* The mark ↓ in each column means that the parameter quantity or calculation amount index is reduced, and the mark * means that the index is the highest in this column)

Table 19.4 shows the ablation experiments based on the self-attention block in WGNet. There are generally two methods to realize the ablation experiment, which are the deletion method and the replacement method. The idea of the deletion method is to delete a certain network structure in the model one by one, such as a certain neuron, a certain layer or a certain feature, while keeping other network structures unchanged, and then compare the performance changes of the model before and after the deletion operation. The idea of the replacement method is to replace a certain network structure in the model with other network structures, such as different activation functions, different convolutional networks, etc., and then compare the performance changes of the model before and after the replacement operation. In order to verify the effectiveness of WGNet, this paper chooses the replacement method, which is to replace the window-SelfAttn and Global-SelfAttn in the network WGNet with a common convolutional network structure. Four comparative experiments show that when the self-attention block is completely replaced by a common convolutional structure in WGNet, the accuracy rate drops by 4.2% compared with the baseline, indicating that the improved self-attention structure can effectively extract the feature information of the image.

Table 19.4 Ablation Experiments

Model	window-SelfAttn	Global-SelfAttn	#Params/MB	mAP/%
Baseline	✓	✓	5.8	72.1
Model-A	✓	×	6.4	70.4
Model-B	×	✓	6.2	69.8
Model-C	×	×	6.8	68.2

19.4.3 Visual Analysis of Inference

We select complex scene images for detection in the COCO validation set. Figures 19.3 and 19.4 show the detection of key points being occluded and multi-person dense, respectively, comparing with the detection results of HRNet [9], our model detection results are basically consistent with them, indicating that our network still has high detection accuracy.

Fig. 19.3 Part occlusion detection

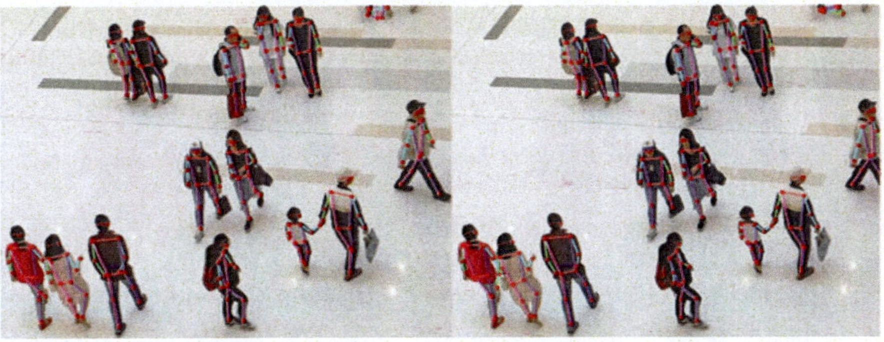

Fig. 19.4 Intensive multi-person detection

19.5 Conclusion

To address the problems of a large number of parameters, computational complexity and long detection time when mainstream human pose estimation networks perform inference, we improve the self-attention module and construct a lightweight model that performs well in the computing process. After training on the COCO2017 dataset for single-image inference, achieve a real-time effect of 24ms, and in practical tests, it can accurately identify, the locations of various key points of the human body in practical test. Further research is needed on how to reduce the number of parameters, computational complexity, and subsequent engineering optimization for deployment to mobile, while ensuring accuracy.

References

1. Lv, X.Y. et al.: Analysis of gait characteristics of patients with knee arthritis based on human posture estimation. BioMed Res. Int. 7020804–7020804 (2022)
2. Amir, N., Ahmad, J., Kibum, K.: Automatic human posture estimation for sport activity recognition with robust body parts detection and entropy Markov model. Multimedia Tools Appl. 1–34 (2021)
3. Liu, J.F.: 3d human pose estimation and action recognition for human-robot interaction. Master's thesis, School of Electromechanical Engineering Guangdong University of Technology (2021)
4. Huu, P.N., Ngoc, N.T., Thien, P.N.: Proposing posture recognition system combining MobilenetV2 and LSTM for medical surveillance. IEEE Access **10**, 1839–1849 (2021)
5. Toshev, A., Szegedy, C.: Deeppose: human pose estimation via deep neural networks. Proceedings of the IEEE Conference on Computer Vision and Pattern Recognition, pp. 1653–1660 (2014)
6. Tompson, J., Jain, A., LeCun, Y., et al.: Joint training of a convolutional network and a graphical model for human pose estimation. Adv. Neural Inf. Process. Syst. **27**, 1799–1807 (2014)
7. Newell, A., Yang, K., Deng, J.: Stacked Hourglass Networks for Human Pose Estimation, pp. 483–499. European conference on computer vision, Springer, Cham (2016)

8. Xiao, B., Wu, H., Wei, Y.: Simple baselines for human pose estimation and tracking. Proceedings of the European Conference on Computer Vision (ECCV), pp. 466–481 (2018)
9. Sun, K., Xiao, B., Liu, D., et al.: Deep high resolution representation learning for human pose estimation. Proceedings of the IEEE/CVF Conference on Computer Vision and Pattern Recognition, pp. 5693–5703 (2019)
10. Yang, S.Q., Wen, J.J., Fan, J.J.: Ghost shuffle lightweight pose network with effective feature representation and learning for human pose estimation. IET Comput. Vision **16**(6), 525–540 (2022)
11. Xu, D.N., et al.: LDNet: Lightweight dynamic convolution network for human pose estimation. Adv. Eng. Inf., 54 (2022)
12. Li, Y.P. et al.: Human pose estimation based on lightweight basicblock. Mach. Vis. Appl. **34**(1) (2022)
13. Vaswani, A., Shazeer, N., Parmar, N., et al.: Attention is all you need. Adv. Neural Inf. Process. Syst, 30 (2017)
14. Li, Y., Zhang, S., Wang, Z., et al.: Tokenpose: learning keypoint tokens for human pose estimation. Proceedings of the IEEE/CVF International Conference on Computer Vision, pp. 11313–11322 (2021)
15. Yuan, Y., Fu, R., Huang, L., et al.: Hrformer:High-resolution vision transformer for dense predict. Adv. Neural. Inf. Process. Syst. **34**, 7281–7293 (2021)
16. Howard, A.G., Zhu, M., Chen, B., et al.: Mobilenets: Efficient convolutional neural networks for mobile vision applications. (2017). https://doi.org/10.48550/arXiv.1704.04861
17. Sandler, M., Howard, A., Zhu, M., et al.: Mobilenetv2: inverted residuals and linear bottlenecks. Proceedings of the IEEE Conference on Computer Vision and Pattern Recognition, pp. 4510–4520 (2018)
18. Howard, A., Sandler, M., Chu, G., et al.: Searching for mobilenetv3. Proceedings of the IEEE/CVF International Conference on Computer Vision, pp. 1314–1324 (2019)
19. Zhang, X., Zhou, X., Lin, M., et al.: Shufflenet: an extremely efficient convolutional neural network for mobile devices. Proceedings of the IEEE Conference on Computer Vision and Pattern Recognition, pp. 6848–6856 (2018)
20. Liu, Z., Lin, Y., Cao, Y., et al.: Swin transformer: hierarchical vision transformer using shifted windows. Proceedings of the IEEE/CVF International Conference on Computer Vision, pp. 10012–10022 (2021)
21. Yu, C., Xiao, B., Gao, C., et al.: Lite-hrnet: a lightweight high-resolution network. Proceedings of the IEEE/CVF Conference on Computer Vision and Pattern Recognition, pp. 10440–10450 (2021)
22. Hu, J., Shen, L., Sun, G.: Squeeze-and-excitation networks. Proceedings of the IEEE Conference on Computer Vision and Pattern Recognition, pp. 7132–7141 (2018)
23. Dosovitskiy, A., Beyer, L., Kolesnikov, A., Weissenborn, D., Houlsby, N.: An image is worth 16x16 words: transformers for image recognition at scale (2020). arXiv preprint arXiv:2010.11929

Chapter 20
Facial Photo-Guided Head Anatomy Modeling Based on Deep Learning and 2D/3D Shape Prior Model Registration

Meng Wang and Hongkai Wang

Abstract Three-dimensional (3D) models of individualized head anatomy are often used in the preoperative planning of plastic surgery. Most current methods use Computed Tomography (CT) to acquire the head anatomy, but CT scan exposes the patients to X-ray radiation. Alternatively, artificial intelligence-based estimation of 3D head anatomy from the facial photo is a safer and more efficient choice, especially for scenarios that do not require high modeling accuracy of internal structures (e.g., doctor–patient communication and surgery procedure demonstration). We develop a method for constructing a personalized 3D model of complete head anatomy (including skin and internal structures) from the front and side view photos. We first detect key facial landmarks from the photos with a convolution neural network (CNN), then perform 2D/3D registration to morph a statistical shape model (SSM) of complete head anatomy to match the dual-view photos. Our method uses deep learning to achieve accurate facial landmark detection and employ the anatomy shape *prior* to yielding reasonable internal structure estimation. We evaluated the method based on 12 subjects (including 7 males and 5 females). The facial surface reconstruction error was assessed using the 3D surface scan of the subjects as the ground truth. Our method's root mean square error (RMSE) was 3.60 ± 0.49 mm, which was 1.06 ± 0.39 mm lower than the state-of-the-art (SOTA) CNN-based face reconstruction method. Our method also predicts all the internal head structures (bones, muscles, vessels, nerve fibers, fat, glands, and brain structures), which are not provided by the SOTA method. To the best of our knowledge, this is the first study modeling complete head anatomy from facial photos toward clinical applications.

M. Wang · H. Wang (✉)
Faculty of Medicine, School of Biomedical Engineering, Dalian University of Technology, Dalian 116024, China
e-mail: wang.hongkai@dlut.edu.cn

H. Wang
Liaoning Key Laboratory of Integrated Circuit and Biomedical Electronic System, Dalian University of Technology, Dalian 116024, China

R. Kountchev et al. (eds.), *Multidimensional Signals, Augmented Reality and Information Technologies*, Smart Innovation, Systems and Technologies 374,
https://doi.org/10.1007/978-981-99-7011-7_20

20.1 Introduction

Three-dimensional (3D) face model reconstruction has been extensively used in game development, film industry, clinical plastic surgery, etc. After decades of development, current deep learning methods can successfully reconstruct personalized 3D face models based on a single front-view photo [1–3]. To make up for the lack of depth information of the single view, multi-view photo reconstruction methods are also proposed [4–6] with improved reconstruction accuracy in 3D space.

Despite the progress in photo-based face model reconstruction, current methods only achieve visually similar modeling of the facial outer look. The accuracy of face models reconstructed from photos cannot meet the requirements of clinical surgery, and quantitative assessment of face reconstruction accuracy is also lacking in the current studies. Moreover, besides the reconstruction of the face surface, successful surgery planning requires modeling internal anatomical structures including multiple head anatomical structures, including bones, muscles, vessels, nerve fibers, fat, glands, and brain structures.

Tomographic imaging devices are usually used to acquire both external face shape and internal structure anatomy, such as Computed Tomography (CT) and Magnetic Resonance Imaging (MRI). However, tomographic imaging is tedious and expensive, patients may not accept this option at the preoperative communication stage. In contrast, facial photos are much more efficient to acquire. Therefore, the estimation of a 3D personalized head anatomy model based on facial photos or optical surface scanning can significantly simplify the preoperative image acquisition procedure and can be useful for scenarios that do not require high modeling accuracies, such as surgery procedure demonstration and approximate surgery outcome simulation.

To address the above concerns, this paper explores the solution of complete head anatomy modeling based on non-tomographic imaging. We select facial photos as the first choice since optical surface scanning is not suitable for remote doctor-patient communication. We detect the facial landmarks with a deep-learning method and then perform 2D/3D registration to match a deformable statistical shape model (SSM) of full head anatomy to the front and side view photos. Since the SSM was previously constructed based on segmented volumetric CT images, it contains the shape prior to full head anatomy and provides a reasonable estimation of the internal structures based on the external face shape. Our method not only reconstructs a visually similar surface model with the color texture from the personal photo but also provides 3D models of the internal structures to assist preoperative planning of plastic surgery.

20.2 Related Work

With the rapid development of deep learning technology, many monocular face reconstruction methods have been proposed. Various network models were designed to regress the parameters of 3D face geometry models and supervise them according to synthetic data [7–9], pre-calculated 3D Morphable Model (3DMM) fitting [10], RGB images [11–15] and feature tags [16]. To deal with complex facial geometry more flexibly, the method [17, 18] regresses geometric residuals to restore fine-scale details. So far, most deep learning methods have used CNN to reconstruct the shape details of the face from 2D photos. Guo et al. [2] proposed a coarse-to-fine framework to reconstruct fine 3D faces from monocular videos and single images in real time.

Chen et al. [3] proposed a conditional generation countermeasure network (CGAN)-based deep detail network (DFDN), which uses a combination of geometry and appearance loss functions. However, this method cannot process low-resolution images, which limits its application.

A series method based on multi-view reconstruction was recently proposed to take advantage of multi-view photos for more accurate face reconstruction [4, 5]. For example, Wu et al. [4] proposed a multi-view image 3D modeling method based on end-to-end trainable CNN, but their method only focuses on shape modeling without color texture details. Later, Bai et al. [6] introduced traditional multi-view geometry into CNN-based face reconstruction. By imposing multi-view geometric constraints to solve the 3D reconstruction problem, they effectively capture shape details and improve the generalization ability of unknown data. Photometric stereo [19] is also used in [20–22] for facial reconstruction, but this method can only have a good reconstruction effect for images with a face deflection angle within ± 30 degrees away from the frontal view. Therefore, this method is not suitable for side profile photos.

For the modeling of internal head anatomy based on external surface shape, Xiao et al. [23] studied the estimation of bone shape based on facial reconstruction. They proposed a facial bone shape estimation framework through 3D facial reconstruction and deformable shape model, using pre-trauma conventional portraits and post-trauma computed tomography (CT) of the head. However, in his research, only facial bones were estimated, and other important anatomical structures, such as nerves and blood vessels, were missing.

At present, there are several common problems in the current research. First, the 90° profile photo is rarely used, leading to the inaccurate reconstruction of the side profile. Second, most methods only focus on the external surface or bony structure modeling without complete internal anatomy, and they reconstruct the 3D shape without a color texture map, limiting their value in preoperative doctor–patient communications. Last, the current face modeling accuracy cannot meet the precision requirements of plastic surgery, barely any study has evaluated such accuracy quantitatively.

Considering the above problems, this paper proposes a method to register a head anatomy SSM to both frontal and lateral view photos. The registration result provides a personalized 3D model of complete head anatomy with face texture mapped from the color photos, and the face modeling accuracy is also evaluated for the reference of surgery planning.

20.3 Method

Figure 20.1 shows the workflow of the proposed algorithm. Our method composes of three steps. The first step is to input dual-view photos to the CNN to detect 2D facial landmarks. The second step is to initialize the SSM of head anatomy according to the facial landmarks and maps the color texture from the photos to the SSM surface. The third step conducts refined matching of the SSM to the dual-view photos via 2D/3D registration, in which the detected landmarks guide accurate matching of key facial features, and the intensity-based registration of the SSM projection images leads to the accurate matching of the texture details.

20.3.1 Landmark Detection Network

Our method detects 68 landmarks from the frontal view photo and 68 landmarks from the side view photo, using a network structure similar to [6] which solved the problem of 3D face reconstruction from multi-view images with different expressions. We use the 2D landmarks to determine a bounding box for face region clipping. Figure 20.2 shows the definition of front and side face landmarks.

Of the 68 landmarks obtained, 11 face landmarks can be customized, including not only the main feature information parts such as facial pupil and nose wing but also the face contours information parts such as the zygomatic arch and mandibular

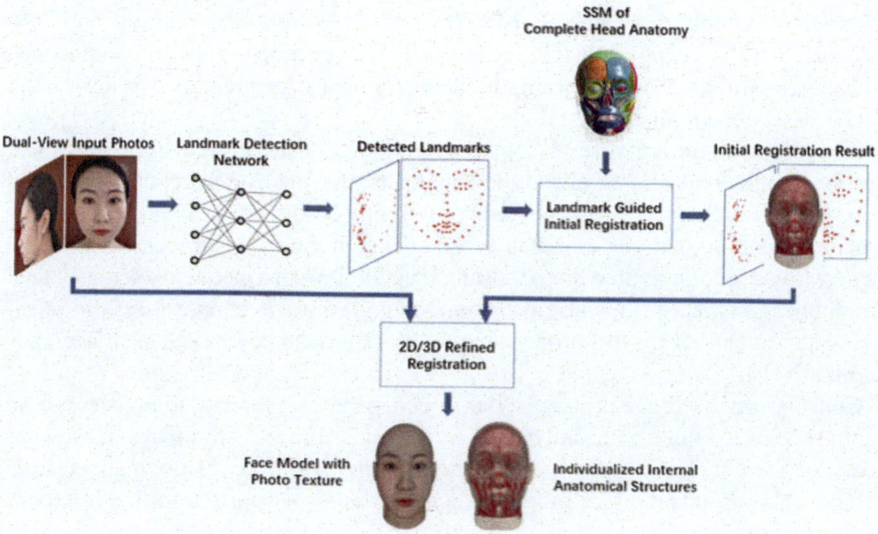

Fig. 20.1 Workflow of the proposed method for head anatomical structure modeling

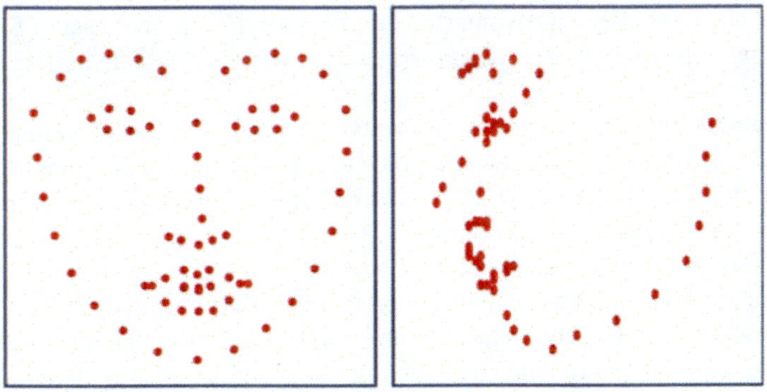
(a) (b)

Fig. 20.2 The front and side face landmarks. **a** The front face landmarks. **b** The side face landmarks

angle, which can effectively extract the face feature information. By using these face landmarks, we can carry out the next step of model initialization.

In addition, the neural network also estimates the model projection angle, which can be used in subsequent 2D/3D registration.

20.3.2 SSM Initialization

The SSM method [24] was used to construct the Atlas based on 65 computer tomography (CT) images (including 46 male and 19 female images). The deformable head atlases (DHAs) include 218 sub-structures of facial muscles, skull bones, vessels, fat, nerves, glands, and brain structures.

The SSM is expressed as a linear combination of average shape plus different shape variation patterns:

$$X = \overline{X} + \sum\nolimits_{i=1}^{n} a_i \Phi_i \tag{20.1}$$

where X is an instance of the shape generated by the model. X represents the shape vector $(x_1, y_1, z_1, x_2, y_2, z_2, \ldots, x_k, y_k, z_k)^T$ containing the 3D coordinates of k mesh vertices. $\overline{X}$ is the average shape vector of all training objects. $\{a_i\}, i = 1, \ldots, n$ are shape parameters that are used as weights for various modes. Different values of $\{a_i\}$ generate different instances of the model. The continuous change of $\{a_i\}$ value causes the continuous deformation of the head anatomical structure to simulate the anatomical changes between subjects.

To use the projection angle obtained in the first step, we need to match the SSM to the coordinate system where the neural network reconstructs the model. The method adopted is to mark 57 3D face landmarks on the corresponding positions of the neural network average model and the average SSM, and apply them to the SSM by calculating the similarity transformation matrix from the landmark set of SSM to the landmark set of the neural network model, to match the SSM to the target coordinate system.

The complete set of head anatomical structures in the SSM [24] is obtained through soft tissue mapping in the standard digital human template. The mapping process uses nonlinear thin plate spline (TPS) transformation, and the control points are the surface vertices on the skin and bones.

The deformation field of the external surface of the head is interpolated to the internal anatomical structure using 3D space interpolation to realize the personalization of the internal anatomical structure and achieve the accurate coincidence between the internal anatomical structure and the external surface model.

20.3.3 2D/3D Registration

The 2D/3D registration algorithm consists of two steps. The first step is to project the SSM mean shape into both views and then perform 2D registration between the projected model image and the facial photo, obtaining the deformation vector field in the 2D projection space. The second step is to back-project the 2D deformation vector to 3D space and to obtain the deformation of the 3D SSM. Figure 20.3 shows the principle of the back-projection.

Where, O is the position of the viewpoint, F_0 is the position of the 2D point before deformation in the world coordinates system on the model photo, F_1 is the position of the 2D point after deformation in the world coordinates system on the model photo, M_0 is the 3D point corresponding to the 2D pixel point of the model photo before deformation, and M_1 is the 3D point corresponding to the two-dimensional pixel point of the model photo after deformation.

Let the distance between O and F_0 be m, the distance between O and M_0 be n, and the distance between F_0 and F_1 be p, then the distance between M_0 and M_1 can be calculated according to the side length relationship of similar triangles, as shown in (20.2).

$$|M_0M_1| = \frac{n \times p}{m} \tag{20.2}$$

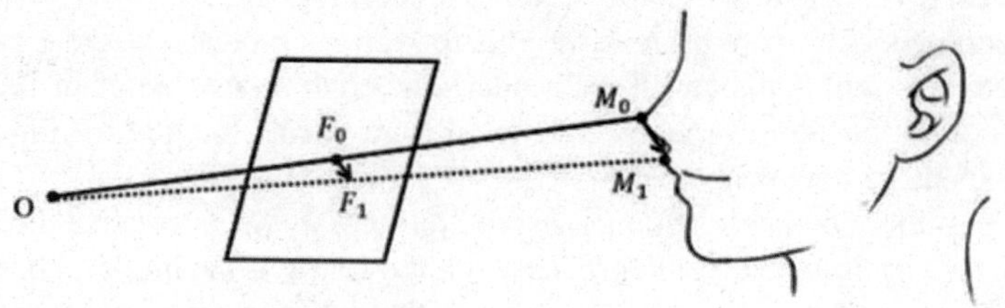

Fig. 20.3 Schematic diagram of the similar triangle

Let the coordinates of F_0 be (X_0, Y_0, Z_0), the coordinates of F_1 are (X_1, Y_1, Z_1), and the unit direction vector of $\overrightarrow{F_0F_1}$ is shown in (20.3).

$$\vec{p} = \frac{\overrightarrow{F_0F_1}}{|F_0F_1|} = \frac{(X_1 - X_0, Y_1 - Y_0, Z_1 - Z_0)}{\sqrt{(X_1 - X_0)^2 + (Y_1 - Y_0)^2 + (Z_1 - Z_0)^2}} \tag{20.3}$$

According to the distance between M_0 and M_1 in (20.2) and the unit direction vector of $\overrightarrow{F_0F_1}$ in (20.3), the coordinates of M_1 can be obtained. Because $\overrightarrow{M_0M_1}$ and $\overrightarrow{F_0F_1}$ have the same unit direction vector, M_1 coordinates are shown in (20.4).

$$M_1 = M_0 + \vec{p} \times |M_0M_1| \tag{20.4}$$

Using the 2D/3D registration algorithm, the front photo is registered to the initial model obtained in the second step to obtain model F, and then the lateral photo is registered to model F to obtain model L. The final reconstructed model is obtained by smoothly deforming model L onto the initial model.

Because the neural network accurately predicts the landmark recognition in the front photo, the thin-plate splines (TPS) transform is adopted for 2D/2D matching of front photos. The TPS transform formula is defined as follows,

$$T_\mu(x) = x + Ax + t + \sum_{x_k^{fix}} c_k G(x - x_k^{fix}) \tag{20.5}$$

where $G(r)$ is the basic function, and c_k are the coefficients corresponding to each landmark. Calculate the coefficients c_k and the elements of A and t according to the landmark displacements $d_k = x_k^{mov} - x_k^{fix}$.

Based on the front reconstruction results, lateral photo registration is carried out. Because the neural network lacks side view training images and does not perform robust enough for recognizing the side view landmarks, the lateral photo 2D/2D registration uses B-splines transform. B-splines registration is a nonlinear deformation registration based on pixel intensity, which serves as a good alternative when registration based on landmarks is not feasible.

20.4 Result and Analysis

Our test set includes 12 young subjects (7 males and 5 females) with an average age of 25 years. The average modeling time of our method is 51.56 s per subject. To evaluate the external face shape reconstruction error, we use a 3D optical scanner to acquire the ground truth surface point cloud of each subject's face. We use average surface distance root mean square error (RMSE) as an error indicator. The calculation formula of RMSE is as follows:

Table 20.1 RMSE comparison (unit: mm)

	Method	
	Ours	Bai et al. [18]
Mean	3.60	4.66
Std	0.48	0.78
Max	4.51	6.42
Min	2.80	3.24

$$RMSE = \sqrt{\frac{1}{N}\sum_{i=1}^{n}(Y_i - f(x_i))^2} \tag{20.6}$$

where Y_i is the accurate value and $f(x_i)$ is the estimated value.

We compare the facial reconstruction error of our method with one of the state-of-the- (SOTA) methods [6], which also art uses CNN to reconstruct facial models from multi-view photos. They [6] combined the traditional optimization based on multi-view geometry with modern neural network to design a new non-rigid multi-view stereo (NRMVS) optimization framework and solved the 3D face reconstruction problem of multi-view images with different expressions, but lacks the refined 2D/3D registration step. Input the test photos to [6] and our method respectively to calculate the RMSE between the reconstructed model and the ground truth. We compared the mean, standard deviation (STD), maximum, and minimum. Table 20.1 shows the comparison results.

It can be seen from Table 20.1 that the model surface error reconstructed by our method is 3.60 ± 0.49 mm, and the model surface error reconstructed by [6] is 4.66 ± 0.78 mm. The facial reconstruction error of our method is lower than that of [6], which proves that the method combining the 2D/3D registration method and neural network is effective and stable. In addition, it should be noted that possible facial expression difference between the photo and the surface point cloud scan also affects the error calculation. Therefore, we may expect when lower reconstruction error if there is no expression difference in ideal cases.

Figure 20.4 shows partial reconstruction results and error heat maps. From the reconstruction results, we can see that our method can well restore the real 3D head model from the photos. It can be seen from the heat maps that our model has high accuracy in facial reconstruction.

From the semi-transparent rendering of the complete head anatomy, one can see that the personalized modeling of all the anatomical structures can be achieved with our method.

We also conducted a qualitative comparison between our method and several popular SOTA deep learning methods for photo-based face modeling from the aspects of whether the reconstructed model includes complete head anatomy and whether it has colored face texture. Table 20.2 shows the comparison of qualitative between our method and other methods [4] uses the novel self-supervised view alignment loss to establish a dense correspondence between different views, and merge the

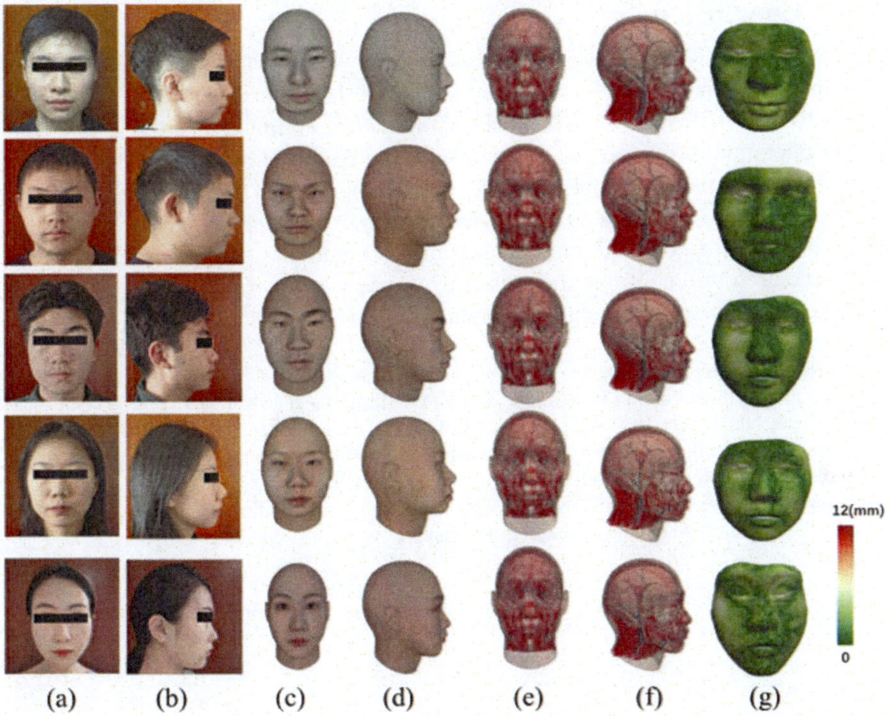

Fig. 20.4 Reconstruction results and error thermodynamic diagram. **a** Facial front photos. **b** Facial side photos. **c** Front display of skin model. **d** Side display of skin model. **e** Front display with head anatomical structure model. **f** Side display with head anatomical structure model. **g** Error heat map

multi-view geometric constraints into the network [5] uses self-supervised training based on a multi-frame video [6] optimizes 3D face shape by explicitly enhancing the consistency of multi-view appearance It is obvious that our model is the only one with both complete anatomy modeling and face texture mapping.

Table 20.2 Comparison of qualitative

	Method			
	MVF-Net [16]	FML [17]	NRMVS [18]	Ours
Complete head anatomy	×	×	×	√
Colored face texture	×	√	×	√

20.5 Conclusion

We combine deep learning-based facial landmark detection and 2D/3D SSM registration to reconstruct an individualized 3D model of complete head anatomy from a front photo and a side photo. The experimental results show that our method can improve the accuracy of model reconstruction compared with the SOTA deep learning methods. To the best of our knowledge, this is the first study providing both a textured facial modal and a complete internal anatomical model based on facial photos. Our future research will continue to assess the accuracy of internal structure modeling using head CT images as the ground truth.

Acknowledgements This work was supported in part by the National Key Research and Development Program Number 2020YFB1711500, 2020YFB1711501, and 2020YFB1711503, the general program of the National Natural Science Fund of China (Number 81971693, 61971445), the funding of Dalian Engineering Research Center for Artificial Intelligence in Medical Imaging, Hainan Province Key Research and Development Plan ZDYF2021SHFZ244, the Fundamental Research Funds for the Central Universities (Number DUT22YG229), the funding of Liaoning Key Lab of IC & BME System and Dalian Engineering Research Center for Artificial Intelligence in Medical Imaging.

References

1. Jackson, A.S., Bulat, A., Argyriou, V., Tzimiropoulos, G.: Large pose 3D face reconstruction from a single image via direct volumetric CNN regression. 2017 IEEE International Conference on Computer Vision (ICCV), pp. 1031–1039 (2017)
2. Guo, Y., Zhang, J., Cai, J., Jiang, B., Zheng, J.: CNN-based real-time dense face reconstruction with inverse-rendered photo-realistic face images. IEEE Trans. Pattern Anal. Mach. Intell. **41**(6), 1294–1307 (2019)
3. Chen, A., Chen, Z., Zhang, G., Zhang, Z., Mitchell, K., Yu, J.: Photo-realistic facial details synthesis from single image. 2019 IEEE/CVF International Conference on Computer Vision (ICCV), pp. 9428–9438 (2019)
4. Wu, F., Bao, L., Chen, Y., Ling, Y., Song, Y., Li, S., et al.: MVF-Net: Multi-view 3D face morphable model regression. 2019 IEEE/CVF Conference on Computer Vision and Pattern Recognition (CVPR), pp. 959–968 (2019)
5. Tewari, A., Bernard, F., Garrido, P., Bharaj, G., Theobalt, C.: FML: Face model learning from videos. 2019 IEEE/CVF Conference on Computer Vision and Pattern Recognition (CVPR), pp. 10804–10814 (2019)
6. Rahim, J.A., Bai, Z., Cui, Z., Liu, X., Tan, P.: Deep facial non-rigid multi-view stereo. 2020 IEEE/CVF Conference on Computer Vision and Pattern Recognition (CVPR), pp. 5849–5859 (2020)
7. Genova, K., Cole, F., Maschinot, A., Sarna, A., Freeman, W.T.: Unsupervised training for 3D morphable model regression. 2018 IEEE/CVF Conference on Computer Vision and Pattern Recognition, vol. 1,2, pp. 8377–8386 (2018)
8. Richardson, E., Sela, M., Kimmel, R.: 3D face reconstruction by learning from synthetic data. 2016 Fourth International Conference on 3D Vision (3DV), pp. 460–469 (2016)
9. Sela, M., Richardson, E., Kimmel, R.: Unrestricted facial geometry reconstruction using image-to-image translation. 2017 IEEE International Conference on Computer Vision (ICCV), vol. 2, pp. 1585–1594 (2017)

10. Tran, A.T., Hassner, T., Masi, I., Medioni, G.: Regressing robust and discriminative 3D morphable models with a very deep neural network. 2017 IEEE Conference on Computer Vision and Pattern Recognition (CVPR), vol. 1,2, pp. 1493–1502 (2017)
11. Deng, Y., Yang, J., Xu, S., Chen, D., Jia, Y., Tong, X.: Accurate 3D face reconstruction with weakly-supervised learning: from single image to image set. 2019 IEEE/CVF Conference on Computer Vision and Pattern Recognition Workshops (CVPRW), pp. 285–295 (2019)
12. Tewari, A., Zollhfer, M., Garrido, P., Bernard, F., Kim, H., Perez, P., Theobalt, C.: Self-supervised multi-level face model learning for monocular reconstruction at over 250 Hz. 2018 IEEE/CVF Conference on Computer Vision and Pattern Recognition, vol. 2,4,6,7, pp. 2549–2559 (2018)
13. Tewari, A., Zollhfer, M., Garrido, P., Bernard, F., Kim, H., Perez, P., Theobalt, C.: MoFA: Model-based deep convolutional face autoencoder for unsupervised monocular reconstruction. 2017 IEEE International Conference on Computer Vision (ICCV), vol. 2, pp. 3735–3744 (2017)
14. Tran, L., Liu, X.: Nonlinear 3D face morphable model. 2018 IEEE/CVF Conference on Computer Vision and Pattern Recognition, pp. 7346–7355 (2018)
15. Tran, L., Liu, X.: On learning 3D face morphable model from in-the-wild images. IEEE Trans. Pattern Anal. Mach. Intell. **43**(1), 157–171 (2021)
16. Sanyal, S., Bolkart, T., Feng, H., Black, M. J.: Learning to regress 3D face shape and expression from an image without 3D supervision. 2019 IEEE/CVF Conference on Computer Vision and Pattern Recognition (CVPR), vol. 2, pp. 7755–7764 (2019)
17. Tran, L., Liu, F., Liu, X.: Towards high-fidelity nonlinear 3D face morphable model. 2019 IEEE/CVF Conference on Computer Vision and Pattern Recognition (CVPR), vol. 2, pp. 1126–1135 (2019)
18. Tran, A.T., Hassner, T., Masi, I., Paz, E., Nirkin, Y., Medioni, G.: Extreme 3D face reconstruction: seeing through occlusions. 2018 IEEE/CVF Conference on Computer Vision and Pattern Recognition, vol. 2, pp. 3935–3944 (2018)
19. Woodham, R.J.: Photometric Stereo: a reflectance map technique for determining surface orientation from image intensity. Proceedings of Spie, vol. 155, pp. 136–143 (1979)
20. Georghiades, A.S.: Recovering 3-D shape and reflectance from a small number of photographs. Proceedings of the 14th Eurographics Workshop on Rendering Techniques, pp. 230–240. Leuven, Belgium (2003)
21. Klaudiny, M., Hilton, A.: High-detail 3D capture and non-sequential alignment of facial performance. 2012 Second International Conference on 3D Imaging, Modeling, Processing, Visualization & Transmission, pp. 17–24 (2012)
22. Roth, J., Tong, Y., Liu, X.: Adaptive 3D face reconstruction from unconstrained photo collections. IEEE Trans. Pattern Anal. Mach. Intell. **39**(11), 2127–2141 (2017)
23. Xiao, D., Lian, C., Wang, L., Deng, H., Lin, H., Thung, K., et al.: Estimating reference shape model for personalized surgical reconstruction of craniomaxillofacial defects. IEEE Trans. Biomed. Eng. **68**(2), 362–373 (2021)
24. Chen, Z., Qiu, T., Huo, L., Yu, L., Shi, H., Zhang, Y., et al.: Deformable head atlas of Chinese adults incorporating inter-subject anatomical variations. IEEE Access **6**, 51392–51400 (2018)

Chapter 21
Design of Water Ecological Cleaning Robot Based on Raspberry PI and OpenCV Visual Recognition

Zhichong Chen, Min Zhang, Hanyang Liu, and Shujian Cai

Abstract In this paper, a water ecological cleaning robot in the near water area is designed with Raspberry PI microcontroller as the core control. Through the OpenCV intelligent visual recognition module, the feature points of various items to be cleaned are extracted to establish a model library, which is fitted in the Raspberry PI microcontroller motherboard. This realizes the automatic detection and identification of near-shore environmental garbage such as water moorings and lakes, as well as the completion of fetching and handling work. The crawler chassis structure is adopted in the design, and Water Sensor is installed before and after the chassis to compare the water level of the car to realize the movement control of the car body so that it has a better function of avoiding water.

21.1 Introduction

With a vast territory and a wide area of near-water areas such as beachlands, the widening of the scope of human activities and the unjustified growth of people's tourism demand have accelerated the environmental pollution of this kind of area. Figure 21.1 is a common water pollution in our life. However, the environment of the beach is relatively complex, and the garbage cleaning workload is large. At present, the cleaning work of the beach and other areas is mostly manual cleaning. Today's more open social environment and increasingly competitive market economy have driven the rapid development of the national economy. At the same time, the unbalanced labor supply structure, the new generation of practitioners have higher requirements for the comfort of the working environment, the significant increase of living costs and other factors have resulted in the situation that the social labor cost has increased several times compared with the previous situation [1, 2]. Nowadays, due to the increasing pressure of young people's survival, the diminishing demographic

---

Z. Chen · M. Zhang (✉) · H. Liu · S. Cai
Department of Electrical and Electronic Engineering, Guangdong Technology College, Zhaoqing, China
e-mail: redxiaobing@foxmail.com

R. Kountchev et al. (eds.), *Multidimensional Signals, Augmented Reality and Information Technologies*, Smart Innovation, Systems and Technologies 374,
https://doi.org/10.1007/978-981-99-7011-7_21

Fig. 21.1 Pollution of shoal in near-water area

dividend and the improvement of people's education level, there is a serious shortage of labor resources in the field of sanitation and cleaning. In the contemporary era of rapid development of artificial intelligence, the use of scientific and technological robots to replace heavy, repetitive and dangerous labor is increasing, but there are few cleaning robot products for the field of water mooring in the market at present. For this reason, an outdoor cleaning robot based on OpenCV intelligent vision sensor is developed, hoping to realize the garbage cleaning function of outdoor man–machine cooperation.

According to the above analysis, based on the principle of existing cleaning robots, and aiming at the environmental problems of water pollution, this paper designs a water moor cleaning robot that can well solve the high cost of human resources and realize the cleaning of environmental waste. The robot uses artificial intelligence and image recognition technology, through the automatic scanning and marking of the head and identify the specific distance between each item and the robot, to achieve the object recognition and automatic obstacle avoidance function.

21.2 Robot System Scheme Design

The robot adopts a tracked chassis structure, which is characterized by many degrees of freedom, strong grip ability and obstacle crossing ability, etc. It is equipped with a LN298n DC motor drive module, which increases the flexibility and mobility of the tracked robot and enables the robot to move freely in the beach, river beach and other wetlands [3, 4]. At the same time, a Water Sensor is installed on the front and rear ends of the chassis of the car. When the robot wades into a certain depth, it can carry out self-detection and adjust the direction of the car to keep it away from the water. It is equipped with a camera and ultrasonic module to achieve a certain degree of self-obstacle avoidance and water avoidance protection, so as to prevent the internal circuit from being damaged by water contact. It has a good performance of human–machine work efficiency and safety protection ability [5]. The garbage cleaning function of the robot is realized by the robot arm grabbing garbage and placing it in the collection basket. The robot arm is composed of the steering gear as the main control unit, and it is combined with the visual aid of the head of the trolley platform. OpenCV intelligent vision sensor can realize the dual functions of distance judgment and object recognition, which can better ensure the work of object grasping. Considering that the distance of beach cleaning requires the distance of communication transmission, the robot adopts the transmission mode controlled by Wi-Fi module, which supports the long-distance signal control. When the robot moves, the real-time working picture and identification data can be processed and displayed on the PC screen, achieving the efficient working mode of "one-to-many" machine. At the same time, the operator can also select the working mode and monitor the working state of the robot through the PC terminal. Equipped with an automatic alarm program, when the robot encounters a problem, it can accurately feedback the problem and issue an alarm to effectively remind the staff and reduce the maintenance frequency.

The machine vision module OpenCV on the mobile chassis can realize the flexible rotation of 360° left and right and 210° up and down on the head. This module is an open source, low-cost, high pixel attached machine vision module, whose identification function is completely dependent on Raspberry PI call code model library for identification, realizing the Raspberry PI motherboard as the core, each collocation component is relatively independent, extension module replacement difficulty and cost of low double low advantages, is a preliminary attempt to the concept of electronic building blocks.

According to the above analysis, the robot is mainly composed of robot chassis, peripheral sensor, robot arm, cradle head and Raspberry PI microcontroller motherboard. The robot chassis is crawler structure, which can overcome the movement problem of complex working environment caused by uneven terrain. And the installed peripheral sensor is the car obstacle avoidance function, water avoidance function, identification function and grab function of the auxiliary and strengthen. Its main components are shown in Fig. 21.2.

21.3 System Function Module Hardware Selection and Design

According to the design idea of the system and the actual working environment, the robot should have several modules such as water avoidance function, obstacle avoidance function, identification function and picking function. Among them, the priority of water avoidance and obstacle avoidance function should be higher than that of identifying items to be cleaned. The modules are described as follows.

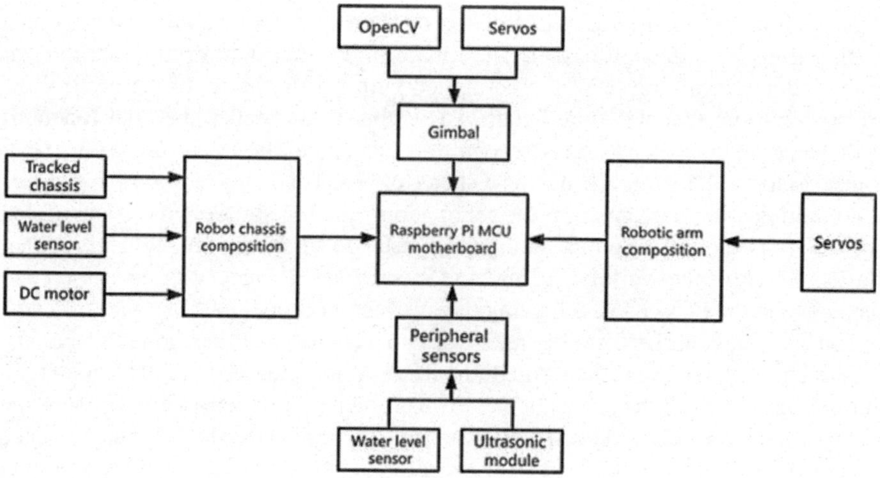

Fig. 21.2 System composition of cleaning robot

21.3.1 *Hardware Selection and Design of Water Avoidance Function Module*

Water Sensor is used to realize the water avoidance function. Specifically, Water Sensor is installed on the front and back ends of the chassis of the car to provide the car with the function of avoiding water in the blind area. The HSV color space of different surfaces, such as water surface, sand surface, mud surface, and grassland, can be compared to realize the difference between water surface and land [6]. When the car moves to a certain area near the water, it will automatically run the water avoidance program. The implementation idea of the program is to compare the data returned by the water level sensor at the front and rear ends of the car. When the water level sensor at any end of the car deviates, the program of the car wading to grab items will be forcibly interrupted. If the water level of one end is high, it will automatically move a certain distance to the low water level end, and then rotate 90° on the right side twice, and then collect the water level height data again for comparison and judgment; Until there is no difference between the water level and the high level, the object to be grabbed will be scanned normally. The function design can improve the ability of the car to grab items and avoid water.

21.3.2 *Obstacle Avoidance and Route Optimization Function Design*

This function uses ultrasonic sensor HC-SR04 and OpenCV intelligent vision sensor to achieve multi-directional, high accuracy of far and short distance obstacle avoidance and can optimize the travel route according to the traveling occasion. Its working principle is that the OpenCV is controlled by the cradle head to scan and identify the left and right rotation of 360° and the up and down rotation of 210°, which can identify both static objects and dynamic living objects such as humans or animals. Ultrasonic sensor can provide assistance to the robot, when the living body or items close to the robot 40 cm range, the robot can be marked according to the ultrasonic hint to the distance between it and the robot, at the same time to modify and adjust the direction of the robot, optimize the cleaning route of the robot, so as to achieve the purpose of dynamic and static obstacle avoidance. At the same time, its function can also be shown in assisting the capture of objects. When there are a large number of obstacles in the environment near the robot and the route needs to be adjusted, the ultrasonic module can conduct ultrasonic scanning of the surrounding environment while the robot is rotating, and then select a road furthest away from the robot and feed it back to the Raspberry PI. With the OpenCV intelligent vision sensor to judge the road, it can better assist the robot to modify the running route and achieve the purpose of optimizing the traveling route. The circuit diagram of ultrasonic module is shown in Fig. 21.3.

21.3.3 *Identify Function Module Hardware Selection and Design*

The core devices used for recognition function are OpenCV camera module and Raspberry PI processor. OpenCV camera is an intelligent visual recognition module with excellent performance and high-cost performance. As a kind of visual sensor, it serves as the peripheral input of Raspberry PI microcontroller motherboard. The working principle of the camera is to first take high-definition photos or videos of various kinds of common articles on the beach and waterfront, extract the features of various articles for summary and generalization, and then convert them into programming languages that can be recognized by the motherboard. Finally, a model library is formed and matched in the Raspberry PI microcontroller motherboard. When the camera recognizes and judges, it needs to call the corresponding programming language in the model library for comparison and judgment, so as to achieve the accurate identification of the items. Then, the identified data will be fed back to the PC side, and the identified items will be framed and the name of the items and the identification accuracy and other information will be displayed. Figure 21.4 is the effect picture of the robot's recognition and display in the laboratory.

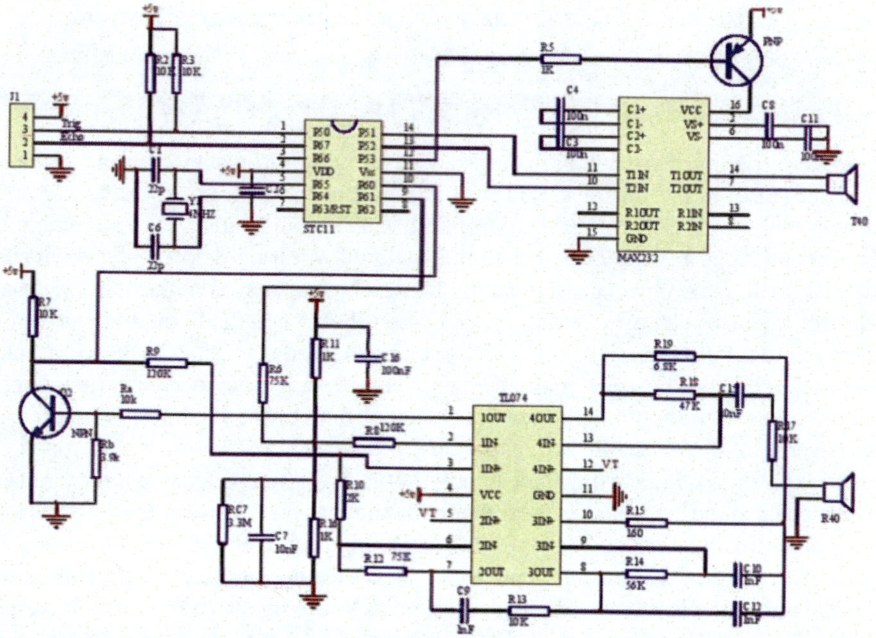

Fig. 21.3 Circuit diagram of ultrasonic module

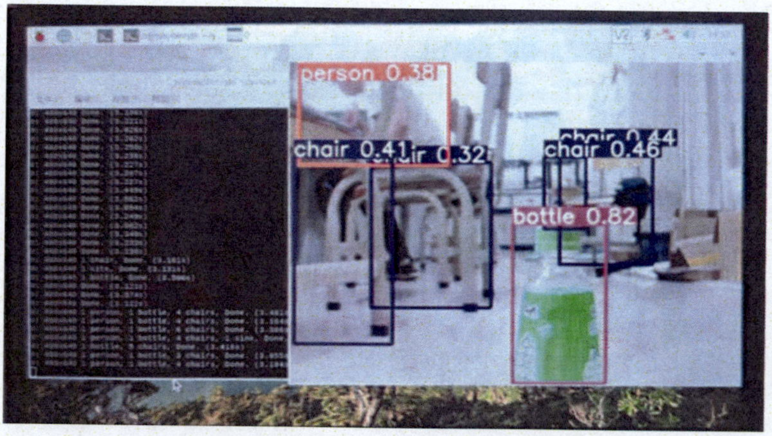

Fig. 21.4 Identification and display effect of OpenCV camera module

Rely on the high precision, high performance, high-cost performance, and more complete image processing library of OpenCV vision sensor, this module can obtain images with higher accuracy than OpenMV vision sensor of the same price, so as to achieve the purpose of accurate recognition.

21.3.4 Hardware Selection and Design of Pickup Function Module

The picking function is realized by the OpenCV intelligent vision sensor assisted by the flexible rotating mechanical arm. Its principle is to control the movement of the mechanical arm and the opening and closing of the mechanical claw through the four TBS2701 digital metal steering gear, and then control the grab of items and place them in the collection basket installed at the rear of the car through the single-chip computer program design. Specifically, the Raspberry PI single-chip microcomputer calculates and processes the data collected and returned by the vision sensor and can control the rotation angle of each steering gear independently, so as to realize the multi-directional rotation of the manipulator. The OpenCV intelligent vision sensor is used to send the identified item data back to the CPU of the Raspberry PI through the serial port and then make the MCU issue a command to control the motor movement to the vicinity of the identified item. At the same time, the OpenCV visual ranging function can accurately locate the actual distance between the object and the robot. After receiving the specific data, the single-chip microcomputer can accurately control the actual moving distance of the robot arm to achieve the purpose of accurate capture. The effects of steering gear circuit diagram and modeling diagram are shown in Figs. 21.5 and 21.6.

Fig. 21.5 Steering gear circuit diagram

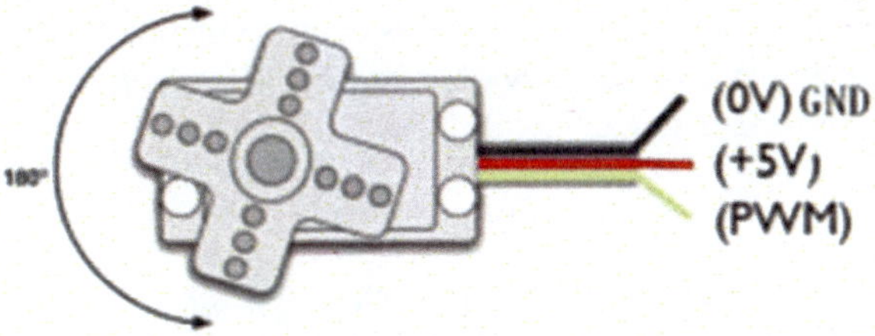

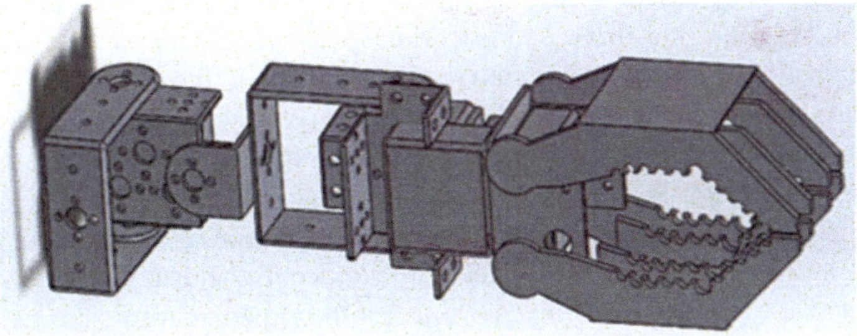

Fig. 21.6 Modeling effect of mechanical arm

21.3.5 *Functional Design of the Main Circuit*

The main circuit of the robot is based on the Raspberry PI microcontroller as the core processor. The main controller uses the BCM2837B0 type CPU, which is composed of 1.4GHZ64-bit 4-core ARM Cortex-A53, USB2.0, and 1G LPDDR2 running memory, meeting the robot's requirements for the main controller [7]. At the same time, Raspberry PI motor drive board as the circuit protection board, combined with Wi-Fi sensor to realize communication processing. LN298 chip can ensure stable voltage output, so as to stabilize the voltage of the whole system. The voltage booster module can provide the voltage required for the operation of the manipulator. Meanwhile, it can protect the Raspberry PI motherboard to avoid the damage of the motherboard chip caused by wrong output Settings during system configuration. The Wi-Fi module can realize the communication between the robot and modern communication devices such as mobile phones and computers and can form a local area network through the Wi-Fi module. The operator can realize graphical window operation on the remote desktop of the PC side and realize the fast real-time control of the Raspberry PI. At the same time, the recognition window and data receiving window of Raspberry PI are presented on the screen of PC in a visual graphical way, which is convenient for remote control [8]. The circuit diagram of Wi-Fi module is shown in Fig. 21.7. The capacitors C9, C12, C164, and C13 in the figure are decoupling capacitors, which can provide a stable power supply and reduce the noise of component coupling to the power end, indirectly reducing the influence of other components on the noise of this component. While the two capacitors C10 and C11 can adjust the frequency of crystal oscillator X1, to make it more accurate, J2 inputs the interface to RUN.

21.4 System Software Flow Design

The software process design of the system is shown in Fig. 21.8. A switch is set on the Raspberry PI motor drive board to run the robot self-test program to check whether each module of the robot can work normally. If the system fails to start normally, the system sends startup failure and fault feedback information to the PC. If it starts normally, check whether there is a stop command. If there is a termination command, the execution ends. If there is no stop command, start the cradle head and ultrasonic module to scan the surrounding environment and drive the robot to move in the direction without obstacles. Obstacle avoidance in the process of operation of the intelligent car to the surrounding environment intelligent identification, detection whether there is a need to clean goods. Since the model library of the items to be cleaned has been stored in the Raspberry PI microcontroller motherboard, when the items to be cleaned are matched, the distance of the items will be measured by the OpenCV intelligent vision sensor and fed back to the microcontroller. Then the microcontroller will issue a moving command to drive the robot to the location of

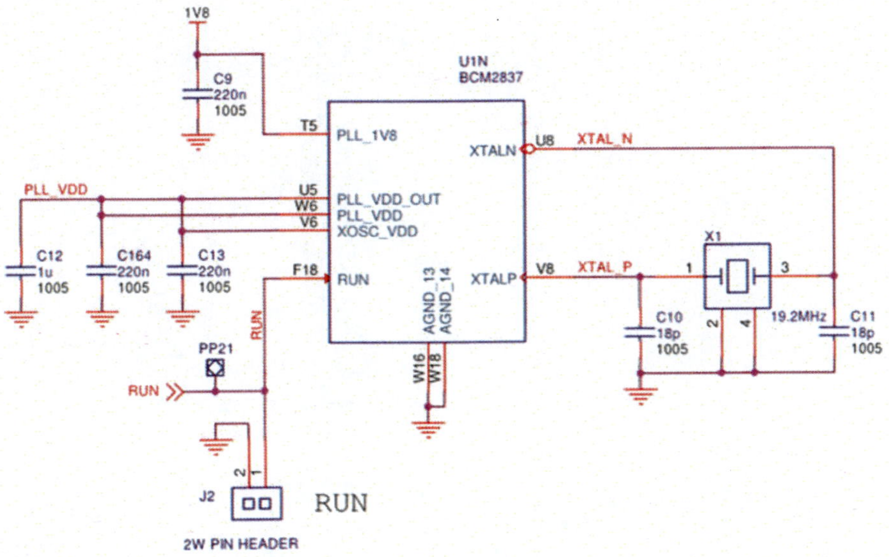

Fig. 21.7 Circuit diagram of Wi-Fi module with Raspberry PI microcontroller

the items. When the car moves within the working range of the manipulator, the MCU assisted by the OpenCV intelligent vision sensor issues a command to start the manipulator to capture the item accurately and put it into the pickup basket at the back of the car. This is a complete operation cycle. After the end of a complete cycle, run the same process again. If items that need to be picked up are found, the same workflow will be maintained for the pickup operation. If no items need to be picked up, the robot will be terminated after receiving the request from the terminal. If no termination request is received, the inspection continues until the item that needs to be picked up is found for cleaning.

In the design, a variety of sensors are used with Raspberry PI microcontroller for application, the main control board has stronger processing capacity, Wi-Fi communication module integrated on the motherboard, can better realize the remote control and monitoring of the robot; The use of OpenCV intelligent vision sensor not only can realize the identification of items but also can realize the identification of the water surface and beach so that the robot has the function of identification, obstacle avoidance and water avoidance ability, at the same time, it will be identified items ranging, improve the accuracy of the robot grasp function; Water level sensor can enable the robot to avoid water autonomously when wading to grab garbage; The tracked chassis of the robot enables the robot to have good mobility in the complex terrain outdoors. It can also wade into water to a certain extent, which improves the cleaning ability of the robot and prolongs the service life of the robot.

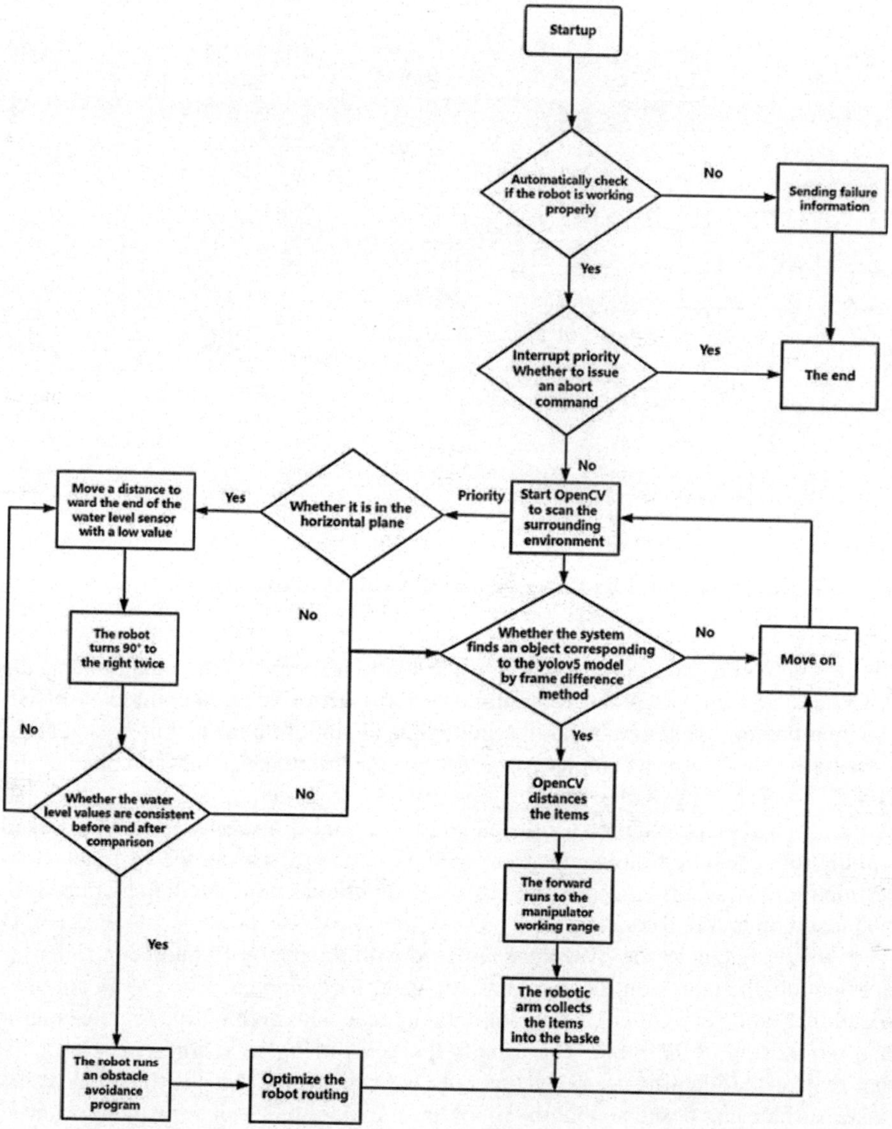

Fig. 21.8 System software flowchart

21.5 Conclusion

The robot is based on Raspberry PI microcontroller in wet areas, such as water mooring, beach cleaning robot, matching with OpenCV intelligent vision sensor to achieve high-precision identification of objects but also uses HSV color difference beach and wave color to achieve automatic water avoidance function, with strong adaptability, wide range of adaptation features. At the same time, it also realizes the efficient working mode of "one to many machines", which can better replace the traditional manpower for the cleaning work of the river and beach. In the current market, there are no robots that have been put into the market that have the ability to treat water beach. The main reason is that the working environment of the river beach and beach in the ecological environment near water is harsh, the difficulty of treatment is much higher than that of indoor surface cleaning, and the reason is that the technology requirements are higher and the research and development cost is higher. The land beach cleaning robot based on Raspberry PI microcontroller designed in this paper has the characteristics of low cost, high practicability, simple operation, etc. At the same time, the robot can adapt to work in harsh environment, can be used as a supplement and substitute for artificial cleaning, and has certain practical value to save human resources' cost and protect the ecological environment.

Acknowledgements This paper is based on the project: 1. National College students' Innovation and Entrepreneurship training program project "A multifunctional integrated cleaning Robot based on water ecological environment protection" (project number: 202113720004); 2. The interim results of the Guangdong Undergraduate University Education and Teaching Reform Engineering construction project "Artificial Intelligence Industry Institute" (Project number: 2020-19). Financial support is provided by the project, and all members involved in the project are thanked for their efforts to complete the project.

References

1. Jiang, Y.: Environmental regulation, environmental regulation competition and employment: an analysis based on inter-provincial spatial durbin model. J. Guizhou Univ. Finance Econ. **05**, 79–89 (2017)
2. Zhang, B.B.: Promoting common prosperity through boosting employment in capacity and quality——take addressing the structural problems in the labor market as the key. Rev. Ind. Econ. **02**, 168–185 (2022)
3. Chen, Y.Q., Wu, Z.L., Li, Y.Y., Cheng, W.L.: Research and design of patrol inspection robot system for circulating water aquaculture workshop. Mod. Inf. Technol. **6**(10), 116–120 (2022)
4. Zhang, H.Y.: Analysis of the structure and stability of a large and highly adaptable hexapod robot. Jilin University (2021)
5. Cai, S.J., Zhang, M., Chen, Z.C., Kang, Z.W.: Design of water park cleaning robot based on STM32 single chip microcomputer. Autom. Instrum. **37**(08), 54–57 (2022)
6. Wang, C., Wan, Z.J., Zhou, Y.J., Liu, Y.H.: Research on garbage classification algorithm based on improved ResNet-50. Intell. Comput. Appl. **12**(10), 184–188 (2022)
7. Wang, Y.F., Zhang, Y.H., Huang, X., Sun, T., Lin, J.P., Zhang, G.Y.: Design of mobile intelligent monitoring system based on raspberry Pi. Shanxi Electron. Technol. **05**, 7–9 (2022)
8. Wang, T., Huang, X.R., Fu, W.B.: Design of multi-function detection inspection vehicle based on STM32 single-chip microcomputer. Electron. World **02**, 208–209 (2020)

Chapter 22
Fragrant Pear Target Identification and Positioning Based on Deep Learning

Peng Zhou and Bingyu Cao

Abstract In the rapid development of artificial intelligence technology, scholars around the world based on the research results of various stages, the relevant technical theories are applied to various fields, not only comprehensively improve the level of social and economic development but also create a digital and intelligent social environment. Especially for the agricultural field, the traditional artificial picking method has been unable to meet the needs of people, it is easy to harm the safety of people and property during the work, the time cost of picking work is too high, so under the development trend of artificial intelligence, it is necessary to focus on the research of picking machines. This paper aims to explore the application value of target recognition and positioning for fragrant pear, by discussing deep learning-centered research on target recognition and positioning based on a comprehensive understanding of the development status of modern agricultural picking work, in accordance with the fundamental principles of deep learning.

22.1 Introduction

Although China has vast geographical resources, due to the rapid population growth rate, relatively weak agricultural production infrastructure, the development level of science and technology is too old-fashioned, the agricultural industrial structure is relatively simple, and the agricultural production efficiency is too low, resulting in an increasing number of rural migrant workers, which will not only aggravate the phenomenon of farmland abandonment but also aggravate the phenomenon of rural migrant workers. There will also be social problems such as left-behind elderly people and left-behind children. In the face of this development situation, Chinese scholars have proposed the all-round penetration and application of artificial intelligence technology in their research to create smart agriculture consistent with the development of the new era, which can not only bring more development opportunities for agriculture

P. Zhou (✉) · B. Cao
Xinjiang University of Science and Technology, Xinjiang, China
e-mail: zp1695670892@163.com

R. Kountchev et al. (eds.), *Multidimensional Signals, Augmented Reality and Information Technologies*, Smart Innovation, Systems and Technologies 374,
https://doi.org/10.1007/978-981-99-7011-7_22

but also attract more enterprises, social institutions, and outstanding talents to participate in agricultural economic construction. At present, agricultural construction and development are mainly divided into the following types: First, green agriculture refers to the combination of agriculture and ecological environment, in a coordinated development at the same time, the real realization of sustainable development goals; secondly, continuously optimize the technical level of modern agricultural construction and regard biophysical factors as practical operating objects, which can not only increase the amount of agricultural production but also reduce the negative impact of pesticide chemical products on the ecological environment. Ultimately, the utilization of cutting-edge technological advancements facilitates the realization of automation and mechanization objectives [1–3].

Nowadays, the most common object detection algorithm is based on deep learning. This kind of theoretical technology is mainly divided into two kinds: on the one hand, the candidate region is regarded as the detection center. This approach applies the detection algorithm to generate a series of candidate boxes and treats these candidate boxes as detection samples, and the rest of the convolutional neural network to efficiently classify the samples, on the other hand, from end to end target detection. This approach turns targeting into a regression problem and does not generate candidate regions. From the perspective of practical application, the principle of the two methods is not consistent, and the application performance is different, the former is more accurate, and the latter is faster. Combined with authoritative PASCAL VOC data sets, we can see that object detection performance has stabilized over the past few years, the most representative of which is the integrated system, which can organically combine high-level context and low-level images. According to the region-centered convolutional neural network algorithm proposed by Ross B. Gurshick of the University of California, Berkeley, Chen Xiaoning et al. mainly studied the process and application effect of small target recognition in complex scenes. The final experimental results show that the average detection accuracy of this method is higher than that of the original model [4].

In essence, this method is very simple to operate, and the accuracy is constantly increasing compared to traditional algorithm applications. This approach effectively combines two key insights: On the one hand, large-scale convolutional neural networks can be applied in a bottom-up order to facilitate localization during the experiment; on the other hand, when there is too little labeled training data, the pre-processing of auxiliary tasks should be supervised, and then the specific fields should be slightly adjusted to optimize the overall application performance. In this study, Fu Miaomiao et al. mainly compared and analyzed deep learning and transformer object detection algorithms based on the region-centered fast convolutional neural network algorithm proposed by Ross B. Gurshick. The final experimental results demonstrate that both target detection techniques are currently at the forefront of research and represent an inevitable trend in future target detection development.

By learning from SPPNET algorithm structure, convolutional neural network is used to classify and process object proposals. Compared with the traditional working mode, fast RCNN can not only accelerate the training speed but also improve the accuracy of detection. Compared to SPPNET, VGG16 with fast RCNN training is 10 times faster, and the practical application technique is more accurate. However, from the perspective of overall application, there is no scientific treatment of the candidate box problem concerning positive and negative samples [5–7].

After comprehensively studying the applied algorithms and theoretical knowledge mastered by domestic and foreign researchers, this paper mainly analyzes the technical process of target recognition and target location according to the application advantages of convolutional neural network, and then verifies and analyzes the experiment of Xiangli target recognition and location, in order to provide an effective basis for the technological development in the agricultural field in the new era.

22.2 Methods

22.2.1 Convolutional Neural Network

As one of the most representative algorithms for deep learning, convolutional neural network was first studied in the 1980s and 1990s. In the twenty-first century, with the continuous improvement of deep learning theory, convolutional neural network algorithms have also been widely used, usually appearing in the field of natural language processing, computer vision and so on. Combined with the analysis of CNN structure diagram shown in Fig. 22.1, it is known that the overall module mainly includes input layer, output layer, and multiple hidden layers. Among them, the input layer is mainly used to process multi-dimensional data, and the input features need to be standardized, which can further improve the learning efficiency and concrete performance of the algorithm; The overall structure of the output layer is consistent with the output of the latent deficit neural network in the traditional sense. For example, when dealing with image classification problems, the output layer can use the planning index or logical function to obtain the classification label bucket. In the process of identifying objects, the output layer can be designed to output object classification, size, center coordinates, etc.; the hidden layer contains three common structures of full connection layer, pooling layer and convolution layer, among which the latter two are one of the unique forms of convolutional neural network algorithm [8].

22.2.2 RCNN

Region-centered convolutional neural network (RCNN) is the initial result of applying the convolutional neural network to the target detection problem. According to the cumulative experience of current practical research, the convolutional neural network itself has unique application performance and is more efficient in solving the

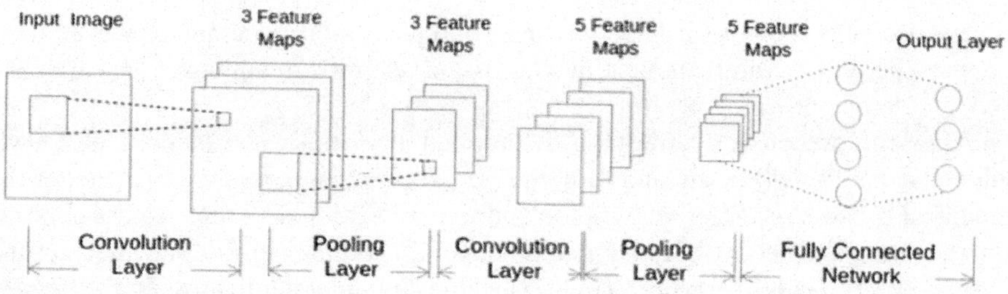

Fig. 22.1 Structure diagram of CNN

target detection problem, so it is the focus of scholars from all over the world. Generally, the convolutional neural network algorithm performs target detection in three steps: first, select candidate regions; second, extract important features; and finally, classify and achieve boundary regression. From a general point of view, RCNN operation is more intuitive, and its purpose is to accurately locate and merge to complete the classification, and reduce the number of experimental analyses.

After the accumulation of RCNN and rapid RCNN, foreign scholars have once again proposed a faster RCNN in their research. By comparing the two, it can be found that the latter will integrate all operations in a certain network in the process of feature extraction, classification processing and border selection, which is helpful to improve the comprehensive performance of practical application and speed up the overall detection speed. In order to effectively solve the problems of RCNN and fast RCNN, faster RCNN introduces a candidate region network (RPN), which can finally reduce the practical application cost by sharing the basic features of images. Combined with the analysis of the RCN algorithm process as shown in Fig. 22.2, RPN can predict the target boundary and target score of each position at the same time, and RPN is trained end-to-end to form a higher quality candidate region and facilitate technical detection and analysis [9–11].

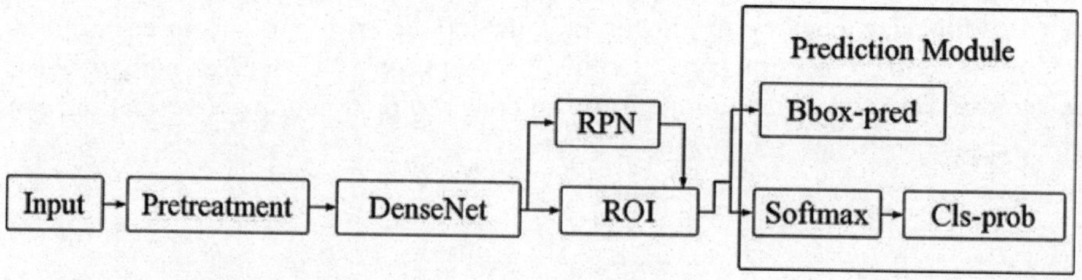

Fig. 22.2 Operation flow of faster RCNN algorithm

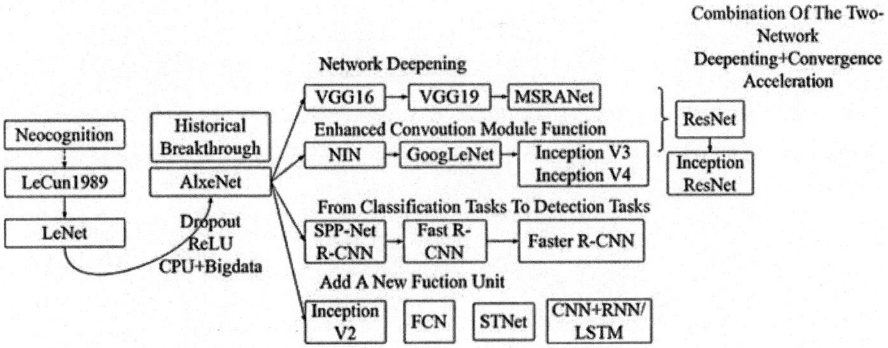

Fig. 22.3 ZF network framework

22.2.3 Shared Convolution Layer

In order to design faster RCNN with different shared convolutional layers and extract interested features, different hierarchies of faster RCNN model, ZF network, VGG16 network and ResNet network should be integrated together to truly realize the iterative training research [12, 13]. Among them, the framework of ZF network is shown in Fig. 22.3. In the shared convolution layer of faster RCNN, only the convolution part is needed.

VGG 16 network structure is shown in Table 22.1. VGG 16 network structure should strictly abide by the basic principles of fairness and justice, accurately judge whether the network depth will affect the accuracy of the model, and at the same time, ensure the consistency of the convolution layer configuration. The size of the convolution kernel is 33, the step size is 1, and the filling value is 1. There are five maximum pooling layers, the kernel size is 22, and the step size is 2. The overall design contains three fully linked layers, where the first two have 4096 channels, while the last one contains 1000 output neurons, implicitly with the RELU non-linear activation function behind it.

22.3 Result Analysis

22.3.1 Experimental Design

In this paper, after understanding deep learning and convolutional neural network algorithms, the Mask R-CNN fragrant pear detection model with DenseNET as the core is used. The DenseNET network structure refers to a brand new deep network structure. The calculation formula is shown as follows:

$$x_l = H_l([x_0, x_1, ...x_{l-1}]) \tag{22.1}$$

Table 22.1 VGG16 network structure diagram

ConvNet configuration					
A	A-RN	B	C	D	E
11 weight layers	11 weight layers	13 weight layers	16 weight layers	16 weight layers	19 weight layers
Input (224 × 224RGB *image*)					
Conv 3-64	Conv 3-64 LRN	Conv 3-64 Conv 3-64	Conv 3-64 Conv 3-64	Conv 3-64 Conv 3-64	Conv 3-64 Conv 3-64
Maxpool					
Conv 3-128	Conv 3-128	Conv 3-128 Conv 3-128	Conv 3-128 Conv 3-128	Conv 3-128 Conv 3-128	Conv 3-128 Conv 3-128
Maxpool					
Conv 3-256 Conv 3-256	Conv 3-256 Conv 3-256	Conv 3-256 Conv 3-256	Conv 3-256 Conv 3-256 Conv 1-256	Conv 3-256 Conv 3-256 Conv 3-256	Conv 3-256 Conv 3-256 Conv 3-256 Conv 3-256
Maxpool					
Conv 3-512 Conv 3-512	Conv 3-512 Conv 3-512	Conv 3-512 Conv 3-512	Conv 3-512 Conv 3-512 Conv 1-512	Conv 3-512 Conv 3-512 Conv 3-512	Conv 3-512 Conv 3-512 Conv 3-512 Conv 3-512
Maxpool					
PC-4096					
PC-4096					
FC-4096					
FC-1000					
Soft-max					

In the above formula, X_1 represents the output of layer 1 and H_1 represents the nonlinear transformation. In this network structure, the input of each layer is the feature graph generated by its previous layer, so for the overall network, there are a total of L (L + 1) /2 connections, where L represents the number of network layers.

In this paper, DenseNET 121 (k = 32, growth rate) is used as the basic network extraction part. The Feature block includes the convolution layer, regularization layer, ReLU layer, and pooling layer. In the convolution layer, 7 × 7 convolution kernel is used, and the step size is 2. In addition, the same padding method is selected to realize the image convolution operation, and then the convolution results are input to the BN layer and ReLU activation function layer, and the pooling operation of the feature map is realized by means of 3 × 3, step size is 2, padding = 1. The network structure of DEnseNet is shown in Fig. 22.4 [14, 15].

In the Mark R-CNN model, ResNet152 is used for feature extraction network area. During the operation of the model, the practice speed is relatively slow. Therefore, ResNet network should be used to replace the DEnseNet network during the

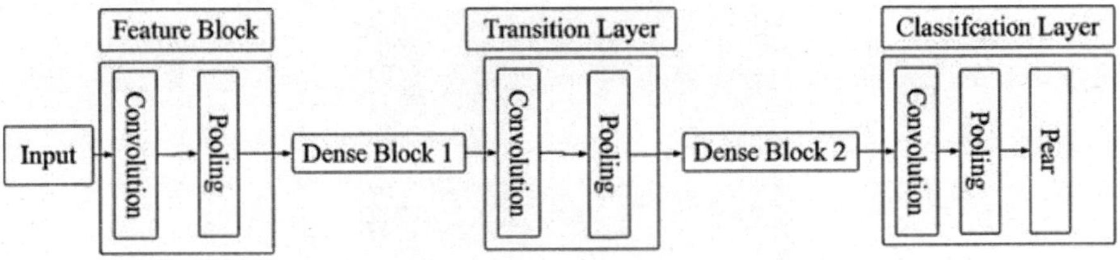

Fig. 22.4 Network structure of DEnseNet

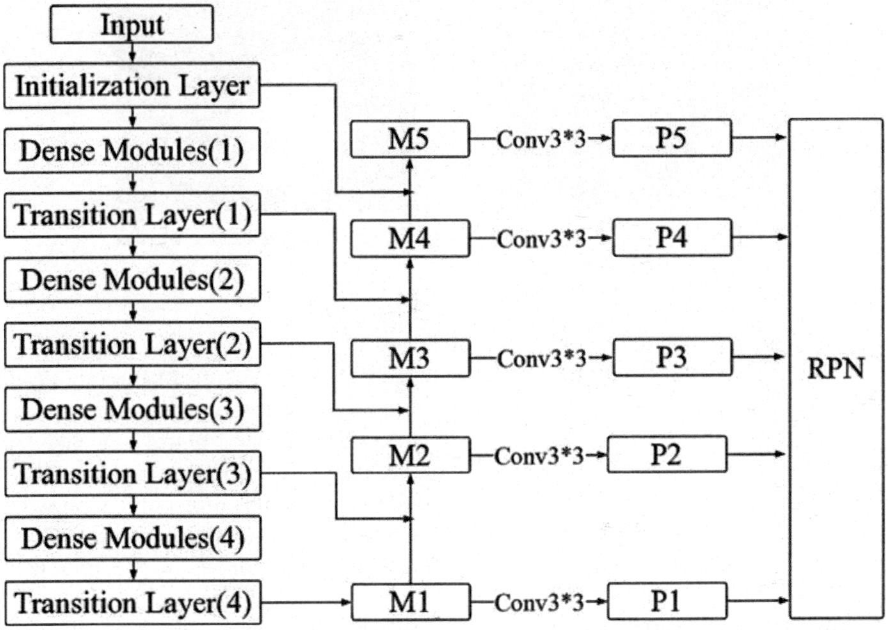

Fig. 22.5 Structure diagram of DEnseNet network combined with multi-scale

experiment, so as to improve the running speed of the network model. Among them, the combination of DEnseNet network and FPN is the feature extraction part of the overall network. The specific structure is shown in Fig. 22.5:

22.3.2 *Experimental Results*

In this experimental analysis, the experimental environment mainly uses the deep learning framework of TensorFlow, and the hardware facilities choose Intel (R) Core (TM) i7-8750H CPU@2.2 GHz six-core CPU and 16 GB memory. In order to further improve the model detection effect, 9600 pieces of fragrant pear data with the size

Fig. 22.6 Comparison and effect diagram

of 400 × 300 were combined for pre-training, and then the pictures of fragrant pear in natural environment were used for re-training. Finally, after data enhancement, 4500 pieces of data set were obtained. Consider 3500 as the training set, 500 as the validation set, and 500 as the test set. Among them, the initial learning rate was 0.001, the quantitative setting was 0.9, the regularization attenuation coefficient was 0.0001, and the compression rate was 0.5. A total of 105 iterative analyses were completed. If the error probability of the verification set continues to rise, then the training should be terminated. Combined with the analysis of the improved model, it can be seen that compared with the ResNet model, the recognition speed of each image increases by 5fps, which proves that the improved target detection model can further improve the detection effect. Combined with the analysis of the comparison effect diagram shown in Fig. 22.6, it is known that compared with the ResNet model, the recognition speed of each picture has increased by 5 fps, so it is proved that the improved target detection model can further improve the detection effect.

22.4 Conclusion

To sum up, as the theory of modern artificial intelligence technology is becoming more and more perfect, the construction and development of intelligent agricultural society have become the focus of our attention. For fruit picking, a variety of intelligent picking machines have appeared in the market at present, which are mainly used to identify and locate basic features such as color and shape of fruit. In this paper, the deep learning-based research on target recognition and positioning of fragrant pear is mainly operated by convolutional neural network algorithm, which can not only improve the identification and positioning efficiency of fragrant pear but also guarantee the fruit quality of fragrant pear. From a long-term developmental perspective, this paper investigates the Mask-CNN model as a representative multi-task learning network capable of performing essential operations such as game target classification, position box regression, and target pixel-level segmentation. Therefore, in future research and development endeavors, it is crucial to enhance the model's speed while maintaining accuracy without compromising its performance for real-time target recognition. Ultimately, this advancement will significantly contribute to fruit picking and agricultural development.

Acknowledgements Autonomous Regional Youth Fund Project: Research on Identification and Positioning of Fragrant Pear in Natural Environment Based on Deep Learning (2022D01B148).

References

1. Yue, C.H., Wang, X., Gong, J.L., et al.: Research on underwater target acoustic recognition and location technology based on deep learning. Digit. Ocean Underw. Offense Def **4**(6), 6 (2021)
2. Zhao, H., Wang, A., Fan, Y.: Research on the identification and evaluation of aerobics movements based on deep learning. Sci. Program. (Pt.14), 6433260.1–6433260.10 (2021)
3. Luo, D.: Detection and fine localization of technical feature objects in basketball based on deep learning algorithm. J. Gansu Union Univ. (Natural Science Edition) **035**(006), 108–112 (2021)
4. Chen, X.M., Zhao, J.: Research on small object recognition in complex scene based on deep learning. Electron. Technol. Softw. Eng. **000**(010), 110–112 (2021)
5. Wang, F., Cui, D.D., Li, L.: Target recognition and localization algorithm of picking robot based on deep learning. Electron. Meas. Technol. (020), 44 (2021)
6. Fu, M.M., Deng, M.L., Zhang, D.X.: Object detection algorithm based on deep learning and transformer. Comput. Eng. Appl. **59**(1), 37–48 (2023)
7. Wang, M.F., Li, Z.M.: Research on port moving target recognition technology based on deep learning. China Water Transp. (10), 2 (2022)
8. Ren, H., Zhu, H.Q.: Research on object orange recognition method based on deep learning. Comput. Age **1**, 5 (2021)
9. Cao, B.Y., Xing, W.X., Zhou, P., et al.: Identification and detection of fragrant pear based on transfer learning faster R-CNN in natural environment. South. Agric. Mach. **53**(21), 4 (2022)
10. Niu, W.Y.: Research on multi-objective vehicle following recognition algorithm based on deep learning. Inf. Technol. Inf. (9), 4 (2022)
11. Yang, W., Yao, T., Geng, X.T., et al.: Application research of infrared target recognition technology for air defense weapons based on deep learning. J. Military Equip. Eng. **001**, 043 (2022)
12. Hu, Q.: Research on key target recognition of smart city based on deep learning. J. Guiyang Univ. Nat. Sci. Edition **16**(2), 5 (2021)
13. Zhang, Y.T., Wang, Z.Y., Fan, Ha. D., et al.: Research on positioning detection technology of parallel robot based on deep learning. Electron. Meas. Technol. (011), 045 (2022)
14. Liu, X.Y.: Research on image object detection algorithm based on YOLO deep learning model. Comput. Program. Skills Maint. (7), 4 (2022)
15. Ezat, W.A., Dessouky, M.M., Ismail, N.A.: Evaluation of deep learning yolov3 algorithm for object detection and classification. Menoufia J. Electron. Eng. Res. **30**(1), 52–57 (2021)

Chapter 23
Online Mini-Game Popularity and Feedback Data Prediction Based on Time Series and BP Neural Network Prediction

Qianyun Huang and Shijie Yang

Abstract In recent years, in order to increase network traffic, online mini-games have accounted for an increasing proportion of the video game market share. For example, the New York Times acquired the online mini-game "Wordle" in order to attract customers, but the online mini-game has the problem of large fluctuations in popularity and large changes in player feedback, and the current research on related issues at home and abroad is still insufficient, resulting in online mini-game companies unable to accurately judge the popularity of the game and player experience and make wrong judgments. This has led to a more serious customer churn. To this end, this article uses the New York Times mini-game "Wordle" player data from January 2022 to January 2023 as an example. An exponential smoothing prediction model and a BP (Back Propagation) neural network prediction model are established to predict the number of players and scores of Wordle, respectively, to realize the prediction of online game popularity and player feedback. The overall fitting error of the final population prediction model does not exceed 3,000 people, and the MAE (Mean Absolute Error) of the fitting error of each of the seven outputs of the score prediction model does not exceed 5. Experimental results show that the two models used in this paper have high feasibility for the prediction of popularity and player feedback of online mini-games.

23.1 Introduction

With the development of science and network technology, the popularity of video games remains high, in a wide variety of video games, online games because of its no need to download and low difficulty of operation convenience in the huge video game market to seize a place, but due to the simple development, a variety of online games

Q. Huang (✉) · S. Yang
Hua Zhong Agriculture University, Wuhan, China
e-mail: 13971333617@163.com

R. Kountchev et al. (eds.), *Multidimensional Signals, Augmented Reality and Information Technologies*, Smart Innovation, Systems and Technologies 374,
https://doi.org/10.1007/978-981-99-7011-7_23

are produced in large quantities almost every day, so online games generally have the problem of game popularity and player feedback changes. If an online mini-game wants to maintain its popularity and ensure stable traffic revenue for a long time, it needs to accurately predict the change in the number of players and player feedback of the game, so as to adjust its game strategy to improve the user experience and maintain the popularity and traffic of the game.

The prediction of the number of players is a typical process of predicting future data based on the data that has existed in the past, and there are many classical methods that can be used in this process, such as multiple linear regression and time series forecasting, especially the holt linear trend exponential smoothing model in time series forecasting is very suitable for solving the prediction problem of one-dimensional data with obvious linear trend; But for the prediction of player game data feedback, because it is associated with the complexity of human thinking and behavior, it is used to mimic the structure of the human brain neural network BP. Neural network predictive models are well suited to solve the prediction problem of these data.

This paper mainly focuses on the time series characteristics of the change of the number of online mini-games and the complexity of feedback data and predicts the changes and feedback data of online mini-games based on holt linear trend exponential smoothing model and BP neural network prediction model, and takes the user data from January to December 2022 of the New York Times game "Wordle" as an example to complete two prediction of player numbers and player feedback on March 1, 2023.

23.2 Analysis of the Current Status of the Study

Time series forecasting model is a forecasting method that uses the past change law of data to infer the possibility and trend of future changes, so it is very suitable for solving the forecasting problem of one-dimensional data.

Aiming at the application of time series prediction method, Meng Zhiqiang et al. used a hybrid prediction model based on time series to predict the number of book borrowings, and the experimental results show that the time series algorithm can be effectively used to predict the borrowing data of university libraries [1]. In order to accurately predict the evolution of heavy pollution in haze, Wang Ying et al. proposed a long short-term memory and multivariate linear regression and used the model to predict the $PM_{2.5}$ concentration in Xi'an, Shaanxi Province, China [2]. Cibicong et al. constructed a new ARIMA (Autoregressive Integrated Moving Average model)-LSTM (Long Short Term Memory) combination model in the form of nonlinear combination, which predicted three financial time series such as Shanghai Securities Composite Index, and used the ARIMA model, LSTM model and linear combination model as the control model, and finally found that the accuracy of the prediction results of the constructed nonlinear combined prediction model was generally better than that of the single prediction model of the control group [3]. In addition, time series forecasting models are also used in agricultural product sales [4–9] and public health [10–14]. From the perspective of the application of time series forecasting methods by experts and scholars, most of them are applied in the fields of finance, health and public safety, but few articles have been published in the fields of online mini-game popularity changes.

Neural network prediction model is a kind of application similar to the brain synaptic connection structure for information processing mathematical model, it is simulated on the basis of human understanding of their own brain tissue combination and thinking mechanism, and BP neural network is a kind of multi-layer feedforward neural network based on error reverse propagation algorithm, BP algorithm is the most successful neural network learning algorithm so far, so it is very suitable for solving the data prediction problems associated with human complex thinking and behavior.

Aiming at the application of neural network prediction method, Zhang Liping et al. established a BP neural network R&D cost prediction model based on the data of 280 companies listed on the Science and Technology Innovation Board as learning samples, and conducted simulation tests on the network, and finally found that BP neural network prediction has better fitting effect and accuracy than traditional regression prediction [15]. Aiming at the problem of stock price fluctuation prediction, Wu Feng et al. proposed a stock price trend prediction model based on the neural network of the LSTM and found that the model has better effect in stock price trend prediction than the existing model [16]. Wang Shunan et al. screened out the main influencing factors affecting channel dredging according to the measured hydrological data and channel dredging mechanism from 2016 to 2018 and established a high-precision prediction model of BP neural network for deepwater channel dredging in the Yangtze River estuary under the action of multiple impact factors, and the research results verify that the model has high channel siltation prediction ability and spatial distribution prediction accuracy [17]. In addition, neural network predictive models also have many applications in the manufacturing [18–20] and physics industries [21–23].

Most of the domestic scholars' research on time series prediction and neural network prediction are distributed in the fields of finance, enterprise, and security, while the research on online mini-game popularity prediction and player feedback prediction [24, 25] is relatively lacking, but the online mini-game popularity and feedback prediction and the prediction in the financial field have obvious homogeneity, so this paper aims at the research status of online mini-game popularity and feedback prediction. A time series model and a BP neural network model were used to predict the number of online mini-game players and feedback data.

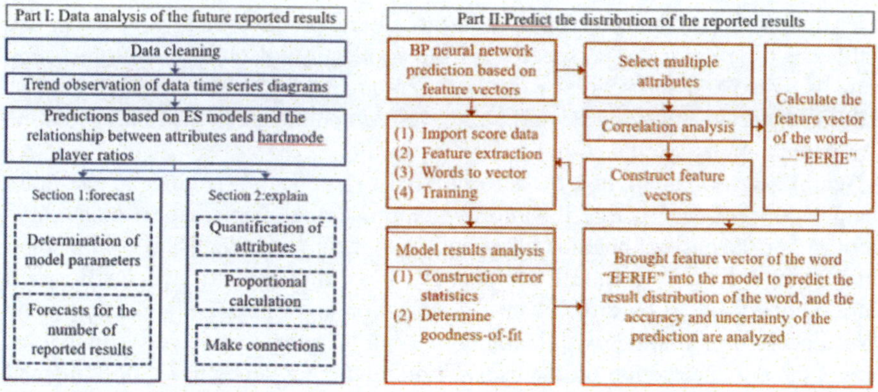

Fig. 23.1 Study the framework diagram

23.3 Research Methodological Settings

23.3.1 Research Framework

After collecting the data required for the experiment, in this paper is supposed first to preprocesses the data and to analyze the trend of the time series graph composed of the dataset, and then to use the time series forecasting model to fit and predict the time series in order to obtain the predicted value of the future player number report results. Secondly, a BP network prediction model is built for the player score distribution results according to the constructed feature vectors, and then is used the collected data set to train the model, counts the model error, calculates the goodness-of-fit of the model, and finally predicts the score distribution of the specified words to obtain the prediction results (Fig. 23.1).

23.3.2 Data Collection and Preprocessing

In order to study the prediction of the popularity of online mini-games, this article collects data such as the number and score of "Wordle" game users and scores from January 7 to December 31, 2022, published by the New York Times on Twitter, and conducts this data a stationarity test, and finally the shuffled data passed the test, indicating that it could be analyzed for time series.

Matlab was used to write the code to visualize the number of reported results by time, as shown in Fig. 23.2. From the figure, it can be concluded that the number of players has been increasing with a large slope since January 2022 until it reached a peak on February 2, 2022, and then it gradually shows a decreasing trend, and the decreasing trend is getting slower and slower. At the same time, it was possible

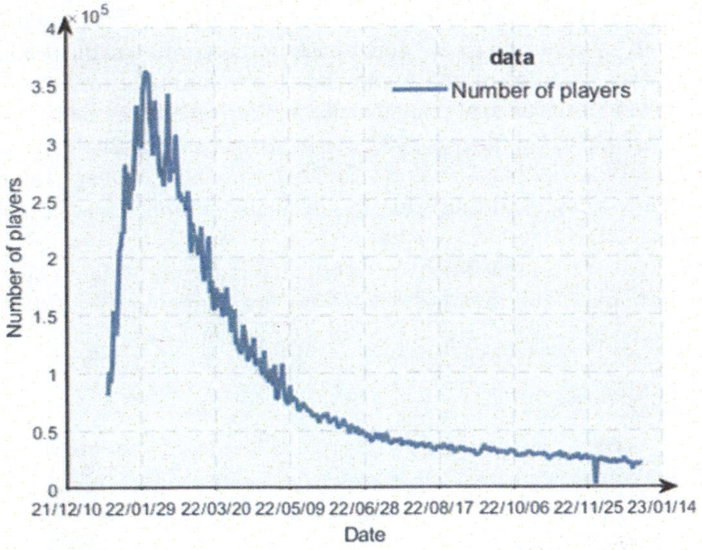

Fig. 23.2 Data time series diagram

to visually identify two outliers of the reported quantities, November 30, 2022 and December 25, 2022, by taking the arithmetic mean of the quantities on the 2 days before and after them for substitution. Similarly, outliers in the difficulty mode scores were handled, in addition to checking and correcting four incorrect words in the data set to facilitate the follow-up process.

The time series diagram for this set of data is shown in Fig. 23.2.

23.4 Case Analysis

Wordle is a popular puzzle currently offered daily by the New York Times, its player base is huge. Many (but not all) users report their scores on Twitter, this article uses these data to predict the popularity of the game, which is the number of players who report the score, and the score data that players have feedback.

23.4.1 Explanation of Changes in Number of Scores

In this problem, we try to construct an exponential smoothing model with data on the number of players reported by past players and obtain a predicted value of the number of players reported by players in the future. Before building the time series model, a smoothness test needs to be performed first. Since the complete sequence of number of reported results did not pass the smoothness test, the data from February 2, 2022 and onwards were taken for the smoothness test according to the peak position, and the series was obtained to be smooth and able to build the time series model. Here, this article focuses on analyzing the effect of the time factor on it. The change from January 7, 2022 to February 2, 2022 was fitted using a straight line, and the data curve for February 2, 2022 and beyond was fitted using holt's linear trend method.

In the fitting process, the fitting error was calculated for nine cases in steps of 0.1, as shown in Fig. 23.3.

From Fig. 23.3, we get that when the threshold parameters' value is taken as 0.5, the error is minimum, and then use Matlab for programming to get the result as Fig. 23.4.

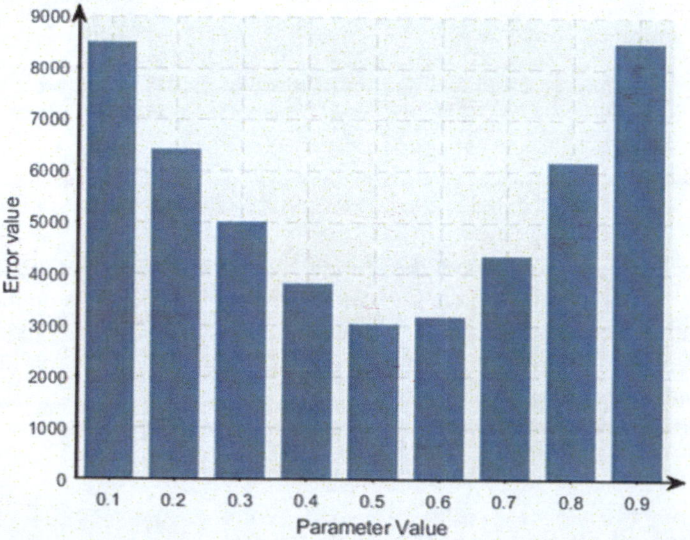

Fig. 23.3 Fitting error under different parameters

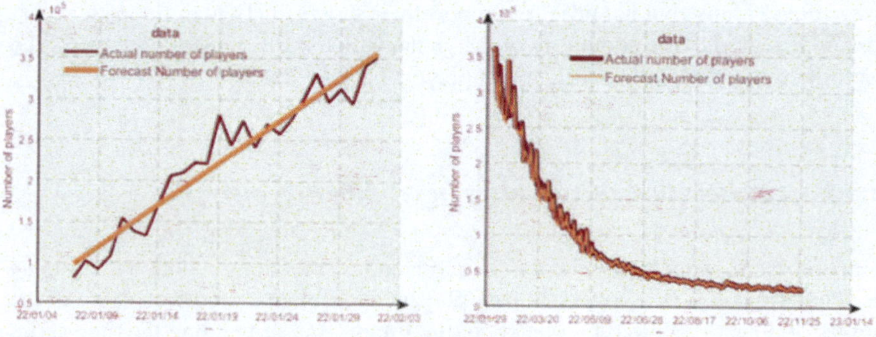

Fig. 23.4 Model fitting

As shown in Fig. 23.4, the fitted values of the two segments of data are similar to the actual value curves, and the surface fitting is good. Based on the rising trend of the previous period, starting from January 7, 2022, the results of the straight-line fit obtained on February 2, 2022, when t = 26, the fitted equation is followed as Eq. (23.1):

$$\widehat{y}_{26+m} = 87378.70 + 10363.04m \tag{23.1}$$

The variable m in the formula represents the number of days between the start date of the fitted data set January 7, 2022 and represents the predicted number of game players after the interval m.

Further refitting the curve to the latter segment while starting on January 1, 2023, yields the linear trend equation when t = 61, the fitted equation is followed as Eq. (23.2):

$$\widehat{y}_{61+m} = 26035.99 - 64.70m^2 \tag{23.2}$$

The variable m in the formula represents the number of days between the start date of fitting the data set and the second segment function, and y represents the predicted value of the number of game players after the interval m.

Since the upper and lower intervals of the predicted values are needed to ensure the accuracy of the prediction, the confidence interval of the above model is used to represent the uncertainty of the prediction results. A wide confidence interval indicates a large uncertainty of the prediction result, while the opposite indicates a high reliability of the prediction result. Here, a 95% confidence interval is used to represent the uncertainty of the prediction result, assuming that the predicted value is y^, the confidence interval is [l, u], and MSE (mean-square error) is the standard deviation of the prediction error, the calculation formula is followed as the Eq. (23.3):

$$\begin{aligned} l &= \widehat{y} - 1.96\sqrt{MSE} \\ \mathrm{u} &= \widehat{y} + 1.96\sqrt{MSE} \end{aligned} \tag{23.3}$$

The MSE represents the mean squared error of the number of player fit, with 1.96 multiple of the error as the predicted interval length.

The forecast values on March 1, 2023 and their upper and lower bounds are obtained in Table 23.1.

Table 23.1 Player count forecast results for March 1

Parameters	Value
Upper bound	22,135.19
Predicted value	18,337.22
Lower bound	14,539.25

From the prediction results, the predicted value of the number of "Wordle" players on March 1, 2023, predicted by the model in this paper, is 18337, which is not exceeded 2,2135 people, with a minimum of 1,4539, and there is a 95% chance that the number of gamers will fall within the range [14539, 22135].

23.4.2 *Player Score Predictions*

23.4.2.1 Word Feature Extraction

How to know the player's feedback data for a level of the mini-game, the player's score can be said to be the most direct way, but if you want to predict the player's score in a given level in the future through the player's score of the previous level, then it is crucial to find the connection between the level and the score because the difficulty of the level will directly affect the player's score, so in the questions studied in this article, we first need to summarize the factors that may affect the player's score in the words every day. In this paper, we argue that the distribution of reported results will be inextricably linked to some features of the words, so before predicting the distribution of reported results, we first perform feature extraction on the words. The extracted features are shown in Table 23.2.

Because the dataset we used already gives a summary of the scores of all players who reported game scores in previous levels, that is the percentage that guessed the word in one try, two tries, three tries, four tries, five tries, six tries, or could not solve the puzzle (indicated by X). Therefore, we only need to express the connection between the above five word characteristics and the different score proportions of players by establishing a prediction model to achieve the prediction of the player's score at a given level.

Table 23.2 Five characteristics of the extracted word

Number	Feature extraction
Feature 1	Using the data of word frequency obtained above as one of the feature vectors of the word
Feature 2	The information entropy was further calculated using the word frequencies as the second feature vector for the words
Feature 3	The frequencies of the initial letters in the 359 words in the dataset are counted and assigned to each word according to its initial letter
Feature 4	The frequencies of the 26 letters of all words in the dataset are summed over the five letters of each word to obtain a feature that measures the total letter frequency of the word
Feature 5	The features are classified according to the presence or absence of repeating letters within the word, with the presence of repeating letters marked as 1 and the absence of repeating letters marked as 0

Fig. 23.5 Score distribution predicts results

Scores	Forecast Percentage
1	0.0768
2	2.2735
3	16.3272
4	33.8418
5	29.6796
6	15.0233
X	2.7777

23.4.2.2 Prediction of Player Score Results

First, we use the data of the training set part to train the parameters of the BP neural network model. When a certain training effect is achieved, we next used the model to predict the distribution of attempts for the word EERIE on March 1, 2023.

The score predictions are shown in Fig. 23.5: It can be seen that the percentage of successful attempts in one or two attempts is low, but the percentage of correct guesses of the word EERIE after four attempts reaches more than 50%, and the results are approximately tending to a normal distribution.

23.4.2.3 Evaluation of the Model

Subsequently, the accuracy and error of the model were tested in some way. In this paper, the BP neural network model was defined to pass the test set data and thus obtain the predicted value of a set of dependent variables as $Z = \left\{\widehat{z}_1, \widehat{z}_2, \ldots, \widehat{z}_n\right\}$, and the true value of the test set dependent variables as $Z = \{z_1, z_2, \ldots, z_n\}$. Then the statistics about the predicted and true values could be constructed to evaluate the goodness of the model, and two error evaluation indicators—RMSE (Root Mean Square Error) were chosen to evaluate the model.

The results for RMSE are calculated from the above formulae as follows.

Figure 23.6 shows the values of the metrics used to evaluate the goodness of fit of the data. The RMSE of the data fit can be used to reflect the goodness of fit of the model. The smaller the RMSE indicates that the higher the goodness of fit, so this paper chose to use RMSE to measure the goodness of fit of the model used this time. The indicators not only reflect the strength of fit of the model to a certain extent but are also an important reference for the accuracy of the prediction results. The table above shows that the maximum values of the data indicator do not exceed 5, respectively, and the mean value of the error is only 3.46, which is magnitude away from the player data with a sum of 100. Therefore, we are confident that our model fits the data well and that the model is sufficient to support our future predictions.

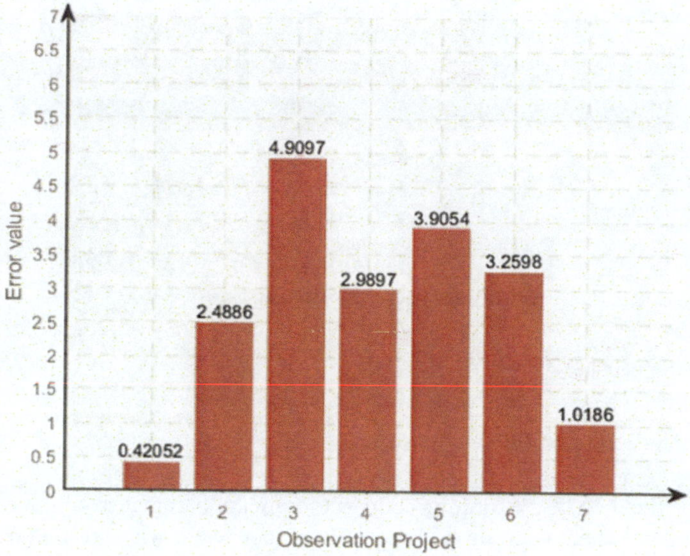

Fig. 23.6 Error test

23.5 Conclusion

In this paper, an exponential smoothing prediction model and a BP neural network prediction model are established based on the player score dataset of Wordle mini-game, and the number of players who report scores and the proportion of score distribution in the future of the game are predicted, and the final prediction result is relatively accurate and the prediction effect of the model is also better. Therefore, the prediction of online mini-game popularity and player feedback based on exponential smoothing model and B-P neural network model is feasible. However, in view of the small and obvious linear trend of the dataset used in this paper, further research is needed in the expansion of the dataset and the improvement of model scalability.

References

1. Li, J., Meng, Z.Q.: Prediction and analysis of university library lending data based on time series algorithm. Inf. Sci. **40**(11), 133–138+147 (2022)
2. Wang, Y., Ran, J.Y., Zhang, J., et al.: Early warning of heavy pollution process of smog in heating season in Xi'an from 2015 to 2020 based on deep time series feature fusion. Chem. Ind. Eng. Progress **41**(10), 5685–5694 (2022)
3. Ci, B.C., Zhang, P.Y.: Financial time series forecasting based on ARIMA-LSTM model. Stat. Decis. **38**(11), 145–149 (2022)
4. Duan, Q.L., Zhang, L., Wei, F.F., et al.: Aquatic product price prediction model and verification based on time series GA-SVR. Trans. Chin. Soc. Agric. Eng. **33**(01), 308–314 (2017)

5. Li, G.Q., Xu, S.W., Li, Z.M., et al.: Research on short-term forecasting method and model of agricultural product market price——prediction based on time series model. J. China Agric. Univ. **16**(02), 172–178 (2011)
6. Wang, R.Z., Zhang, X.S., Wang, M.H.: Agricultural product price prediction based on signal decomposition and deep learning. Trans. Chinese Soc. Agric. Eng. **38**(24), 256–267 (2022)
7. Li, Z.M., Xu, S.W., Cui, L.G., et al.: Forecasting research based on dynamic chaotic neural network: a case study of potato time series prices. Syst. Eng. Theory Pract. **35**(08), 2083–2091 (2015)
8. Zhang, D.B., Zeng, L.L., Ling, L.W.: Integrated prediction model for price decomposition of agricultural futures based on VMD-ELM. Operat. Res. Manag. **32**(01), 127–133 (2023)
9. Liu, X., Liu, J.T., Li, J.L., et al.: Egg price prediction in Beijing based on seasonal decomposition and long-term short-term memory. Trans. Chinese Soc. Agric. Eng. **36**(09), 331–340 (2020)
10. Ren, J.H., Xu, J., Yang, H.Y.: Application of ARIMA and holt-winters index smoothing model in prediction of tuberculosis epidemic trend in Henan Province. J. Zhengzhou Univ. (Med. Sci.), **57**(06), 756–760 (2022)
11. Guo, Z.J., Gong, H., Zhou, L.J.: Application of SARIMA model and holt-Winters index smoothing method in predicting the incidence of pulmonary tuberculosis in Jiangsu Province. Dis. Surveill. **37**(08), 1042–1047 (2022)
12. Liu, Q., Liu, L., Zhang, Z., et al.: Etiological characteristics and time series prediction of hand, foot and mouth disease in Dali. Chin. J. Dis. Control **27**(05), 574–580 (2023)
13. Luan, P.X., Xiao, J.H., Chen, X., et al.: Comparison of prediction of new occurrences of swine fever month based on gray model and ARMA model. Trans. Chin. Soc. Agric. Eng. **27**(12), 223–226 (2011)
14. Li, B.K., Wang, J.J., Wu, S., et al.: Prediction of the incidence of malaria in northern Anhui based on remote sensing technology and time series analysis. Chin. J. Dis. Control **21**(03), 291–294 (2017)
15. Zhang, L.P., Cheng, Y., Wang, Y.C., et al.: R&D cost prediction of enterprises on the science and technology innovation board based on BP neural network. Friends Account. **704**(08), 74–81 (2023)
16. Wu, F., Xie, C., Ji, S.P.: Stock price trend prediction based on dual-stream LSTM neural network. J. Appl. Sci. **41**(02), 344–358 (2023)
17. Wang, S.N., Gu, F.F., Li, J.H., et al.: Prediction of silt reclamation in deepwater channel of Yangtze River estuary based on BP neural network. Yangtze River **54**(02), 206–213 (2023)
18. Ji, X.M., Hou, M.L., Wang, L., et al.: Modeling and application of deformation resistance model of medium and thick plate based on machine learning. Acta Metall. Sin. **59**(03), 435–446 (2023)
19. Ren, L., Jia, Z.Z., et al.: Data-driven industrial intelligence:status quo and prospects. Comput. Integr. Manuf. Syst. **28**(07), 1913–1939 (2022)
20. Yuan, Y., Zhang, Y., Ding, H.: Key technologies of industrial artificial intelligence and its application status in predictive maintenance. Acta Autom. Sinica **46**(10), 2013–2030 (2020)
21. Huang, L., Li, L.X., Cheng, Y., et al.: Short-term wind power prediction based on SAM-WGAN-GP. J. Solar Energy **44**(04), 180–188 (2023)
22. Mounir, N., Ouadi, H., Jrhilifa, I.: Short-term electric load forecasting using an EMD-BI-LSTM approach for smart grid energy management system. Energy Build. 88 (2023)
23. Limouni, T., Yaagoubi, R., Bouziane, K., Guissi, K.: Baali El Houssain. Accurate one step and multistep forecasting of very short-term PV power using LSTM-TCN model. Renew. Energy 205 (2023)
24. He, D.F., Luo, J., Shu, X.X.: Autonomous networked vehicle time delay feedback prediction cruise control. J. Jilin Univ. (Eng. Sci.) **51**(01), 349–357 (2021)
25. Tang, X.M., Deng, L., Yu, J.M., et al.: Output feedback prediction control of network control system based on interval 2 T-S fuzzy model. Acta Autom. Sinica **5**(03), 604–616 (2019)

Part III
IT-based Applications

Chapter 24
Mileage Pile Detection for Vehicle-Borne Video

Han Liu, Ronggui Ma, Gaoli Cheng, Xufeng Li, and Juan Du

Abstract To realize easy positioning and convenient operation for road maintenance, proposed the mileage pile detection and classification of vehicle-borne video for pile number recognition. This method lay the foundation for subsequent mileage pile number recognition. Firstly, Contrast Limited Adaptive Histogram Equalization (CLAHE) image enhancement enhances the contrast between the mileage pile plates and background. And then, YCrCbC simplifies redundant pixel information; and Adopting the Maximally Stable Extremal Region (MSER) roughly locates mileage pile plates area. Thirdly it adapts Histograms of Oriented Gradients (HOG) extracts the pile plate features; Lastly it uses Support Vector Machine (SVM) to classify pile plates category. The experiment results indicate that classification accuracy of 100M is 86.73% and 1 km classification accuracy is 89.77%, which provides important practice for highway maintenance.

24.1 Introduction

Mileage piles detection and classification has become important parts for the road intelligent positioning system. The automatic mileage piles location includes two aspects: mileage piles detection and classification. The static mileage piles images detection does not conform to the actual scene. Therefore, the mileage pile detection based on the video is more suitable for the actual task scenario. Video object detection in traffic field mainly includes the optical flow method, frame difference method, Beijing difference method and Gaussian mixture model method. When the camera shakes and the road environment illumination changes, the detection error of the above method is greatly large.

H. Liu (✉) · R. Ma
School of Information and Engineering, Chang'an University, Xi'an, China
e-mail: 2020024011@chd.edu.cn

G. Cheng · X. Li · J. Du
Shannxi High-Speed Mechanical Engineering Company, Xi'an, China

R. Kountchev et al. (eds.), *Multidimensional Signals, Augmented Reality and Information Technologies*, Smart Innovation, Systems and Technologies 374,
https://doi.org/10.1007/978-981-99-7011-7_24

Detection based on target feature is more suitable for mileage pile detection, which includes the color feature [1], geometric feature [2], texture feature [3] of pile plates, and the machine learning detection method based on multiple features [4–6]. The color feature detection separates pile plates by threshold in color spaces. RGB color space separates fast, but vulnerable to light distortion; while Hue Saturation Value (HSV) and Hue Intensity Saturation (HIS) color space can avoid illumination. The three components (chroma, saturation brightness) of HSV color space have low correlation. The one component change does not influence the other two components on human vision, but its spatial nonlinearity has a huge calculation amount. The shape feature detection is too large a calculation. Therefore, the color feature is used to roughly classify pile categories, and then the shape feature is used to optimize the classification results to improve classification precision, but has low robustness in complex natural background images.

Liu [7] proposed a multi-view landmark detection system based on cascade detector, SW-Gaussian model, and shape context matching methods. The SW-Gaussian model extracts strong color and is matched by shape context matching method, which has high detection accuracy. Chen [8] combined HSV color space with affine transformation to correct the geometric shape, which also is used to reduce the projection distortion and improve accuracy.

Nowadays, machine learning based on image features is popular because of high accuracy and strong robustness. Kiran [9] combined SVM based on the color segmentation technology of enhanced hue and saturation product. Soendoro [10] combined CIElab hue and shape arc based on color and used linear cSVM to detect binary images. Liao [11] used SVM to train classification and used genetic algorithm to realize adaptive optimization of shape classification, which has the fast speed and strong adaptive ability. Most of methods are aimed at the warning, prohibition, indication pile plates as a test sample, and few people study the mileage pile. Therefore, this paper proposes a mileage pile detection and classification method based on the Hidden Markov Model (HMM)speed prediction mechanism. The overall research steps are shown in Fig. 24.1.

1. Image preprocessing: using CLAHE image enhancement to enhance the contrast between the mileage pile plate and background; and then MSER to locate the pile plate area, and then the crop pile plate area.
2. HOG extracts the character feature of the cropped pile plate area.
3. The extracted feature sets as the SVM input data to realize the pile images classification.
4. HMM is introduced to overcome the redundant frame problem, and establish a target detection model based on speed estimation. The 100-m (100 M) pile and 1000-m (1 km) pile detection are reasonably called to improve detection efficiency.

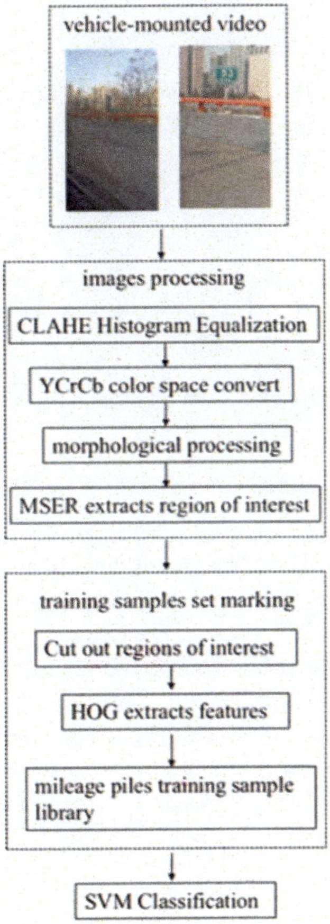

Fig. 24.1 Mileage pile detection based on vehicle-borne video

24.2 Mileage Pile Target Detection

24.2.1 Mileage Pile Video Acquisition

The data acquisition equipment is shown in Fig. 24.2. Camera is placed on the right vehicle roof front with a magnet and active connection. Camera selection is from Shendu Technology. Imaging resolution is 2248*2048, pixel dimension is 3.45 μm*3.45 μm, and focal length is 3.5 mm. While the vehicle driving, the mileage pile enters the camera's scope at time T and leaves the camera's scope at time T + 1. At the time T + 1, the vehicle position is approximately parallel to the mileage pile, and the distance between the vehicle and the mileage pile is the shortest, and the mileage pile image quality is high. Therefore T + 1 is the time stamp of the key video frame. The GPS data acquisition equipment is assembled directly below the video acquisition equipment, and it is fixed at the co-pilot position. The GPS antenna has signal enhancement function to prevent multipath effects.

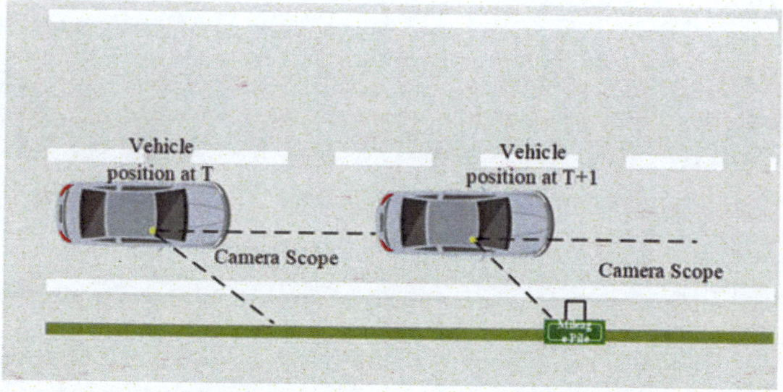

Fig. 24.2 Data acquisition equipment

24.2.2 Mileage Pile Positioning

Figure 24.3 shows the mileage pile images acquired from the Xi 'an Ring Expressway (G3002) and the Yukun National Expressway (G85). The 100M pile is a small round circle plate. The 1 km pile plate big hexagon with two right angles. The above pile plate information is position information that road maintenance personnel can directly obtain. The vehicle-borne mileage pile image is affected by the driving environment, the road environment, which seriously affects detection accuracy. This paper proposes preprocessing technology as follows:

(1) CLAHE image enhancement [12] is used to reduce the color distortion, image detail loss, and the blurring of pile plate edge.
(2) MSER detect pile plate area. Expansion of morphological closing operation is performed to expand the boundary of the connected domain, connecting the two adjacent connected domains. And then erosion operation to decrease the expansion area of the connected domain boundary caused by the expansion operation. Last calculating the pixel width and height of the entire plate area, and the pile number area is cropped using the height and width as a rectangle.

After the CLAHE histogram equalization, the detailed information of the 1km piles and 100M piles is enhanced, and the background noise is suppressed, making the pile plate area stand out in complex background, as shown in Fig. 24.4c. YCbCr sets a threshold based on the road edge brightness to segment the pile plates area, as

Fig. 24.3 Road sign images of different angles and scale

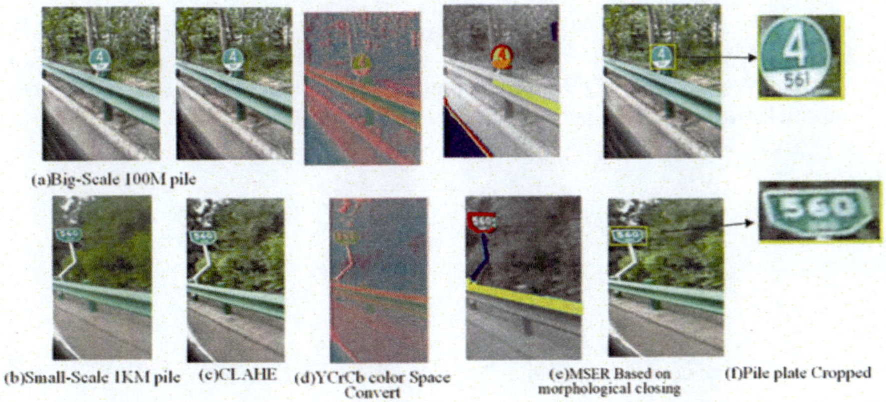

Fig. 24.4 Pile number detection

shown in Fig. 24.4d. MSER detects the pile plate area, and adopts the morphological closing operation to calculate the width and height of the character area, as shown in Fig. 24.4e. Finally, according to the width and height of the area, crop the pile plate area, to reduce the image size and image pixel information for the pile number recognition, as shown in Fig. 24.4f.

24.2.3 Mileage Pile Classification Based on HOG + SVM

The mileage pile images resolution collected is not high and the image scale is greatly changed, so it is difficult to obtain fine texture information. HOG can greatly describe the pile plate edge, as well as the insensitivity to illumination and translation invariance [13, 14]. This paper adopts SVM to realize pile plates classification. The SVM with small dimension of the classification sample and small amount of calculation has great advantages.

24.2.3.1 HOG

HOG extracts features by calculating and counting the histogram of gradient directions in the local image. HOG focus on the shape information distribution. The extraction steps are follows:

(1) GAmma correction to reduce local illumination changes. Considering that most of the photos are taken during the day, the image brightness is high, and the GAmma is set to 0.8;
(2) Divide the image into several cells, and calculate the gradient direction map of each cell;

(3) Merge the gradient direction map into blocks, and calculate the HOG feature of each block;
(4) Normalize the block HOG eigenvectors to reduce the interference of noise. The normalization function is

$$\vartheta = \frac{\vartheta}{\sqrt{\|\vartheta\|_2^2 + \varepsilon^2}} \tag{24.1}$$

$$\vartheta = \frac{\vartheta}{\sqrt{\|\vartheta\|_1 + \varepsilon}} \tag{24.2}$$

ε is a minimal constant, preventing the denominator from being 0; $\|\vartheta\|_1$、$\|\vartheta\|_2^2$ is *L1*、*L2* norm of ϑ.
(5) Generate the HOG eigenvectors and slide the block on the image with a grid unit. The blocks features are combined to obtain the HOG features.

For example, every 4*4 pixels form a cell, 1*1 cells form a block. There are 9 features in each cell, and 9 features in each block. The detection window step size is set to 4 pixels, with 3 scanning windows in the horizontal direction, and 7 scanning windows in the vertical direction. The 60*60 pile image has 3*7*9 = 189 features. The HOG feature extraction result is Fig. 24.5. HOG utilizes the pixel gradient direction to reflect the contour information, and is insensitive to shape and illumination.

24.2.3.2 SVM Image Classification

SVM image classification reduces the complex calculations, and achieves better classification speed and quality. This paper chooses the radial basis kernel function to the SVM classification algorithm. The extracted HOG feature sets have some noise points, which is prone to the problem of decreased classification accuracy. Therefore, constructing the edge pixel classification hyperplane.

(1) Construct SVM classification function

$$\frac{1}{2}\|w\|^2 + C\sum_{j=1}^{N}\xi_j \tag{24.3}$$

w is the hyperplane weight, C is the penalty function, and ξ_j is the slack variable.
(2) Converted SVM classification function into a convex optimization problem:
$y_i(w * x_i + b) \geq 1 - \xi_i$, and $\xi_i \geq 0$, i = 1, 2, 3 ⋯ , N

$$\min_{w,b,\xi}\frac{1}{2}\|w\|^2 + C\sum_{i=1}^{N}\xi_i \tag{24.4}$$

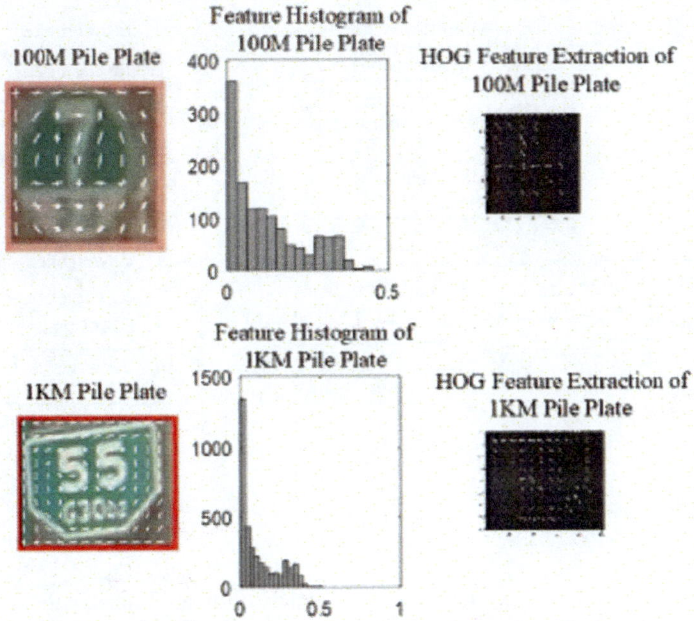

Fig. 24.5 HOG feature extract of pile plate

The solution of the convex optimization is the weight of the classification hyperplane

$$w = \sum_{i=1}^{N} a_i^* y_i x_i \tag{24.5}$$

(3) When the optimal solution a_i^* satisfies $0 \le a_i^* \le C$, The offset of the SVM hyperplane is

$$b^* = y_i a^* (x_i * y_i) \tag{24.6}$$

The optimal hyperplane for SVM classification is

$$w^* * x + b^* = 0 \tag{24.7}$$

24.2.4 Target Prediction Based on Hidden Markov Model

Video processing found that mileage pile images occupy only less than 10% of the video. Frame-by-frame video recognition leads to a great increase in the amount of

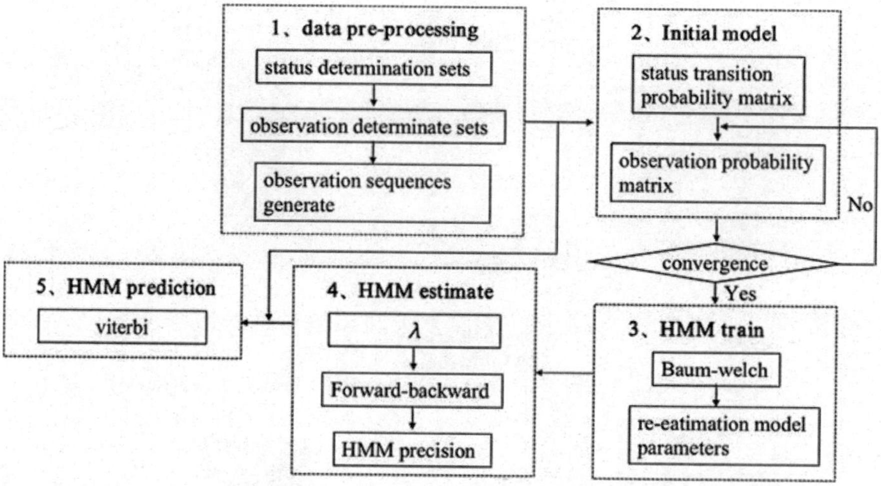

Fig. 24.6 Sliding windows mile pile forecasting based on HMM

meaningless calculation to reduce unnecessary detection and improve detection efficiency. Therefore, it proposes a predictive sliding time window mileage pile detection algorithm, which can flexibly allocate the detection algorithm of the kilometer pile and hundred-meter pile.

The timestamp sequence of the vehicle-borne video is used as the observation sequence. To accurately detect the mileage pile and improve detection efficiency, we propose a target prediction scheme based on HMM [15]. Figure 24.6 shows the prediction steps, and is mainly composed of five steps:

(1) Divide the whole video frame time series into four parts: state set, observation set, observation sequence and state sequence;
(2) The observation sequence set as input data and initial model parameter λ;
(3) Adopt Baum-Welch algorithm to train HMM model parameter λ;
(4) Test set to evaluate the accuracy of the HMM model parameters using Forward–Backward algorithm;
(5) The trained HMM model, combined with the observation sequence of the preprocessing process to forecast frame of the mileage pile using Viterbi.

24.3 Experimental Results and Analysis

7769 images were collected on the G85. There are 1,165 1 km pile plate images and 6,604 100M pile plate images. To verify the effectiveness of the image classification method proposed, it uses G3002 pile plate images for HOG + SVM classifier training, and then uses the weights to classify the G85 pile plate images and compared with

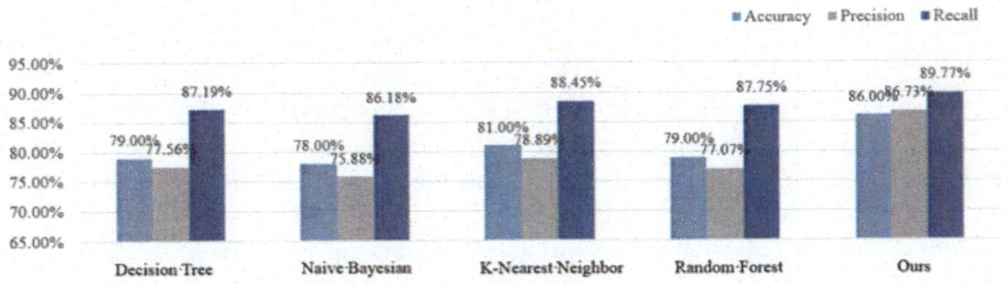

Fig. 24.7 Classification results of classifier

the decision tree classifier, Naive Bayesian classifier, K-Nearest Neighbor classifier and Random Forest classifier.

The classification results are evaluated by the Accuracy, Recall and Precision, as shown in the formula (24.8), (24.9), (24.10), TP is the prediction sample and is a 100M pile plate, and TN is the prediction sample and is a 1 km pile plate. FP indicates that the sample is a 100M pile plate but is classified as the 1 km pile plate, and FN indicates that the sample is 1 km pile plate but is classified as the 100M pile plate.

$$\text{Accurary} = \frac{TP + TN}{TP + FP + TN + PN} \tag{24.8}$$

$$\text{Precision} = \frac{TP}{TP + FP} \tag{24.9}$$

$$\text{Recall} = \frac{TP}{TP + FN} \tag{24.10}$$

Formulae (24.9) and (24.10) show that FP and FN are 100M piles and 1 km piles that were failed to be classified respectively. The larger the Precision and Recall, the fewer the pile images are incorrectly classified, and lesser the number of piles that are correctly classified. The classification result is shown in Fig. 24.7.

The Decision Tree and Random Forest classifier were trained on the G3002 road piles, and then the training weights were used to classify the G85 road piles. The classification accuracy is poor. The naive Bayesian classifier is based on the Bernoulli model, which has a poor classification effect on blurred images and images with small changes in foreground and background colors, where many unrecognizable images appear. The K-Nearest Neighbor classification requires a large calculation quantity and is the most time-consuming method. It needs to calculate the distance from each pixel point to each center K, which requires a large calculation quantity and is not suitable for the deployment of vehicle-borne equipment. But the accuracy is better. The HOG + SVM classifier proposed is superior to the above traditional machine learning methods in classification accuracy.

24.4 Conclusion

To solve the problem of the maintained area positioning under the Satellite denial, this paper builds a road information database through the identification of 1 km piles and 100M piles in expressway video, to realize the maintained area position for the road management department, which improves the accuracy and efficiency of road maintenance and management. The following conclusions were obtained from the experiment:

CLAHE image enhancement plays an important role in the mileage pile plates extraction, YCbCr highlights the region of interest (ROI), and reduces data storage to improve efficiency; while MSER can quickly locate the pile plate area in a complex background. Later, the SVM image classifier is based on HOG feature extraction to realize 1 km pile plates and 100M pile plates classification. Its performance has been very well. Its classification precision reaches, classification recall reaches 89.77%, which indicate 100M classification accuracy is 86.73% and 1 km classification accuracy is 89.77%.

References

1. Chen, C., Wu, W.: Color pattern recognition with the multi-channel non-zero-order joint transform correlator based on the HSV color space. Optics Commun. **244**(1–6), 51–59 (2005)
2. Sharma, G., Singh, L., Gautam, S.: Automatic facial expression recognition using combined geometric feature. 3D Res. **10**(2), 14(2019)
3. Chen, H., Xie , Z.G., Lu, H.: License plate location method combining color and edge texture. Mod. Electron. Tech. **41**(21), 67–70+75 (2018)
4. Bente, T.F., Szeghaimy, S., Fazekas, A.: Detection of lanes and traffic signs painted on road using on-board camera. In: Proceedings of 2018 IEEE Internation Conference on Future IoT Technologies (Future IoT), pp. 1–7. IEEE (2018)
5. Liu, C., Chang, F., Chen, Z.: Traffic sign detection based on regions of interest and HOG-MBLBP features. J. Electron. Inf. Technol. **38**(5), 1092–1098 (2016)
6. Namdev, N., Agrawal, S., Silkari, S.: Recent advancement in machine learning based internet traffic classification. Procedia. Comput. **60**, 784–791 (2015)
7. Liu, C.,Chang, F., et al.: Multiview road sign detection via self-adaptive color model and shape context matching. J. Electron. Imaging **25**(5), 051202 (2016)
8. Chen ,Y., Ye, F., Xiao, F.: Detection and recognition of traffic signs based on HSV vision model and shape feature. J. Jianghan Univ.(Nat. Sci. Ed) **44**(2), 119–125 (2016)
9. Kiran, C.G., Prabhu, L.V., Abdu, R.V.: Traffic sign detection and pattern recognition using support vector machine. In: Proceedings of 2009 Seventh International Conference on Advances in Pattern Recognition, pp. 87–90. IEEE, Kolkata (2009)
10. Soendoro, D., Supriana, I.: Traffic sign recognition with color-based method, shape-arc estimation and SVM. In: Proceedings of the 2011 International Conference on Electrical Engineering and Informatics, pp. 1–6. IEEE, Bandung (2011)
11. Liao, L.C., Jiang, X.H., Zou, F.M., et al.: An automatic recognition approach for traffic congestion states based on traffic video. J. Highw. Transp. Res. Dev. **8**(2), 72–80 (2014)
12. Ma, J.Y., Ye, B., Wang, S.M.: Vein image enhancement algorithm based on CLAHE and multi-scale detail fusion. Semicond. Optoelectron. **41**(05), 738–742 (2020)
13. Xu, S.P., Li, L.Y., Jia, J.Y., et al.: AR-assisted mobile rice disease intelligent analysis and identification system based on HOG-SVM. J. Gr. **42**(03), 454–461 (2021)

14. Liu, D.Q., Chen, Z.J., Xu, Y., et al.: Research on composite SVM algorithm for imbalanced data classification. Comput. Appl. Res. **35**(04), 1023–1027 (2018)
15. Zucchini, W., Macdonald, I., Alliot, P.: Hidden Markov models for time series: an introduction using R. Aust. N. Z. J. Stat. **53**(2), 265–266 (2011)

Chapter 25
Exploration of Future Temperature Analysis Based on ARIMA Time Series Model and GA-BP Neural Network Prediction Model

Tangliang Wang, Yong Jiang, and Mengzhu Liu

Abstract The persistent ultra-high temperature caused by global warming effect has sounded the alarm for the development of human society. In order to improve these problems and make some constructive suggestions, this paper focuses on the main factors affecting global temperature using time series and GA-BP neural network prediction models to investigate the trend of global average temperature. We first preprocess the data using mean interpolation, and then compare the significance of the global mean temperature change using Kruskal–Wallis test with multiple independent samples concluding that the global temperature increase triggered by March 2022 does not exceed the temperature of the past decade. Then, using a time-series prediction model and a GA-BP neural network, temperatures in 2050 are predicted to be 15.275, 14.863 and 14.862,15.660, and 15.204, 15.178, based on global average annual temperature data from previous years and taking into account various factors such as carbon dioxide, solar radiation intensity and fossil fuel combustion. The above model shows that the predicted temperatures reach 20 °C only at 2758, 2912 and 2860, which implies that the global average temperature will not reach 20 °C from 2050 to 2100. Finally, we compare the performance of the above models by determining the coefficients and find that the prediction based on the GA-BP neural network which is the most accurate.

T. Wang (✉)
Collage of Mechanical Engineering, Chongqing University of Technology, Chongqing, China
e-mail: 1677702204@qq.com

Y. Jiang
College of Science, Chongqing University of Technology, Chongqing, China

M. Liu
Chongqing Intellectual Property School, Chongqing University of Technology, Chongqing, China

R. Kountchev et al. (eds.), *Multidimensional Signals, Augmented Reality and Information Technologies*, Smart Innovation, Systems and Technologies 374,
https://doi.org/10.1007/978-981-99-7011-7_25

25.1 Introduction

Climate change is one of the important issues of global concern today. Due to the interaction of human activities and natural factors, the earth's temperature continues to rise, leading to global climate change. As the trend of climate change continues to intensify, the prediction of future temperatures has become particularly important. Accurate prediction of future temperature can help the government and social decision makers make corresponding policies and plans, which is conducive to protecting the environment and coping with the impact of climate change.

Time series analysis is a commonly used forecasting method, which can analyze and model time series data to predict future development trends. The ARIMA(Autoregressive Integrated Moving Average mode) model is one of the most commonly used models in time series analysis. It can differentiate and stabilize time series data to obtain a stable model for forecasting. However, the prediction accuracy of the ARIMA model is affected by many factors, such as the nonlinear characteristics and complexity of the data, so it is necessary to further improve its prediction accuracy.

A neural network is a computational model capable of simulating the human nervous system, with adaptive learning capabilities and nonlinear processing capabilities. BP neural network is a commonly used neural network model that can be used for time series forecasting. However, the training process of the BP neural network requires a lot of computing resources and time, and it is easy to fall into the problem of local minimum, resulting in unstable prediction accuracy.

The summer of 2022 will see continuous high temperatures in many parts of the world, with some regions reaching temperatures above 50 degrees. In order to analyze whether the temperature increase in March 2022 is greater than the temperature increase in any of the previous decades, we need to collect data for the missing years, use imputation of means to fill in the missing values, and use multiple independent sample tests to analyze the temperature change from March 2022 to October 2022. The next step is to characterize past and future average temperatures using a cyclical prediction model based on existing global annual average temperatures, but future temperature trends can often only be predicted from global annual average temperatures and not with high accuracy, so we introduce various factors such as CO_2, solar radiation intensity, fossil fuel burning, etc. The global annual average temperature is predicted by the GA-BP neural network, using a cycle prediction model and a GA-BP neural network prediction model to predict and analyze the year 2050 [1], 2100 and the time when the temperature reaches 20 °C. Finally, the most accurate model was determined by reviewing the relevant literature, the accuracy of the test set and the periodicity of the prediction results.

25.2 Data Preprocessing

This paper collects global annual mean temperature data from 1850–2022 from the Berkeley Earth Climate Database as supplementary data. Since there are many null values in the data, which may cause deviations in the prediction of future global warming, and removing null values may affect the accuracy of the model's predictions, we use the mean imputation method to use the mode of the average temperature of each city as its substitute value. Considering the variety of factors affecting temperature, we collected four sets of data from the Berkeley Earth Climate Database and Our World in Data on carbon dioxide emissions, solar radiation intensity [2], fossil fuel burning, and ocean temperature as supporting materials for our prediction models.

25.3 Model Construction and Solving

25.3.1 Kruskal–Wallis Test

We used the Kruskal–Wallis test of multiple independent samples to divide the data into 11 groups in 10-year groups, and the temperature in January 2022 was numbered as the parent sequence 12.

Descriptive statistics and tests of normality were obtained for the mean temperature variables, together with the mean and standard deviation phase separation data which were used to test the normality of the data. The mean temperature sample in this paper is $N < 5000$, and we believe that the S-W test will be more accurate through analysis. From the calculation results, it can be seen that the P value of the significance S-W test is $0.053 > 0.05$ [3], so the significance data does not meet the normal distribution at the level, and the next test of multi-independent sample Kruskal–Wallis can be performed.

Through the Kruskal–Wallis test results, it was found that there was a significant difference in average temperature between the different groups, and the difference range was moderate.

The results of the Mann Whitney U test for independent samples of the grouped variable and the parent sequence number 12, as well as the sample size, statistic, and effect measure Cohen's d values. According to the average temperature comparison group of 1.0 and 2.0, the P value of the test result was $0.00^{**} < 0.05$, so the statistical results were significant, indicating that there was a significant difference in the average temperature between 1.0 and 2.0; the difference range Cohen's d value was:$6.648 > 0.8$, and the difference range was very large [4].

The results showed that the P value of the 2–11 groups was less than 0.05, and the statistical results were significant, and the difference range was Cohen's $d > 0.8$, and the difference range was large. The P value of group 12 (2011–2021) was 0.175 >

0.05, and the statistical results were not significant, and the difference range Cohen's d value was 0.368, which was small.

25.3.2 *Periodic Forecasting Model*

25.3.2.1 Time Series Forecasting

Overview

Time series models are mainly used as short-term predictions for the future which belong to the trend forecasting method, which can effectively predict the future changes of variables.

Introduction to the Model

An autoregressive model can be expressed as:

$$x_t = a_1 x_{t-1} + a_2 x_{t-2} + \cdots + a_p x_{t-p} + \mu_t \tag{25.1}$$

where: $a_1, \cdots, a_p$ are the model parameters. μ_t is the white noise with value 0 and variance σ.

When $\mu_t = \epsilon_t + \beta_1 \epsilon_{t-1} + \cdots + \beta_q \epsilon_{t-q}$ holds, (25.1) is the moving average of order q and ϵ_{t-q} represents the white noise series 1.The ARIMA model can be obtained:

$$\begin{aligned} x_t = a_1 x_{t-1} + a_2 x_{t-2} + \cdots + a_p x_{t-p} + \epsilon_t + \beta_1 \epsilon_{t-1} \\ + \cdots + \beta_q \epsilon_{t-q} \end{aligned} \tag{25.2}$$

where: $\beta_1, \cdots, \beta_q$ denote the regression coefficients. $\epsilon_{t-1}, \cdots, \epsilon_{t-q}$ denote the error terms of the time series data at time $t - q$.

Model Testing

Based on the AIC information criterion system to automatically find the optimal parameters, the model result is the ARIMA model (0,1,4) test table and based on 1 differential data, the model formula is as follows:

$$\begin{aligned} y_t = 0.008 - 0.338 * \epsilon_{t-1} - 0.321 * \epsilon_{t-2} - 0.133 * \epsilon_{t-3} \\ + 0.148 * \epsilon_{t-4} \end{aligned} \tag{25.3}$$

The following figure represents the raw data plot, model fit value, and model prediction value for this time series model (Fig. 25.1).

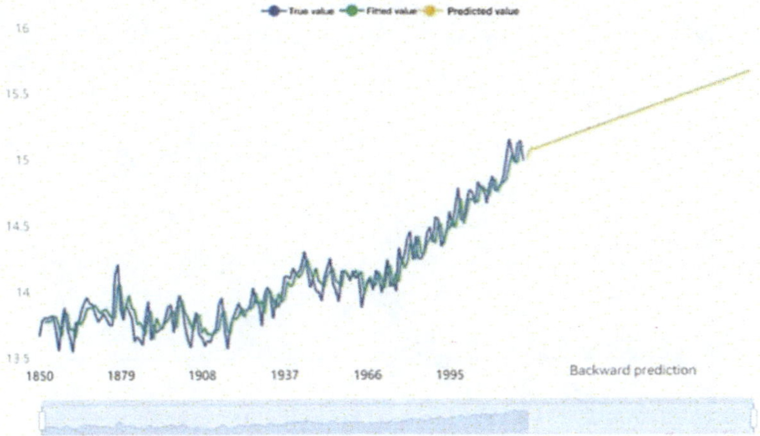

Fig. 25.1 Time series plots

25.3.2.2 Cyclical Fluctuation Forecasting

Trending Average Temperature

Through Fig. 25.2, it can be seen that there are long-term trends and short-term oscillations in the global average temperature series, in which the warming trend is 1850–2020, 1911–2008 and 1976–2008, and the global average temperature increase is 0.44 °C, 0.73 °C and 1.7 °C per hundred years, respectively. Subtract the corresponding trend value from the annual average temperature to obtain the deviation temperature series, and through analysis, we can obtain the three warm plateons in 1870, 1940 and 1988 and the two cold troughs in 1910 and 1970 [5], which shows that the change law of the quasi-65-year cycle fluctuation is alternating between cold and warm periods, and there are ten-year cold and warm periods, quasi-21 years, quasi-65 years, quasi-115 years and quasi-200 years cycle change combination, which can be predicted in the absence of external conditions interference. The global average temperature and time (year) show cyclical changes.

Cyclical Fluctuation Exploration

In time series analysis, empirical mode decomposition and wavelet transform methods are usually used to express the local spectral characteristics of signals. In this paper, several main periodic fluctuations are obtained by wavelet transformation of the deviation temperature series, and the period of fluctuations is fitted, and then the whole sequence is fitted by the cosine function of the specific fluctuation form, so that the sum of squared deviations of the fit is minimized.

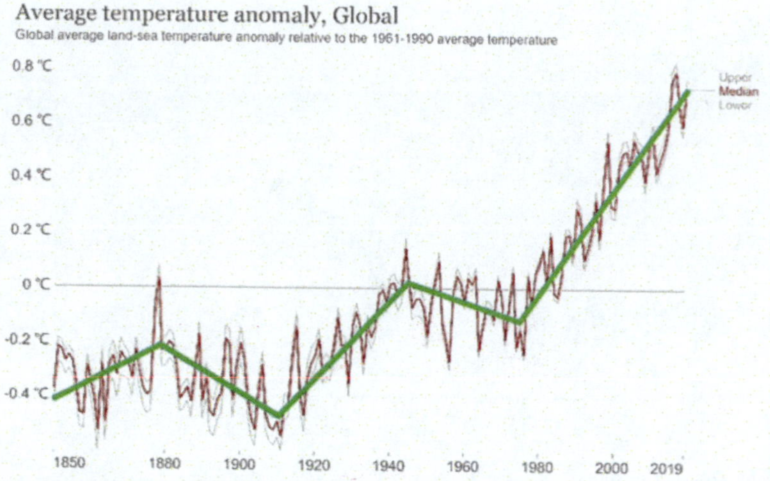

Fig. 25.2 Deviation temperature sequence

Quasi-21 years:

$$y = 0.55\cos(0.296t + 10) \tag{25.4}$$

Quasi-65 years:

$$y = 0.111\cos(0.098t + 4.70) \tag{25.5}$$

Quasi-115 years:

$$y = -0.109\cos(0.035t + 39) \tag{25.6}$$

The past 171 years have only seen a 10-year positive phase superposition phenomenon (wave peak superposition) between the twenty-first century [6]. The warm plateau (peak of global temperature rise) of these three periodic fluctuation functions is the result of the positive superposition of three periodic fluctuations in 1880, 1940 and 2020, and 2020 in the 10-year inter-annual warm period. These three periodic fluctuation functions were used to simulate the coldest troughs (trough superposition) around 1910 and 1970 [7].

Considering the long-term trends of the three equations and the linear superposition of the three cyclical fluctuations, the global mean temperature is expected to drop from a warm plateau between the twenty-first century until 2035 (0.22 °C) and reach another warm plateau of 0.58 °C in 2068. These two cold and warm periods are consistent with the predicted results of the time series.

25.3.3 GA-BP Neural Network Predictive Model

The results obtained by directly using the annual average temperature data may be influenced by natural environmental factors with uncertainty when making predictions. We use the GA-BP neural network method to predict global temperature by considering CO_2 emissions, solar radiation, fossil fuel combustion and ocean temperature as the main influencing factors [8].

Model Introduction

GA-BP Neural network prediction model combines genetic algorithm with BP neural network, and makes up for the shortcomings of neural network sensitivity to the initial value and slow convergence speed through MATLAB software modeling and simulation operations [9].

Step 1: Population initialization, determining structure, weights, and thresholds.

Step 2: Calculate individual fitness F:

$$F = k\sum_{i=1}^{n} |y_i - o_i| \tag{25.7}$$

where k is a coefficient, n is the number of output nodes of the neural network, y_i is the expected output value of the ith node of the network, and o is the predicted output value of the ith node [10].

Step 3: Select the operation mode as the fitness ratio method.

$$f_i = \frac{k}{F_i};\ P_i = \frac{f_i}{\sum_{j=1}^{N} f_i} \tag{25.8}$$

where k is a coefficient, N is the number of neural network output nodes, f_i is the expected output value of the ith node, and P_i is the predicted output value of the i_{th} node.

Step 4: The k chromosome a_k nd the lth chromosome a_l perate at the j position crossing:

$$\begin{cases} \alpha_{kj} = \alpha_{kj}(1-b) + \alpha_{ij}b \\ \alpha_{ij} = \alpha_{ij}(1-b) + \alpha_{kj}b \end{cases} \tag{25.9}$$

where b is any number between [0,1].

Step 5: Select a_{ij} to mutate:

$$\alpha_{ij} = \begin{cases} \alpha_{ij} + (\alpha_{ij} - \alpha_{max})f(g), \gamma \le 0.5 \\ \alpha_{ij} + (\alpha_{min} - \alpha_{ij})f(g), \gamma > 0.5 \end{cases} \tag{25.10}$$
$$f(g) = \gamma_2\left(\frac{10g}{G_{max}}\right)$$

where a_{ij} is the i-th individual and j-th gene, a_{max} denotes the upper limit of a_i, a_{min} denotes the lower limit of a_j, γ_2 refers to an arbitrary number [11], g is the current number of iterations, G_{max} denotes the maximum number of evolution, and γ is an arbitrary number between [0,1].

Model Testing

At around 1,000 trainings, the predicted data accuracy reached 85%, but it was far from accurate [12]. After 5,000 trainings, it can be seen from the figure below that the training effect is good, and the accuracy of the prediction data reaches 92.45%, that is, the training is successful and the future global temperature can be predicted Fig. 25.3.

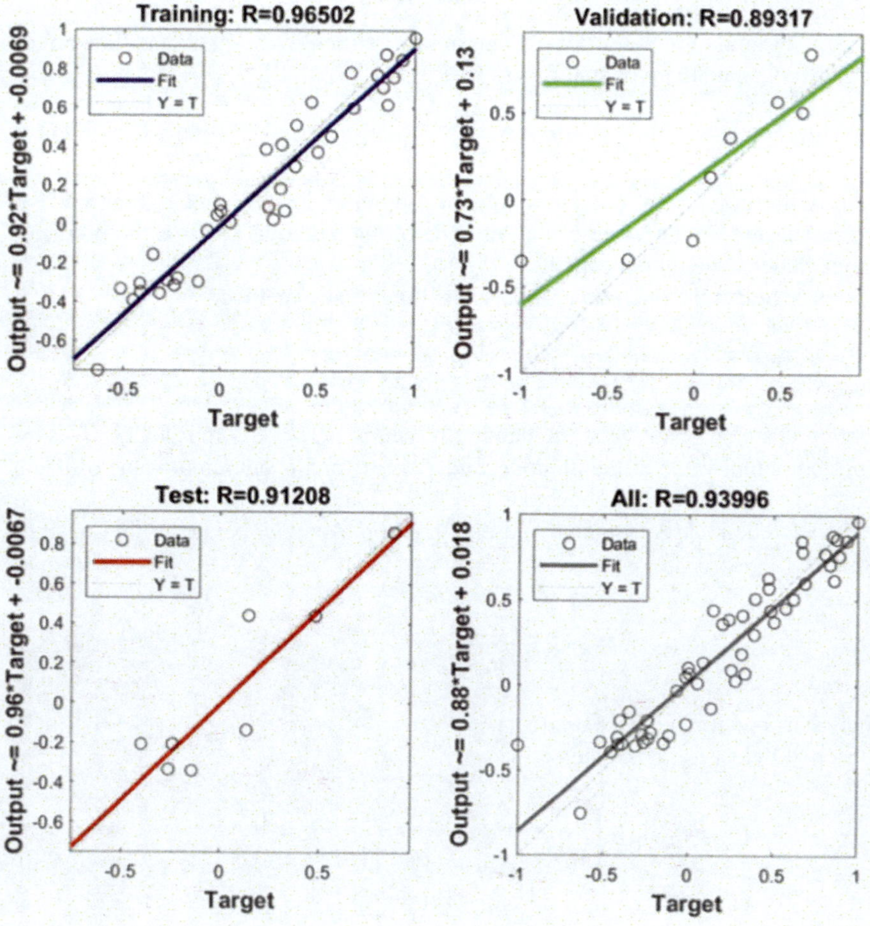

Fig. 25.3 Model training error graph

25.4 Conclusion

At present, the world has a lot of research on the future global temperature change law. The prediction of future global temperature is mainly divided into ordinary prediction models, time-based forecasting models and machine learning forecasting models [13].

Comparing the prediction results of scholars in previous years with the actual temperature results at present, we find that the prediction effect of the model through machine learning is about 10% higher than that of other models, and the usage rate of deep learning, genetic algorithms, neural networks and other models is more common than other models [14].

Through the visual analysis of the prediction results, we find that the global temperature change law of the GA-BP neural network prediction results from 2070 to 2120 is almost the same as the temperature superposition cycle law after the twenty-first century.

It can be concluded that the prediction trend of the GA-BP neural network measurement model is more stable than that of the periodic prediction model, and the nonlinear fitting ability of the GA-BP neural network model is strong, and the weight and threshold of the neural network are optimized by the genetic algorithm, and a large number of training is carried out to make its goodness of fit higher than that of the periodic prediction model.

25.5 Discussion

In order to find the appropriate parameters for this model, models such as periodic fluctuation prediction, time series and GA-BP neural network were selected for global average temperature prediction. In the case of poor model fit, we decisively adopted the lightGBM back-scan model for machine learning, which is highly efficient, has low memory consumption, produces accurate results and ultimately finds the right relationship between temperature and time and latitude. The disadvantage of this model, when using the Manner-Kendall mutation point test function to detect the effect of natural disasters on global temperature, is that it is more complex to detect for larger sample sizes and the results vary to a certain extent.

As the effects of global climate change become more pronounced, weather forecasting has become an integral part of daily life and economic activities. Time series analysis is a technique used to analyze time data, and the ARIMA (autoregressive integral moving average) model is a commonly used time series analysis method, which has been widely used in meteorological forecasting [15]. However, there are some limitations in the ARIMA model, such as the inability to handle nonlinear relationships and complex time series. Therefore, researchers have proposed many other methods, including the GA-BP (Genetic Algorithm-Back Propagation) neural network model.

Future research can explore how to combine the ARIMA model and the GA-BP neural network model to improve the accuracy and reliability of meteorological forecasts. This hybrid model can use the time series analysis ability of the ARIMA model to deal with linear relationships, and use the GA-BP neural network model to deal with nonlinear relationships and complex time series. With this approach, forecasting models will be better able to handle complex weather data and improve forecast accuracy.

Additionally, future research could explore how this hybrid model can be combined with other advanced techniques, such as machine learning and artificial intelligence techniques. These techniques can help predictive models automatically learn complex relationships in meteorological data and improve forecast accuracy and reliability.

Finally, future research could also explore how this hybrid model can be applied to other domains beyond weather forecasting. For example, the model can be used to predict data on financial markets, environmental pollution, and healthcare. These applications will help to better understand and manage complex data sets and improve the accuracy and reliability of decision making.

References

1. Yang, H., Li, X., Qiang, W., Zhao, Y., Zhang, W., Tang, C.: A network traffic forecasting method based on SA optimized ARIMA–BP neural network. Comput. Netw.. Netw. **193**, 108102 (2021)
2. Abhishek, K., Misra, B.B.: Hybrid genetic algorithm and time delay neural network model for forecasting traffic flow. In: 2016 IEEE International Conference on Engineering and Technology (ICETECH), IEEE, pp. 178–183 (2016)
3. Deng, Y., Weng, Z., Zhang, T.: Metaverse-driven remote management solution for scene-based energy storage power stations. Evol. Intell. 1–12 (2022)
4. Ma, X., Jin, Y., Dong, Q.: A generalized dynamic fuzzy neural network based on singular spectrum analysis optimized by brain storm optimization for short-term wind speed forecasting. Appl. Soft Comput.Comput. **54**, 296–312 (2017)
5. Guo, N., Wang, Z.: A combined model based on sparrow search optimized BP neural network and Markov chain for precipitation prediction in Zhengzhou City, China. AQUA—Water Infrastructure. Ecosyst. Soc. **71**(6), 782–800 (2022)
6. Qian, Z., Pei, Y., Zareipour, H., Chen, N.: A review and discussion of decomposition-based hybrid models for wind energy forecasting applications. Appl. Energy **235**, 939–953 (2019)
7. Zhang, H., Tian, Z.: Failure analysis of corroded high-strength pipeline subject to hydrogen damage based on FEM and GA-BP neural network. Int. J. Hydrogen Energy **47**(7), 4741–4758 (2022)
8. Xie, Y., Ishida, Y., Hu, J., Mochida, A.: A backpropagation neural network improved by a genetic algorithm for predicting the mean radiant temperature around buildings within the long-term period of the near future. In: Building Simulation, Vol. 15, pp. 473–492 (2022)
9. Jiang, P., Ma, X.: A hybrid forecasting approach applied in the electrical power system based on data preprocessing, optimization and artificial intelligence algorithms. Appl. Math. Model. **40**(23–24), 10631–10649 (2016)
10. Li, Z., Piao, W., Wang, L., Wang, X., Fu, R., Fang, Y.: China coastal bulk (Coal) freight index forecasting based on an integrated model combining ARMA, GM and BP model optimized by GA. Electronics **11**(17), 2732 (2022)

11. Wang, L., Zhan, L., Li, R.: Prediction of the energy demand trend in middle Africa—a comparison of MGM, MECM ARIMA and BP models. Sustainability **11**(8), 2436 (2019)
12. Xiao, L., Shao, W., Yu, M., Ma, J., Jin, C.: Research and application of a hybrid wavelet neural network model with the improved cuckoo search algorithm for electrical power system forecasting. Appl. Energy **198**, 203–222 (2017)
13. Zheng, D., Semero, Y.K., Zhang, J., Wei, D.: Short-term wind power prediction in microgrids using a hybrid approach integrating genetic algorithm, particle swarm optimization, and adaptive neuro-fuzzy inference systems. IEEJ Trans. Electr. Electron. Eng.Electr. Electron. Eng. **13**(11), 1561–1567 (2018)
14. Hua, L.: Power quality prediction of active distribution network based on CNN-LSTM deep learning model. In: Artificial Intelligence for Communications and Networks: Third EAI International Conference, AICON 2021, Xining, China, October 23–24, 2021, Proceedings, Part I 3, pp. 108–122 (2021)
15. Zheng, W., Peng, X., Lu, D., Zhang, D., Liu, Y., Lin, Z., Lin, L.: Composite quantile regression extreme learning machine with feature selection for short-term wind speed forecasting: a new approach. Energy Convers. Manage. **151**, 737–752 (2017)

Chapter 26
Research on Feature Extraction Based on Time Series Images

Sixin Li, Meiji Zhu, Fusheng Zhu, Qingya Yang, Keke Li, and Yanmei Chen

Abstract The application of mold flux has greatly expanded and improved the continuous casting process and has become an indispensable metallurgical auxiliary material. In order to design a mold flux that is more convenient and better able to perform metallurgical functions, realize the automatic extraction technology of mold flux sequence image features, and study the relationship between temperature, time and phase distribution, it has become the top priority of current research. In this paper, the gray level co-occurrence matrix method is firstly used to analyze the crystallization and melting process of the mold powder in combination with relevant data and literature. Then use the image segmentation algorithm to intercept the central part of the image as the research object, and use the RGB color mode to reflect the color features through the color moment. Finally, the gray level co-occurrence matrix is used to describe the texture features, and the data is visualized and analyzed differently.

26.1 Introduction

In the context of fierce competition in the steel market, continuous casting technology in various countries has been continuously improved, and efforts have been made to improve the yield of steel products, the quality of billets, and the simplification and automation of the production process. At the same time, the mold powder is also continuously improved, and the technical and economic benefits are remarkable. During the continuous casting process, the mold slag thermally isolates the surface of the curved steel, prevents secondary oxidation of the steel, controls heat transfer to reduce heat loss of the steel, provides lubrication for the billet, and absorbs and dissolves floating non-metallic inclusions. Whether mold slag technology can

S. Li · M. Zhu · F. Zhu · Q. Yang · K. Li · Y. Chen (✉)
Guangdong Polytechnic Normal University, Guangdong, China
e-mail: chch1980@163.com

Q. Yang
e-mail: y15026087656@126.com

R. Kountchev et al. (eds.), *Multidimensional Signals, Augmented Reality and Information Technologies*, Smart Innovation, Systems and Technologies 374,
https://doi.org/10.1007/978-981-99-7011-7_26

promote the overall improvement of continuous casting technology is a common concern of continuous casting manufacturers, mold slag researchers and manufacturers, and it is also a very noticeable trend in the development of the contemporary iron and steel industry.

The effect of mold flux on the heat transfer of the mold is mainly realized through the crystallization layer. The thickness of the crystalline mineral phase mold flux film directly affects the heat transfer and lubrication properties of the mold flux. It is important to study the phase distribution of the mold flux in the void between the mold wall and the solidified shell. Therefore, mathematical modeling is required to study the relationship between crystallization temperature, time change and slag melting and crystallization process, so as to provide a basis for the design of mold powder that can play its metallurgical function well. In the SHTT-II melting point testing machine experiment, the staff who observe the crystalline phase state of mold flux need to observe and record the information in the upper left corner of the image one by one, and use the naked eye and experience to judge and identify the key node image.

Obviously, the old method wasted a lot of manpower, and there was a risk of inadvertently missing key information, which also caused problems for further information work during the experiment. There is also an urgent need to implement automatic feature extraction techniques on sequential images of mold flux melting and crystallization. Dynamic changes between adjacent sequence photos, such as crystallizer flow and crystallization, must be investigated and quantified to complete this topic. Use the image segmentation algorithm to intercept part of the image to avoid the error caused by the random mixed color extraction of the entire image in the past, and better reflect the dynamic differences between images, because there are a large number of colors in the image. The background part of the image interferes with the color feature extraction of the mold powder. The texture and color information of the image is taken out.

26.2 Image Capture

Use the image segmentation algorithm to intercept a part of the image, avoiding the error mold powder caused by the random mixed color extraction of the entire image due to a large number of colors in the background part of the image interfering with the color feature extraction. This will more accurately reflect the dynamic differences between adjacent sequences of photographs during mold flux melting and crystallization. The size of the matrix is carefully considered for the improvement of analysis efficiency and accuracy. Smaller images can accelerate computation speed and response time, while also helping to present details and clarity, making it easier to capture relevant information. The cutting effect is as follows Fig. 26.1.

According to the six node images, it can be found that the dynamic differences between adjacent sequence images in the melting and crystallization process of mold slag mainly have obvious different color features and texture features [1]. Therefore,

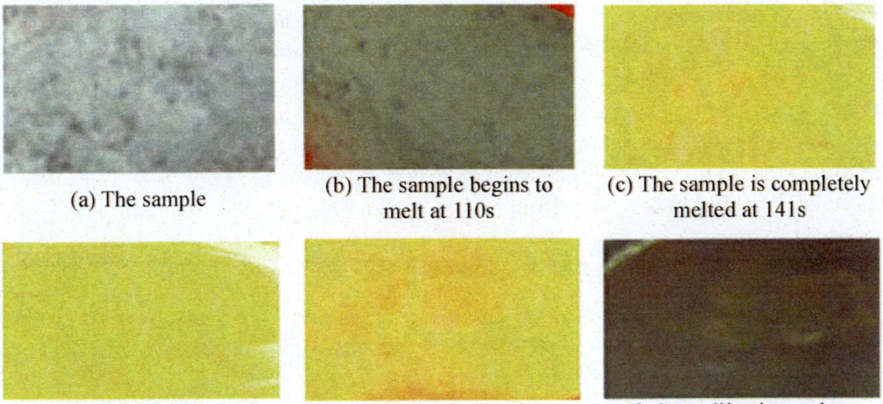

(a) The sample
(b) The sample begins to melt at 110s
(c) The sample is completely melted at 141s
(d) The sample sees steels melting at 149s
(e) Crystallization begins at 150s
(f) Crystallization ends at 671s

Fig. 26.1 Image interception part

the RGB color mode is used to extract color features, and GLCM (Grey-Level Co-occurrence Matrix) is used to extract texture features. Next, the detailed process of image feature extraction will be shown.

26.3 Image Feature Extraction

26.3.1 Color Feature Extraction Using RGB Color Mode

By observing that the color feature has less dependence on the size, direction and viewing angle of the image, that is, it is not easily affected by other external factors. Therefore, color features will be employed first to describe the dynamic differences between adjacent image sequences. The RGB color mode obtains multiple colors by changing the three color channels of red (R), green (G), and blue (B) and superimposing each other. RGB is the color representing the three channels of red, green and blue, and is one of the most widely used color systems at present [2]. RGB generally uses numerical values to represent the brightness of each channel. Each color has 256 levels of brightness for the red, green, and blue channels. The brightness is the weakest at 0 and the brightest at 255. Therefore, the color characteristics of these six node images can be quantified by RGB values.

Quantification of color features has been done. Now in order to study the dynamic difference between adjacent sequential images during the melting and crystallization of mold powder, it is necessary to continue to characterize the color [3]. Color moment is a commonly used color feature description method, which is used to represent the color distribution in an image. In statistics, the first moment represents the mean of the data, the second moment represents the variance of the data distribution, and the third moment represents the offset of the data distribution. Therefore, these three statistical features are used to represent the color distribution of digital images [4]. In this paper, the first-order (μ_i), second-order (σ_i) and third-order (s_i) moments of the three color components of RGB are used to represent the color distribution of the melting and crystallization process images of mold flux. There are nine components in total, with three lower-order moments per color channel. The calculation formula is.

First-order color moments—mean values, reflecting image lightness and darkness.

$$\mu_i = \frac{1}{N}\sum_{j=1}^{N} p_{i,j} \tag{26.1}$$

Second-order color moments—standard deviation, reflecting the range of image color distribution.

$$\sigma_i = \left(\frac{1}{N}\sum_{j=1}^{N} (p_{i,j} - \mu_i)^2\right)^{\frac{1}{2}} \tag{26.2}$$

Third-order color moments with one variance, reflecting the symmetry of the image color distribution.

$$s_i = \left(\frac{1}{N}\sum_{j=1}^{N} (p_{i,j} - \mu_i)^3\right)^{\frac{1}{3}} \tag{26.3}$$

When $= s_i$ 0, the color distribution of the image is symmetric;

When s_i<0, the color distribution is left-biased or negatively biased;

When s_i>0, the color distribution is right-biased or positively biased;

Where p_i and j represent the probability of occurrence of pixels with gray value of j in the i th color channel component, and n represents the gray level of the image. As can be seen from the above formula, the color moment feature of the image consists of nine components (three color components, each of which has three low-order moments). Compared with other color features, the color moment is very simple and time efficient.

The output from Python is a 9-dimensional matrix:

$$\text{Color} = \left[\mu_R, \mu_G, \mu_B, \sigma_R, \sigma_G, \sigma_B, s_R, s_G, s_B\right] \tag{26.4}$$

So, the mean, variance, and offset of the three RGB color components could got from it.

26.3.2 *Texture Feature Extraction Based on Gray Level Co-occurrence Matrix*

The process of extracting and analyzing the spatial distribution of gray levels (or shades) or textures in an image. Gray-Level Co-occurrence Matrix can describe the relationship between pixels in a local or entire area of an image and adjacent pixels or pixels within a certain distance. The matrix can be used to define the grayscale relationship between a certain pixel in the image. Partial image A region or an entire region and adjacent pixels or pixels within a specified distance. However, it cannot directly provide texture distinguishing features, so GLCM is used to calculate nine different statistical properties of texture features, including mean, variance, standard deviation, homogeneity, contrast, dissimilarity, entropy, angular second moment, and correlation properties[5]. Next, we describe and define the symbols that will appear in this article Table 26.1.

Below is the definition of each property and how they are calculated.

(1) Mean

The mean value reflects the regularity of the texture, and the value is small if the texture is chaotic and difficult to describe. Strong regularity, easy to describe, large value.

Also, here "quant k" equals 562 in terms of sample size.

$$\text{Mean} = \sum_{i=0}^{\text{quant}_k} \sum_{j=0}^{\text{quant}_k} p(i, j) * i \tag{26.5}$$

(2) Std.

Variance and standard deviation reflect the measurement of the deviation between the pixel value and the mean value. When the gray level of the image changes greatly, the variance and standard deviation values are larger.

Table 26.1 Notations used in this paper

Symbol	Description
P_{ij}	Probability of occurrence of a pixel with grayscale j in the i-th color channel component of the image
μ_i	Mean value of the i-th color channel component
σ_i	Variance of the i-th color channel component
s_i	Offset of the i-th color channel component
N	Total number of pixels in the image

$$Std = \sqrt{\sum_{i=0}^{quant_k} \sum_{j=0}^{\text{quant}_k} p(i, j) * (i - \text{Mean})^2} \tag{26.6}$$

(3) Contrast

Both the depth of texture and the sharpness of an image are reflected in the contrast. The sharpness of the image and the depth of the texture are reflected in the distribution of the metric matrix values and the magnitude of the local variation in the image. The larger the contrast, the deeper the groove of the texture, and the clearer the effect; on the contrary, the smaller the ratio, the shallower the groove, the more chaotic the effect [6]. This contrasts the brightness of a pixel value relative to its domain pixel value. The brightness of the image fluctuates rapidly. In other words, richer textures and greater contrast make textures stand out.

$$\text{Contrast} = \sum_{i=0}^{\text{quant}_k} \sum_{j=0}^{\text{quant}_k} p(i, j) * (i - j)^2 \tag{26.7}$$

(4) Dissimilarity

When computing the contrast, the weights are exponentially multiplied with the diagonal and the spacing between matrix elements. It can be changed to linear growth, and the difference can be found.

$$\text{Dissimilarity} = \sum_{i=0}^{\text{quant}_k} \sum_{j=0}^{\text{quant}_k} p(i, j) * |i - j| \tag{26.8}$$

(5) Homogeneity

If you measure the local uniformity of an image, non-uniform images have lower values, while uniform images have larger values. In contrast to contrast or dissimilarity, the weight of homogeneity decreases exponentially as the distance of the element value from the diagonal increases.

$$\text{Homogeneity} = \sum_{i=0}^{\text{quant}_k} \sum_{j=0}^{\text{quant}_k} p(i, j) * \frac{1}{1 + (i - j)^2} \tag{26.9}$$

(6) ASM

The texture thickness and regularity of the gray distribution of the image are described by GLCM. The ASM value is small when all GLCM values are closely related; it is large when the values of the matrix elements differ significantly [7]. High ASM values have a rough texture and high energy. Conversely, when the ASM value is low, the texture is fine and the energy is low.

$$ASM = \sum_{i=0}^{\text{quant}_k} \sum_{j=0}^{\text{quant}_k} p(i, j)^2 \tag{26.10}$$

(7) Energy

Energy, which reflects the uniformity and thickness of the texture, is an index to measure the stability of the grayscale change of the image texture, and is the sum of the squares of the elements of the grayscale co-occurrence matrix. High energy values indicate that the texture is now highly stable and changes frequently.

(8) Entropy

Computes the texture randomness (intensity distribution) of an image. Entropy is a seemingly random indicator of how much information an image contains and how complex it is. Entropy is greatest when all co-occurrence matrix values are the same or when pixel values exhibit maximum randomness; thus, entropy values represent the complexity of the grayscale image distribution [8]. The complexity of an image increases as the entropy value increases.

$$\text{Entropy} = -\sum_{i=0}^{\text{quant}_k} \sum_{j=0}^{\text{quant}_k} p(i, j) * \ln p(i, j) \tag{26.11}$$

(9) Maximum probability

The texture feature with the highest likelihood is the most prevalent texture feature in the image. There are 9 outputs in the constructed gray co-occurrence matrix. Means cannot be used to represent either of these results, since they are all matrices. To represent the upper left, upper right, lower left, lower right and center of a 380*800 matrix, five-point sampling and the last five key nodes are used.

In image processing, texture usually refers to the spatial repetition of detailed structures. In order to determine the properties of a texture, we can use many different methods, one of the common methods is to calculate the spatial offset of the image gray level. That is, a grayscale offset matrix can be obtained by calculating the grayscale difference of each pixel in the image. According to this matrix, the characteristic information of texture such as direction interval and amplitude can be calculated.

One approach we take is to use gray co-occurrence matrix (GLCM) to calculate texture features. GLCM is a directional sensitive matrix that records the cumulative probability distribution of gray levels between two pixels in an image region. By calculating some statistics in the specific direction of GLCM, such as energy, contrast, correlation, entropy, etc., the values reflecting the texture characteristics can be obtained. In addition, other texture feature extraction algorithms, such as Gabor filter and wavelet transform, can also be used for practical applications. As for decision rules, they are usually designed according to specific problems. For example, in an image classification task, texture features can be used as input feature vectors, and then the classification models (such as SVM, KNN, etc.) can be used to classify images. In practical applications, more complex decision rules can be designed according to the requirements of specific problems.

From this, we determine the properties of the texture by reflecting the details such as the direction, interval and size of the gray displacement of the image. These results are then applied to the GLCM model Fig. 26.2.

26.3.3 Comparison of "RGB + GLCM" and "HSV + GLCM" Image Feature Extraction Methods

Both "RGB + GLCM" and "HSV + GLCM" based image feature extraction methods are commonly used computer vision techniques. They use GLCM to describe the texture features in an image. The following is a brief description of the advantages of the two methods:

(1) Advantages of the "RGB+GLCM"-based image feature extraction method
First, color information is taken into account: RGB represents the red, green and blue channels of an image, so the color information of the original image can be preserved. Second, wide applicability: RGB is a common image representation and can be applied to a variety of image processing tasks. Third, Intuitive: the RGB color model is easier to understand and interpret for many people.
(2) Advantages of "HSV+GLCM" based image feature extraction method
First, hue, saturation and brightness are taken into account: HSV color model describes the three attributes of hue, saturation and brightness of an image, which is more suitable for expressing the color characteristics of an image. Second, robust to lighting changes: Saturation and lightness in HSV color space are insensitive to lighting changes, so it can better adapt to different lighting conditions. Third, more discriminative in certain scenarios: In certain image classification or target detection tasks, HSV color models can better distinguish between different objects or backgrounds.

In summary, the choice of which feature extraction method to use hinges on the specific application scenario and task requirements. If color information is a crucial factor and a high level of intuitiveness is necessary, "RGB + GLCM" may be a suitable approach. Alternatively, the "HSV + GLCM" method may be more appropriate for tasks focused on image color characteristics, robustness to lighting changes, or better differentiation in certain scenarios. In the case of crystallizer co-solvent sequence image feature extraction, the "RGB + GLCM" method is more practical due to these reasons.

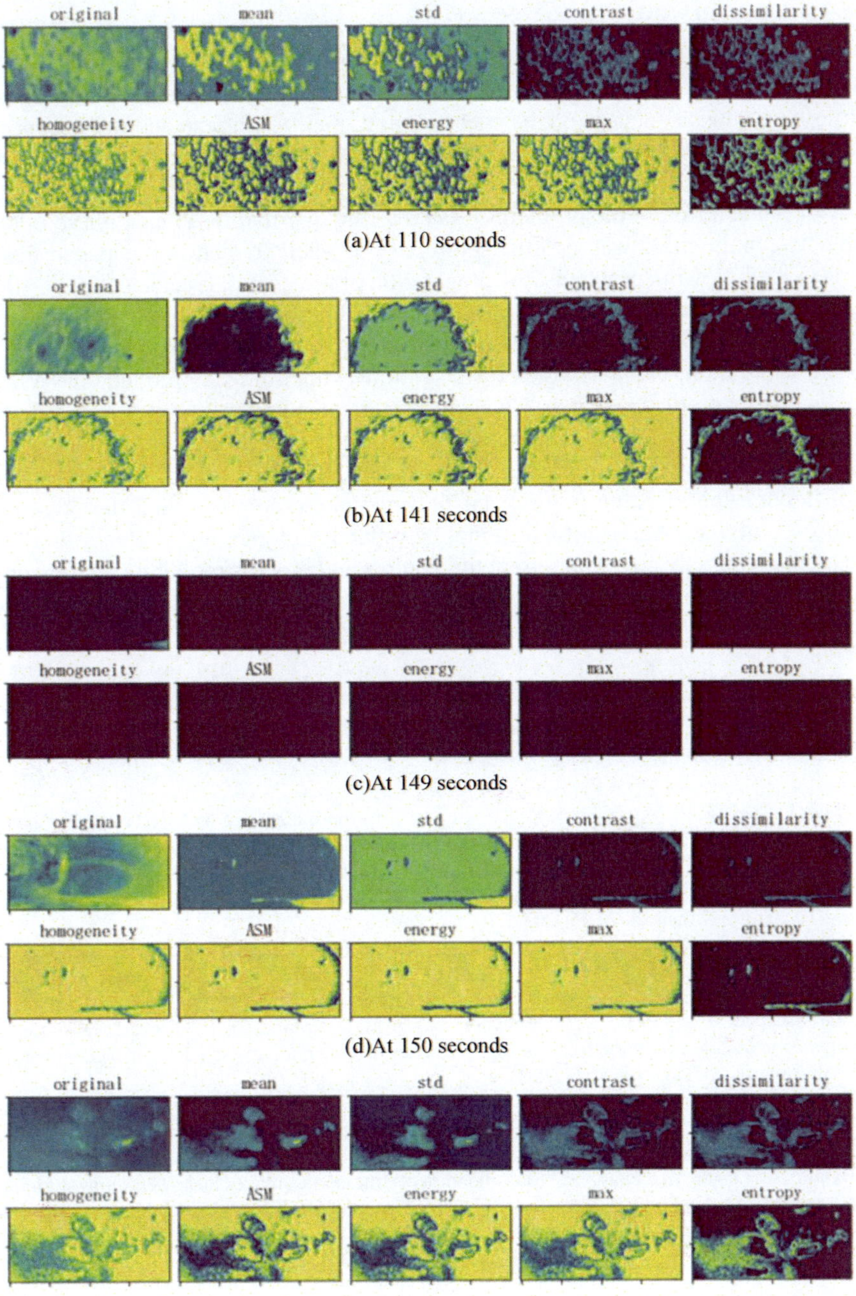

(a)At 110 seconds

(b)At 141 seconds

(c)At 149 seconds

(d)At 150 seconds

(e)At 671 seconds

Fig. 26.2 GLCM results graph

26.4 Conclusion

Mold slag is an important part of the continuous casting process, and its design purpose has always focused on the needs of high-quality, efficient and stable continuous casting production, and the development, production and use of mold slag varieties. However, in the stage of comprehensive upgrading of the continuous casting industry, in addition to the problems related to the distribution of slag phase in the gap between the mold wall and the solidified slab shell studied in this paper, there are also the following problems: the preparation of the materials selected for casting mold flux, and the avoidance of mold slag phase distribution. slag rollers, optimum surface quality to reduce dynamic pressure inside mold powder, overcapacity, production concentration, low awareness R&D, and investment in slag production. When conducting theoretical research, we should fully consider the complex multi-factor effects in actual production, and avoid the deviation caused by excessive isolation of some problems in pursuit of superficial articles. As the decarburization mold flux goes through the grinding process, air becomes more complex, and the presence or absence of water vapor, which cannot be influenced by giving a clear definition, can more complexly affect important time points and process mold flux melting and crystallization. If necessary, parametric variables can be searched more broadly to increase the feasibility and generality of the model.

References

1. Zhao, L., Zhang, P., Ma, X., Pan, Z.: Land cover information extraction based on daily NDVI time series and multiclassifier combination. Math. Probl. Eng. **2017**(pt.12), 1–13 (2017)
2. Wang, H., Wang, G.: The prediction model for haze pollution based on stacking framework and feature extraction of time series images. Sci. Total. Environ. **839**, 156003 (2022). https://doi.org/10.1016/j.scitotenv.2022.156003
3. Yuan, Y., Lin, L., Zhou, Z.G., Jiang, H., Liu, Q.: Bridging optical and SAR satellite image time series via contrastive feature extraction for crop classification. ISPRS J. Photogramm. Remote Sens.Photogramm. Remote. Sens. **195**, 222–232 (2023)
4. Marin Zapata, P.A., Roth, S., Schmutzler, D., Wolf, T., Manesso, E., Clevert, D.A.: Self-supervised feature extraction from image time series in plant phenotyping using triplet networks. Bioinformatics **37**(6), 861–867 (2021)
5. Yan, J., Chen, Y., Zheng, J., Guo, L., Zheng, S., Zhang, R.: Multi-Source Time Series Remote Sensing Feature Selection and Urban Forest Extraction Based on Improved Artificial Bee Colony. Remote. Sens. **14**(19), 4859 (2022)
6. Hatami, N., Gavet, Y., Debayle, J.: Classification of time-series images using deep convolutional neural networks. In Tenth international conference on machine vision (ICMV 2017), SPIE, Vol. 10696, pp. 242–249 (2018)
7. Fahim, S.R., Sarker, Y., Sarker, S.K., Sheikh, M.R.I., Das, S.K.: Self attention convolutional neural network with time series imaging based feature extraction for transmission line fault detection and classification. Electric Power Systems Research **187**, 106437 (2020)
8. Ma, R., Wu, W., Wang, Q., Liu, N., Chang, Y.: Offshore hydrocarbon exploitation target extraction based on time-series night light remote sensing images and machine learning models: a comparison of six machine learning algorithms and their multi-feature importance. Remote Sensing **15**(7), 1843 (2023)

Chapter 27
Research on Human Eyesight Tracking Algorithm Based on Monocular Vision

Haochun Tang, Jindong Zhang, and Jing Yang

Abstract This paper focuses on the optimization and improvement of the line-of-sight tracking algorithm based on monocular vision, and aims to achieve a series of complex functions through image analysis based on line-of-sight tracking technology and more in-depth analysis of monocular images. Therefore, we use the webgaze framework and use face detection models and feature point recognition libraries to analyze the recognition rate and provide further development of the technology.

27.1 Introduction

Current human eye detection implemented in different ways is mainly used in the fields of iris recognition, human eye gaze point extraction, facial expression recognition, and identity verification. However, there are still many problems in the practical application [1]. Existing human eye detection algorithms often need to obtain high-resolution images of localized areas of the human eye or face images in order to accurately extract the contours of the human eye, and need to be pre-processed to determine the localized areas of the face or human eye. In addition, most existing iris recognition and eye detection systems use near-infrared (NIR) light source cameras to capture images, but NIR light sources are weakly affected by natural light under outdoor conditions, which limits the application of eye detection technology to indoor locations. In addition, commercial vision tracking devices (eye-tracking devices) and specialized cameras for human eye detection are often very expensive.

H. Tang · J. Zhang (✉) · J. Yang
College of Software, Jilin University, Changchun 130012, China
e-mail: zhangjindong_100@163.com

J. Zhang
College of Computer Science and Technology, Jilin University, Changchun 130012, China

Key Laboratory of Symbol Computation and Knowledge Engineering of the Ministry of Education, Jilin University, Changchun 130012, China

R. Kountchev et al. (eds.), *Multidimensional Signals, Augmented Reality and Information Technologies*, Smart Innovation, Systems and Technologies 374,
https://doi.org/10.1007/978-981-99-7011-7_27

The study of human eye realization tracking through monocular vision is based on the project characteristics such as the low cost of the research project itself and the lower hardware requirements. In practice acquiring image equipment is often simple and the environment is often random and complex, so in the case of monocular cameras, the cameras have many practical applications. The best way to select the best combination of the face detection model and feature point recognition library can maximize the accuracy rate and lay the foundation for the application of the detection model.

27.2 System Description

The system uses the webgaze framework, which uses a monocular webcam to monitor changes in the human eye in real time, determine if the sight changes, and if it does, the changed to predict the line of sight of the human eye and to mark the location of the human eye in a graphical interface [2].

To help the network, we can provide it with only a portion of the image around the user's eye. This rectangle around the eye can be located using a third-party library. Thus, the first part of the system is shown in Fig. 27.1.

In order to use the vision tracking system on a web page, we need to collect some calibration data first. This is done by asking the user to click on some points on the screen with their mouse, while looking at the same points with their eyes. The program will record both the coordinates of the mouse clicks and the coordinates of the eye regions captured by the camera. These data are then used to train and calibrate the model, which is a machine learning algorithm that learns to map the eye regions to the gaze points on the screen. After the training is completed, the model

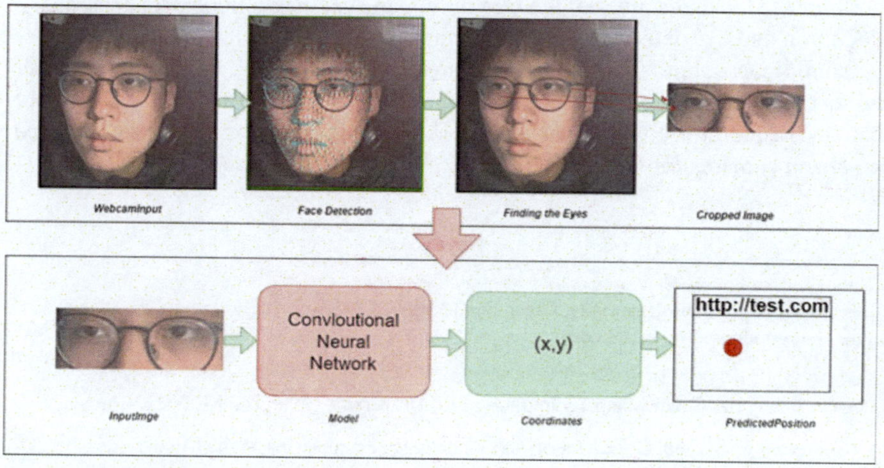

Fig. 27.1 Algorithm flow chart

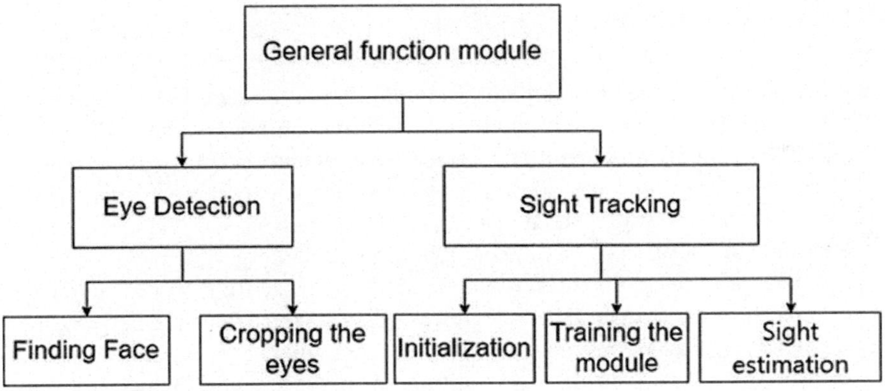

Fig. 27.2 Algorithm design module hierarchy graph

can predict the coordinates on the screen that correspond to where the user is looking at, and highlight them with a dot or a circle. The design hierarchy diagram of this process is shown in Fig. 27.2.

In this program, we are going to introduce the core functions of the vision tracking system, which aims to track and estimate the gaze direction of a human user. The system consists of two main components: eye detection and vision estimation. Eye detection is responsible for locating and extracting the eye region from an image or video frame captured by a camera. Vision estimation is responsible for predicting the gaze point on a screen or a scene based on the eye region.

The design method for eye detection is as follows: First, we use a face tracker to detect and track the face of the user in real time. Then, we crop a rectangular area around each eye and place it in a separate canvas. This step is usually related to the human eye detection model used, which can be based on different techniques such as edge detection, feature extraction, or deep learning.

The design method for vision estimation is as follows: First, we need to initialize the model, which is a machine learning algorithm that learns to map the eye region to the gaze point. To do this, we use a combination of mouse and calibration points on the screen. The user needs to follow the cursor with their eye while pressing the mouse button when they want to record an example. The program will record the position of the human eye, the coordinates of the cursor, and other information for data analysis and processing. Then, it will feed these data to the model for training and calibration. After initialization, the model is ready to predict eyesight and detect human eye.

To use the system, we need to combine the predicted information from vision estimation with the current graphical interface state. The system will output the graphical interface coordinates that correspond to the gaze point of the user. According to these coordinates, we can highlight or select an element on the screen (usually with a dot of a specific color) to indicate where the user is looking at.

27.3 Modules Design

The movement of the line of sight requires monitoring of the relative position of the eye, which is related to a variety of factors: including sitting posture, the amount of head offset, and also the input is captured by the picture of the camera, which is also related to the position of the camera. Therefore, we combine the data of the human 3D coordinate system and dynamic model for analysis and judgment, which can determine the position of the line of sight more accurately.

First we need to select the tracker module in the webgaze framework, the tracker module controls how to detect the face, refresh the browser after the selection, and the face will appear as a green mask.

Next we need to crop in the eyes to cmltracker for example, which not only provides us with the location of the face also provides the location of 70 facial features, and by querying the document, we can select to the exact point we are looking for [3] Fig. 27.3.

We define the eye as a rectangle with contact points 23, 28, 24, and 26, extending 5px in each direction. This rectangle should cover everything that matters.

For model training, we store the captured graphs as a tensor, simply storing all examples in a large training set, but in machine learning it is important to make sure that your model is not just reciting data. This is why we should store some examples in a separate validation set. Then, we can check how the model performs on unseen data and make sure it doesn't overfit the training set. To do this, we split 20% into the validation set [4] Fig. 27.4.

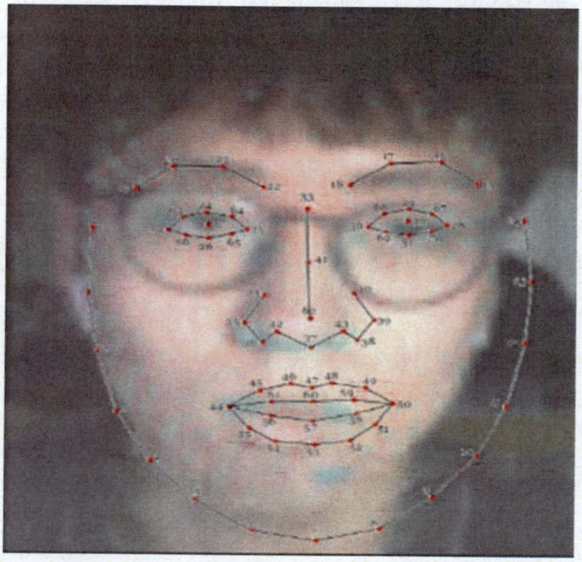

Fig. 27.3 Facial features

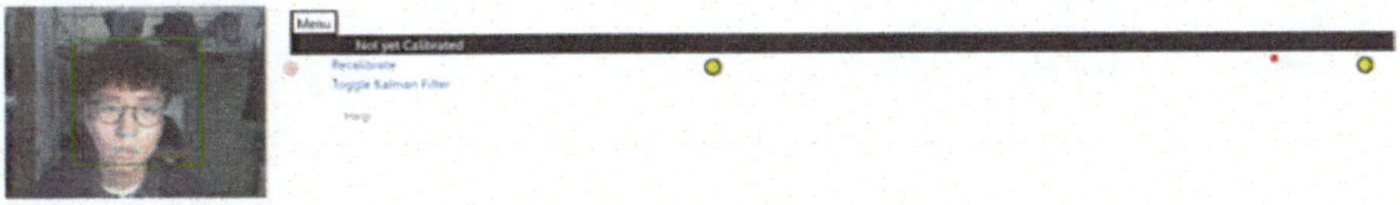

Fig. 27.4 Model calibration

In this model, we aim to build a simple convolutional neural network (CNN) that can predict the gaze point of a user based on their eye region. To implement the CNN, we use TensorFlow.js, which is a JavaScript library for machine learning that provides a Keras-like API for creating and training models in the browser or on Node.js. The network architecture consists of three main layers: a convolutional layer, a max-pooling layer, and a dense layer. The convolutional layer applies a set of filters to the input image to extract features such as edges, shapes, and patterns. The max-pooling layer reduces the size of the feature maps by taking the maximum value in each subregion, which helps to reduce overfitting and computational cost. The dense layer is a fully connected layer that outputs two values: the x-coordinate and the y-coordinate of the gaze point on the screen. To improve the performance and generalization of the network, we also add some additional layers in between: dropout and flatten. Dropout randomly drops out some units in the network during training, which prevents co-adaptation of features and reduces overfitting. Flatten reshapes the 2D feature maps into 1D vectors, which are then fed into the dense layer. To train the network, we use the Adam optimizer, which is an adaptive learning rate algorithm that adjusts the learning rate based on the gradient of each parameter. The training process consists of several steps: data preparation, model initialization, model fitting, model evaluation, and model saving. The details of each step are shown in the following sections Fig. 27.5.

After the model is calibrated, we can use it smoothly to make implementation predictions. In this model we use tracker to get the human eye changes, so we need a function to get the human eye changes, geteyes ().

Algorithm 1: Model Training

Data: detected data $DATA$, elapsed time Te

Result: model information $DATAm$

```
X ← DATA.X;
Y ← DATA.Y;
T ← currentperiod;
while T ≥ Te do                    // Every elapsedTime down a test
    if Available(DATA) then  // In order to filter out invalid
     signals in the data and eliminate or attenuate some of
     the signal changes, so as to achieve the effect of noise
        Model.RECORD(DATA);
    end
    T ← 0;
end
```

Fig. 27.5 Model training process pseudocode

```
Algorithm 1: Model Predict
Data: detected set CAM
Result: highlighted location
while CAM.OPEN() do        // Continuous monitoring while the
  camera is on
    if MOVED(CAM.GET()) then      // if gaze moved then Get
      the eye position image
        image = Crop(CAM.GET());
        // First crop the image
        batchedImage = image.expandDims(0);
        // Create the batchedImage object
        data = DATA(batchedImage);
        // Wrap the batchedImage object in the DATA object
        preditionp = webgazer.getCurrentPrediction(DATA));
        // End the prediction function
        locationl = p.convert();
        // Convert the prediction value to the value of the
           corresponding screen
        Hight(l);
    end
end
```

Fig. 27.6 Model predicition process pseudocode

After cropping and tensorization, we encapsulate the processed data in a data object, use webgazer.getCurrentPrediction(data) to make predictions, and finally highlight the predicted points.

A pseudo-code description of the prediction algorithm is performed, as follows Fig. 27.6.

27.4 Results and Analysis

After the model has been initialized, the framework provides an accuracy calculation tool. The user is required to stare at the calibration point for a certain amount of time, record the predicted position during this time, and record the distance between the predicted position and the calibration position to estimate the accuracy of the line-of-sight prediction.

We used a laptop with a 3060 graphics card to test the program in the Edge browser of Windows 11. Also, we have done several calculations under different environments and light sources, and the accuracy is above 60%, which shows that our model, which uses deep neural networks and eye-tracking data, can estimate the gaze position more accurately than the baseline model, which relies on simple geometric methods. The following is a line graph of the accuracy rate and some experimental results Figs. 27.7 and 27.8.

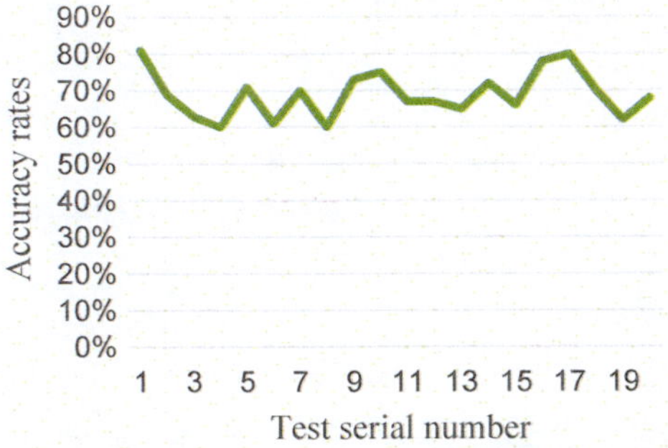

Fig. 27.7 Accuracy statistics

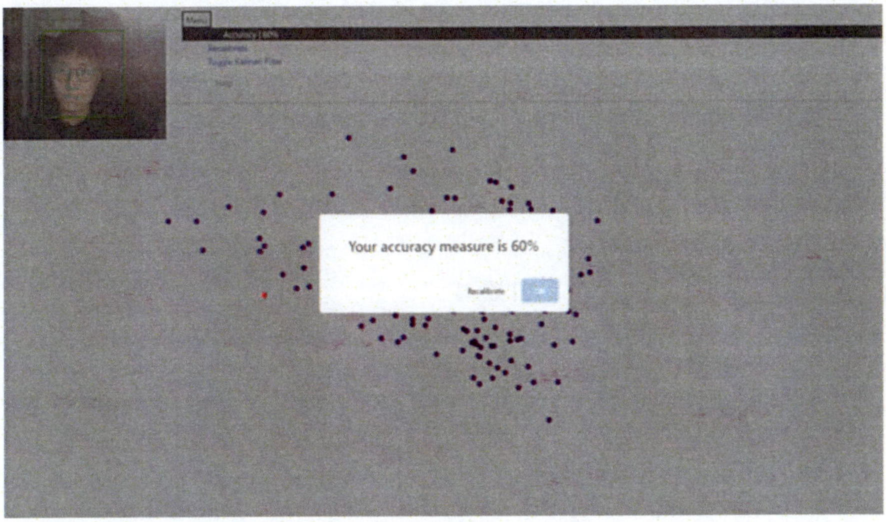

Fig. 27.8 Partial experimental results

We also implemented WebGazer on Search Engines (legacy) on top of the model using the existing features of the framework.

The reliability of the model was again verified by implementing the Ball Collision Game Figs. 27.9 and 27.10.

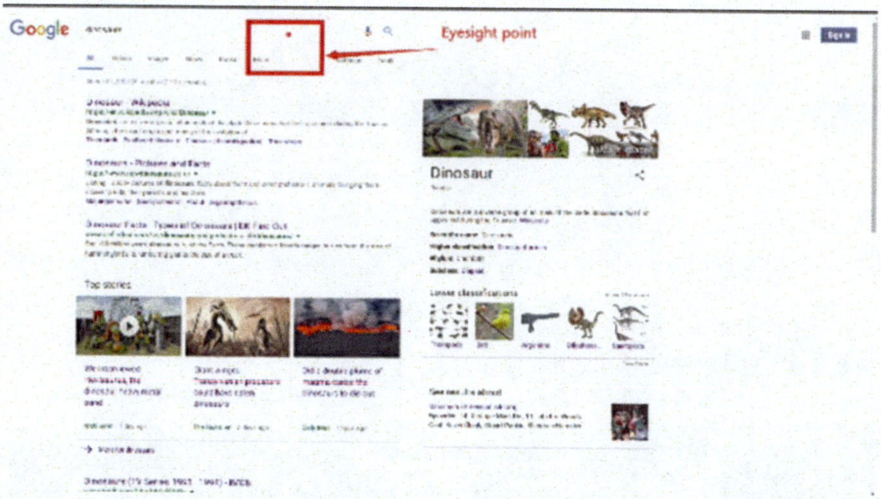

Fig. 27.9 WebGazer on Search Engines

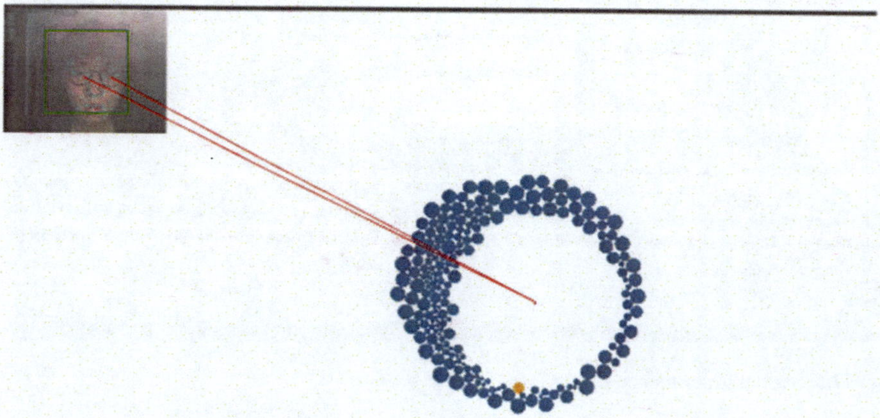

Fig. 27.10 Ball Collision Game

27.5 Conclusion

A human eye monocular line-of-sight tracking system designed based on the webgaze framework is presented. The program implements line-of-sight tracking through a face monitoring model. It is developed using html language, which provides an intuitive and easy-to-use line-of-sight tracking system for users, avoiding the defects of traditional line-of-sight tracking systems that are expensive and not easy to use. It plays a role in the wide application of line-of-sight tracking in real life.

Eye-tracking technology is a technique that reflects human visual attention and cognitive processes by measuring the movement, position, and fixation point of the eyes. It has a wide range of applications in academic and scientific research fields, such as psychology, neuroscience, education, media studies, etc. It can also be used in assistive and rehabilitative fields, such as controlling wheelchairs, robotic arms, and prosthetics.

With the development of artificial intelligence, eye-tracking technology is also constantly innovating and improving, bringing more possibilities and opportunities for human society. Some of the prospective and practical application scenarios include:

Autonomous driving technology: By monitoring the eye state of the driver, the driver's attention, fatigue, emotion, etc. can be judged, thereby improving the safety and intelligence of the autonomous driving system.

Human emotion capture: By analyzing the movement and pupil changes of the eyes, human emotions and psychological states can be identified, thus providing more information and services for fields such as mental health, social interaction, entertainment games, etc.

Classroom attention test: By recording the students' eye fixation points and time, the students' interest and comprehension level of the classroom content can be evaluated, thus providing more effective teaching feedback and guidance for teachers.

In summary, eye-tracking technology is a technique with great potential and value, which can enable us to understand human visual cognition mechanisms more deeply, and also allow us to interact with computers more conveniently. With the advancement of technology and the expansion of applications, eye-tracking technology will bring more benefits and innovations for human society.

Students' interest and comprehension level of the classroom content can be evaluated, thus providing more effective teaching feedback and guidance for teachers.

In summary, eye-tracking technology is a technique with great potential and value, which can enable us to understand human visual cognition mechanisms more deeply, and also allow us to interact with computers more conveniently. With the advancement of technology and the expansion of applications, eye-tracking technology will bring more benefits and innovations for human society.

Acknowledgements This study was supported by innovation and entrepreneurship training program for college students of Jilin University.

References

1. Weisenburger, S., Wallace, G.: Gnss Satellite Line of Sight Detection: EP3940427A1[P] (2022)
2. Zhang, J.D., Xiu, Y.: Image stitching based on human visual system and SIFT algorithm. Vis. Comput.Comput. (2023). https://doi.org/10.1007/s00371-023-02791-4
3. Zhang, J., Zhong, H.: Curve-based lane estimation model with lightweight attention mechanism. Signal Image Video Process **17**, 2637–2643 (2023). https://doi.org/10.1007/s11760-022-02480-0

4. Zhang, J., Pan, D., Zhang, K., Jin, J.Y., Ma, Y.Z., Mai, C.: Underwater single-image restoration based on modified generative adversarial net. Signal Image Video Process **17**, 1153–1160 (2023). https://doi.org/10.1007/s11760-022-02322-z
5. Yang, X., Krajbich, I.: Webcam-based online eye-tracking for behavioral research. Judgm. Decis. Mak. **16**, 1–39 (2021)

Chapter 28
Research and Development of Automatic Leakage Inspection System for Gas Pipeline Based on Internet of Things Technology

Wen Zhou, Kun Mao, Shengbin Hua, Chengwei Huang, YuyuYang, Jie Liu, Jun He, Rongwang Chai, and Jiangang Ye

Abstract According to the existing problems of data reading difficulty and falsification in the process of gas pipeline installation test, a novel automatic leakage inspection system for gas pipelines based on the Internet of Things technology is proposed. The system consists of a detection device, cloud, and client, which can automatically complete the automation process from data acquisition and processing to data visualization. By integrating the gas pressure sensor, gas temperature sensor, and atmospheric pressure sensor, the system can simultaneously detect three kinds of data so that accurate and efficient inspection can be achieved. During the inspection process, the client can view the visualized detection process data and obtain leakage warning information in real time. This paper also proposes an automatic anomaly assessment algorithm based on complexity to realize automatic assessment of the risk status of gas pipelines. The experiment results show that the proposed system can realize automatic leakage inspection. Besides, the proposed system can detect abnormal conditions in time, and the anti-interference ability is better than the traditional judgment method.

W. Zhou · S. Hua · C. Huang · J. Liu · J. He · J. Ye (✉)
Quzhou Special Equipment Inspection Center, Quzhou, China
e-mail: dester19831219@126.com

K. Mao
Quzhou College of Technology, Quzhou, China

YuyuYang
Hangzhou Juwei Technology Engineering Co., LTD, Hangzhou, China

R. Chai
Kaihua County Natural Gas Co. LTD, Quzhou, China

R. Kountchev et al. (eds.), *Multidimensional Signals, Augmented Reality and Information Technologies*, Smart Innovation, Systems and Technologies 374,
https://doi.org/10.1007/978-981-99-7011-7_28

28.1 Introduction

Natural gas as efficient energy has become rapidly popular, and the natural gas pipeline is the main carrier for transporting gas to thousands of households. The natural gas pipeline is an essential part of the urban pipe network in China, while natural gas is combustible and explosive. Improper operation in gas pipeline installation and misuse of natural gas may lead to leakage, casualties, and social instability [1–5]. According to pipeline construction standards, such as CJJ33 [6], CJJ63 [7], and other standards on the gas pipeline tightness or leakage inspection requirements, the precise inspection duration is 24 h. During the inspection, the natural gas pressure, gas temperature in the pipeline, and atmospheric pressure should be measured every hour, which is used to judge whether the leakage inspection is qualified. However, in the actual inspection process, only read the natural gas pressure data from the gas pipeline's pressure gauge and measure the atmospheric pressure while replacing the natural gas temperature data with the ambient temperature data. So, one gas pipeline inspection requires more labor costs, and errors can easily occur in manual data reading and calculation. In order to simplify the process, some installation companies adopt falsification in the data reading process, which shows the root of the hidden danger of accidents. In order to standardize the process of inspection, this paper has developed an automatic system for the leakage inspection of gas pipelines. Through this system, the stability data of the pipeline can be confirmed, the status data of the gas pipeline can be visualized, the abnormal status can be detected in time, and the inspection results can be calculated automatically. When the result exceeds the upper or lower limit, it can send an early warning, which provides a basis for the automatic management of third-party inspection and gas pipeline construction companies.

28.2 Design of Automatic Leakage Inspection System

An automatic leakage inspection system based on Internet of Things technology is designed in this paper in order to automatically complete the pressure resistance and leakage inspection of the natural gas pipeline. The system is composed of three parts: detection device, cloud, and client. The schematic diagram of its composition is shown in Fig. 28.1.

The detection device is responsible for data collection and wireless transmission of the relevant status of the natural gas pipeline, while the cloud is responsible for receiving the data from the detection device. The client is the HMI for users to operate and view the detailed status of the natural gas pipeline.

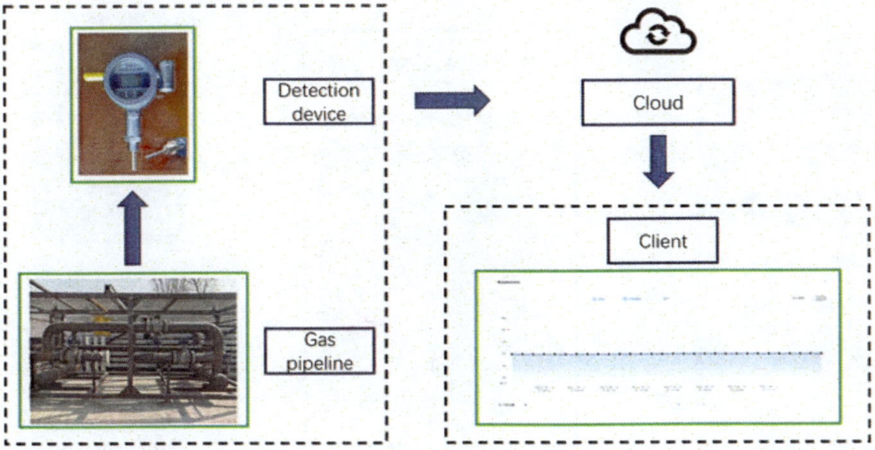

Fig. 28.1 Schematic diagram of the automatic inspection system

28.2.1 Design of the Detection Device

Considering the natural gas pressure and gas temperature inside the pipeline, the atmospheric pressure should be simultaneously recorded during the tightness inspection, the temperature sensor, pressure sensor, and atmospheric pressure sensor are integrated into the hardware, as shown in Fig. 28.2.

Sensor Design

According to the actual environment and the status of the natural gas in the pipeline, the appropriate pressure sensor and temperature sensor are selected. Among them, the pressure sensor used for natural gas pressure detection has a range of 0–1.6 MPa and an accuracy of 0.4, while the atmospheric pressure sensor has a range of 26 kPa-126 kPa, and the temperature sensor is the type of thermocouple with a range of 0–400°C. In order to minimize the external dimension, the probe of the temperature sensor and the pressure hole of the pressure sensor are integrated, as shown in Fig. 28.3.

Operation Display Design

According to different actual requirements of natural gas pipeline inspection, the frequency of data acquisition needs to be adjusted accordingly. Therefore, this paper designs a display screen that can be operated by buttons so that users can directly and manually set and adjust the acquisition frequency of the device at the detection device.

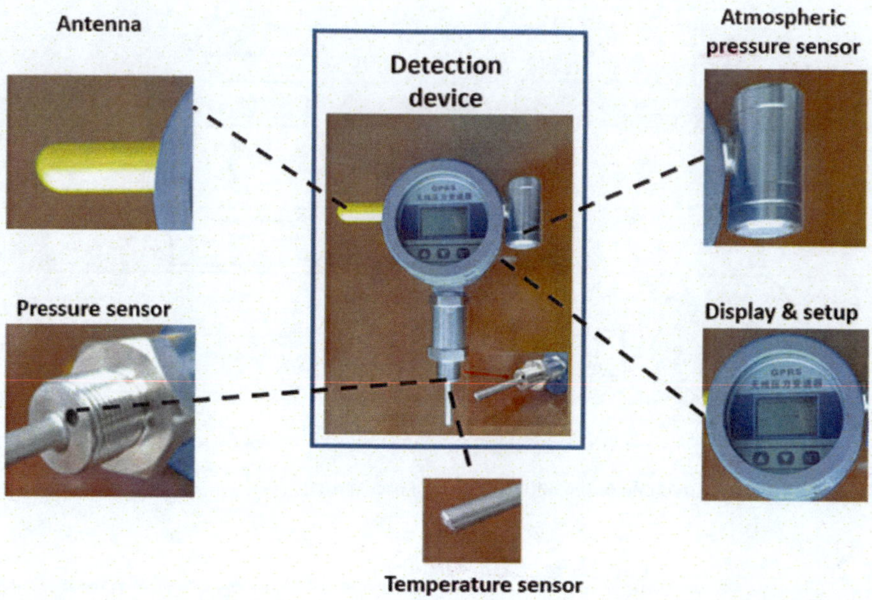

Fig. 28.2 Composition of the detection device

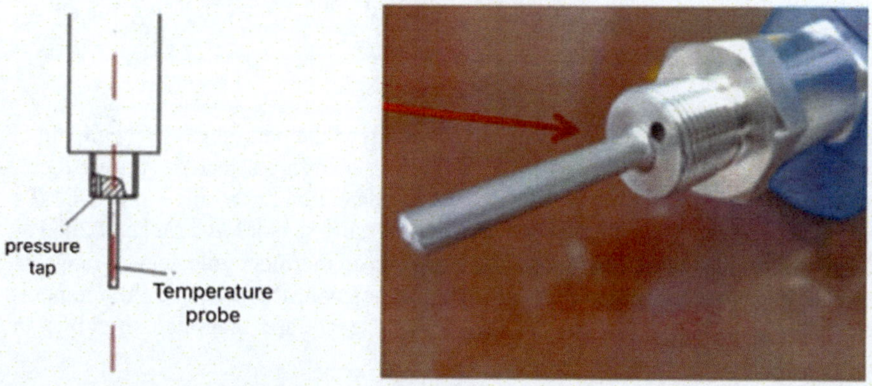

Fig. 28.3 Schematic diagram of temperature probe and pressure hole integration

Wireless Transmission Design

In order to improve the safety of the inspection process, the data is periodically uploaded to the cloud server through 4G, which can realize stable data transmission in remote areas and significantly improve the application fields of the system.

28.2.2 Design of the Cloud

The cloud server of this system uses the *OneNET* cloud platform of China Mobile. The platform supports the adaptation of various network environments and protocol types, realizing rapid access to various sensors and intelligent hardware and meeting platform-level service requirements such as device connection, protocol adaptation, data storage, data security, and big data analysis in the field of the Internet of Things.

The communication protocol between the detection device and the cloud is HTTP, and the transmitted data consists of "*Workpressure*", "*Temperature*" and "*Airpressure*", which respectively represent the natural gas pressure, gas temperature, and atmospheric pressure in the natural gas pipeline. All detection data is uploaded to the cloud server by the detection device and stored temporarily.

28.2.3 Design of the Client

In this paper, C# programming language is used to realize the HMI to the client [8–10]. The framework design structure of the interface is shown in Fig. 28.4. Users log in to the client software through the computer to realize real-time monitoring of the inspection process data, and at the end of the inspection, automatically calculate the conclusion result and visualize the status data.

Login Interface Design

The client uses the login interface with permissions, which must be through the user name and password login to enter the system's main interface successfully. After the user logs in successfully, the system will automatically identify the user's identity information and permissions to display different data and open different operation

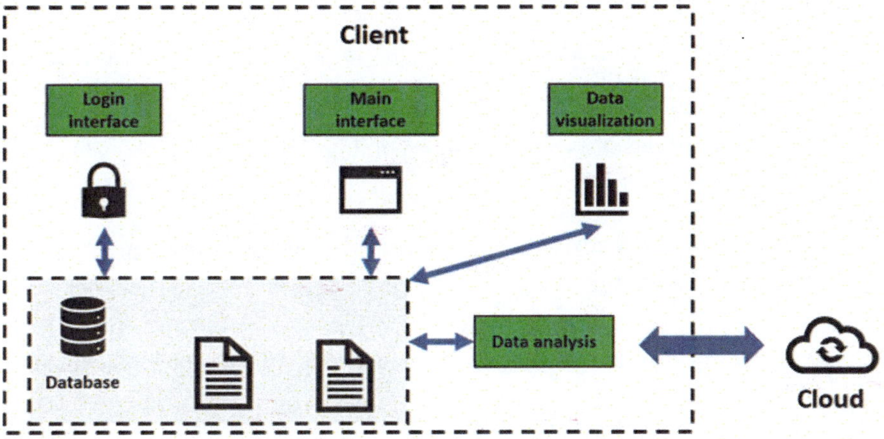

Fig. 28.4 Client software architecture diagram

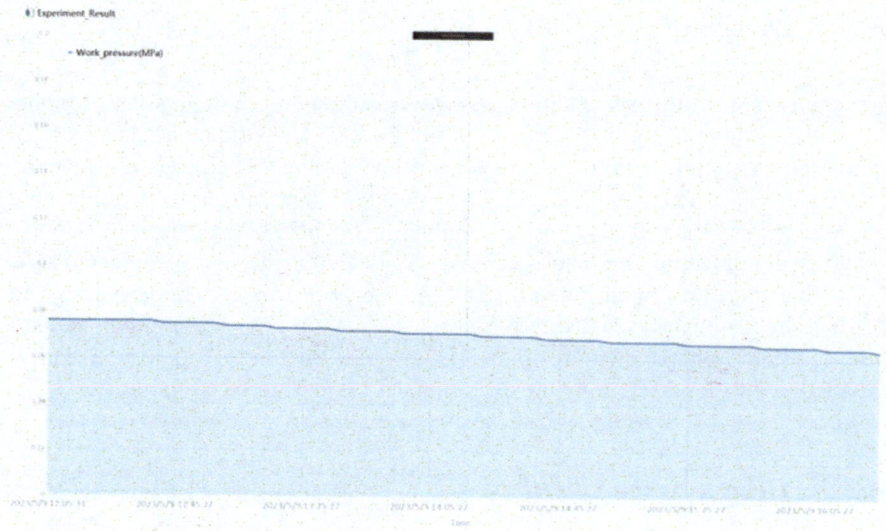

Fig. 28.5 Visualization of real-time data

permissions according to different permissions in the subsequent query and data visualization function.

Main Interface Design

The main interface has the function of query and inspection start-stop. The query module can automatically display the corresponding historical inspection data according to the query conditions entered by the user; The inspection start-stop function can realize the start and stop operation of the complete leakage inspection of the natural gas pressure pipeline.

Data Visualization Design

The data visualization interface can display the inspection process data to users intuitively. The data visualization of this system can be divided into real-time data and decision data. The real-time data curve realizes real-time drawing of all data from the beginning of this inspection, as shown in Fig. 28.5.

28.3 Design of Automatic Evaluation Algorithm for Gas Pipeline Leakage

The traditional gas pipeline leakage inspection applies the following judgment algorithm:

$$\Delta P'=(H_1+B_1)-\frac{273+t_1}{273+t_2}\cdot(H_2+B_2) \tag{28.1}$$

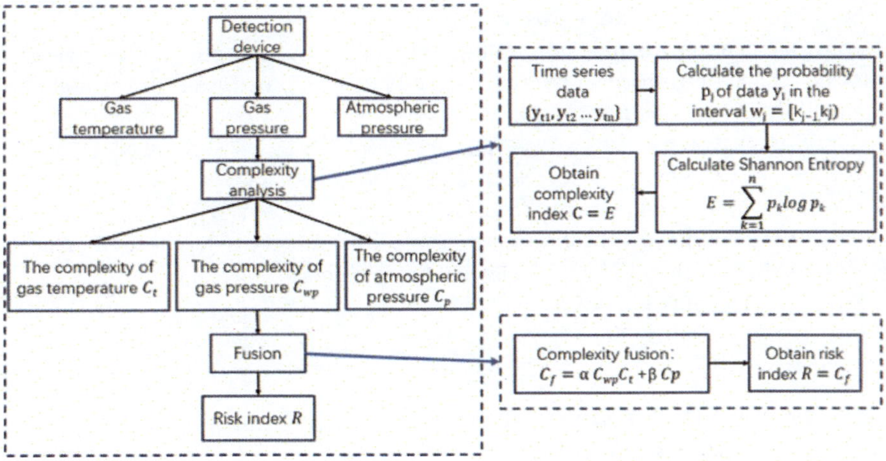

Fig. 28.6 Schematic diagram of the proposed evaluation algorithm

where $\Delta P'$ is the revised pressure drop (unit: Pa), H_1 and H_2 respectively represent the gas pressure at the beginning and end of the inspection (unit: Pa), B_1 and B_2 respectively represent the atmospheric pressure at the beginning and end of the inspection (unit: Pa), t_1 and t_2 respectively represent the gas temperature at the beginning and end of the inspection (unit: °C).

The traditional judgment algorithm only calculates the data at the beginning and end moments, ignoring the data's influence in the intermediate change process on the judgment results. Therefore, this paper proposes a new automatic evaluation algorithm for gas pipeline leakage inspection based on the complexity.

The schematic diagram of the proposed evaluation algorithm is shown in Fig. 28.6, and the detailed process of the proposed automatic evaluation algorithm is as follows:

Step 1: Data Receiving

Receive real-time data from the detection device, including gas pressure, gas temperature, and atmospheric pressure;

Step 2: Complexity Analysis

Step 2–1: Convert the raw data into time series data $\{y_{t1}, y_{t2} \cdots y_{tn}\}$;

Step 2–2: Calculate the probability p_j of the data y_{ti} appearing in the interval $w_j = [Y_{j-1}, Y_j)$;

Step 2–3: Calculate the corresponding Shannon entropy:

$$E = \sum_{k=0}^{n} p_k \log p_k \tag{28.2}$$

Step 2–4: The complexity index corresponding to time series data is obtained $C = E$;

Step 3: Complexity Index Fusion

Step 3–1: The gas temperature complexity index C_t, gas pressure complexity index C_{wp}, and atmospheric pressure complexity index C_p are fused. The calculation formula for the fusion is as follows:

$$C_f = \alpha C_{wp} C_t + \beta C_p \tag{28.3}$$

where α is the internal coefficient of the gas pipeline, β is the external coefficient of the gas pipeline, both of them are adjustable parameters;

Step 3–2: Obtain risk index $R = C_f$;

28.4 Experimental Study of the Automatic Leakage Inspection System

28.4.1 Design of Gas Cylinder for Experiment

In order to validate the proposed automatic leakage inspection system, a gas cylinder is designed, which consists of an upper head, a lower head, and a cylinder body, as shown in Fig. 28.7. The volume of the gas cylinder is 3.5L, the maximum diameter is 210 mm, the thickness is 2.5 mm, and the material is HP295. Two seats are designed on the top of the cylinder body. One is used for threaded assembly with the detection device, the internal thread of M20 × 1.5. The other is used for assembling the valve, the internal thread of G1/2, and the distance between the seats is 100 mm.

In order to ensure the safety of the experiment, the gas cylinder is compressed by V-0.25/8 air compressor with the highest pressure of 0.8 MPa. The compressed gas cylinder is shown in Fig. 28.8. In order to test the sensitivity of the sensor, the thread interface was not completely tightened during the installation, and there were micro leaks. At 12:05:31, the power supply of the device was switched on, with the initial condition of the gas pressure 0.076 MPa. At 16:25:27, the power supply of the device was switched off. After 4 h, 19 min, and 56 s, the gas cylinder pressure was 0.064 MPa, as shown in Figs. 28.8 and 28.9.

28.4.2 Data Visualization Verification

After the experiment, the real-time curves of the gas temperature, gas pressure, and atmospheric pressure in the gas cylinder can be seen in the HMI, as shown in Fig. 28.9. It can be seen from the curves that the gas pressure value is consistent with the reading value of the pressure gauge, indicating that the experiment process is successful.

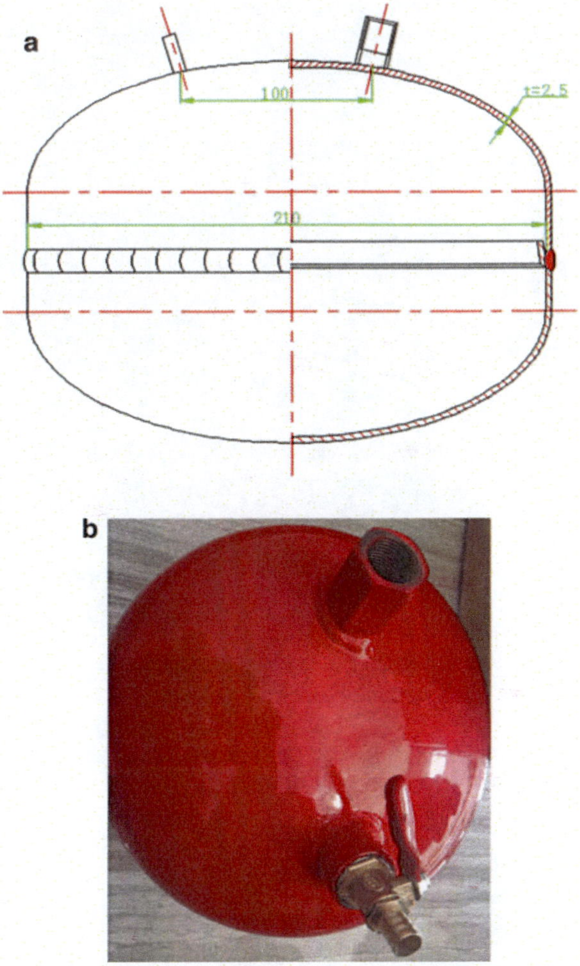

Fig. 28.7 Schematic diagram (left) and physical diagram (right) of the gas cylinder used in the test

28.4.3 Inspection Performance Verification

The complexity index and risk index of the proposed algorithm can be calculated based on the experimental results, where α is 0.7 and β is 0.3. For easy observation, the time of the horizontal coordinate is converted to a number. When the horizontal coordinate is 1, it represents the first moment, i.e., 2023/5/9 12:05:31, as shown in Fig. 28.11.

As shown in Fig. 28.10, the gas pressure gradually decreases monotonically over time, while the atmospheric pressure and temperature maintain a stable status, with occasional slight fluctuations. It can be seen from Fig. 28.11, after analysis by the algorithm proposed in this paper, the blue curve shows a continuous rise, indicating that the gas pressure value continues to have anomalies; The red curve shows an

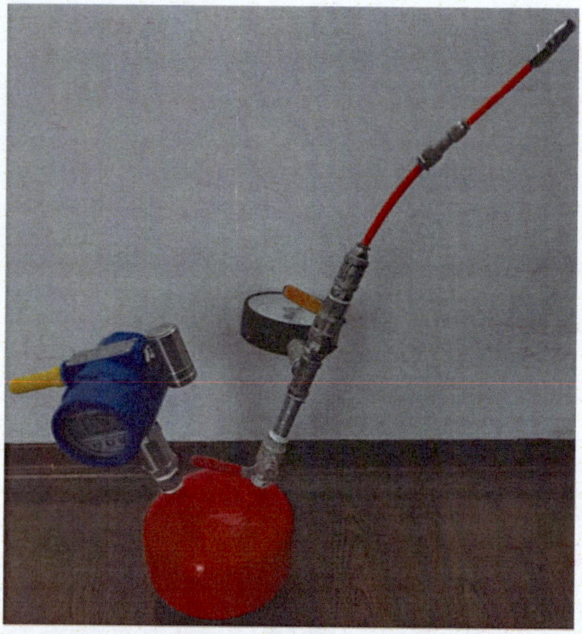

Fig. 28.8 Cylinder for experiment and detection device

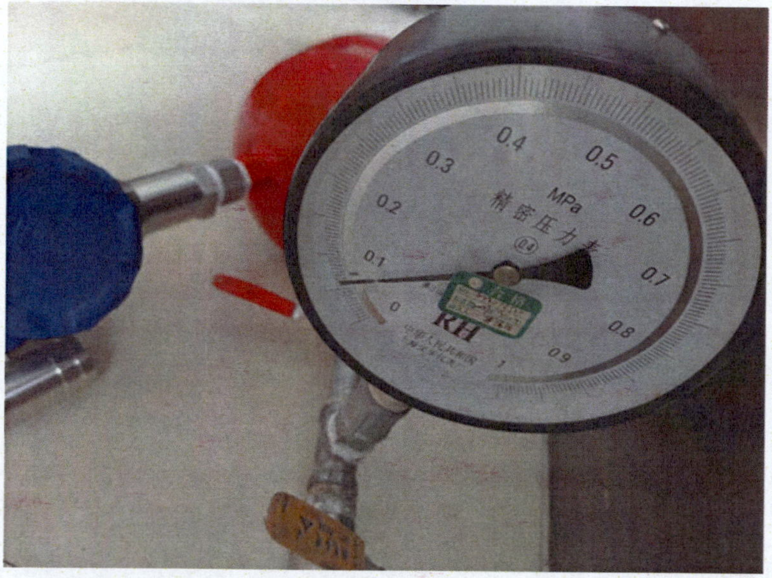

Fig. 28.9 Cylinder pressure after experiment

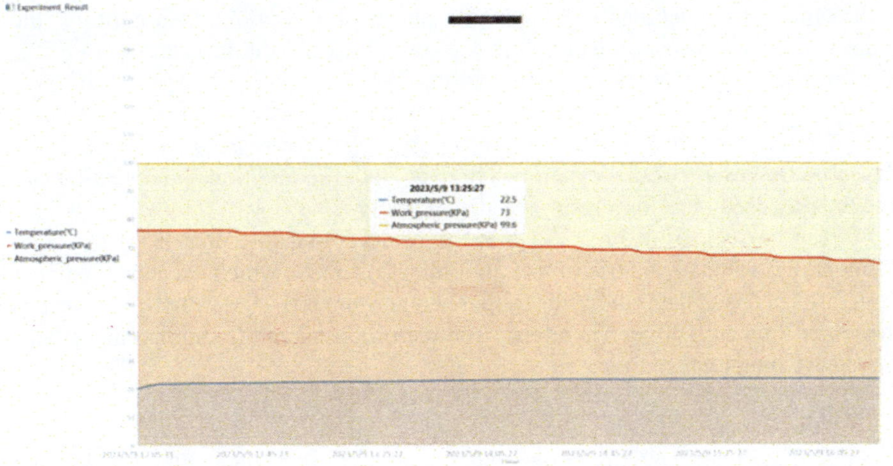

Fig. 28.10 Real-time curves of gas temperature, gas pressure and atmospheric pressure

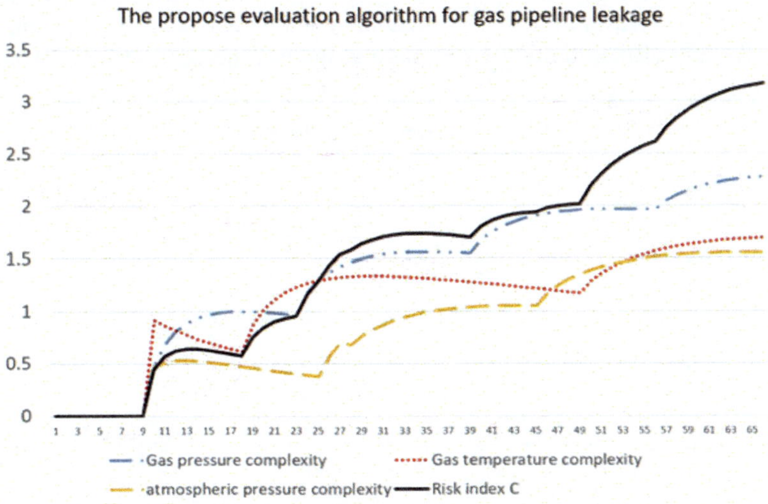

Fig. 28.11 The complexity index and risk index of the proposed algorithm

inevitable fluctuation, but the overall trend is still rising, indicating that the gas temperature has a certain degree of irregular fluctuation in the whole process. The yellow curve is similar to the gas temperature complexity index curve, reflecting that atmospheric pressure also fluctuates. The black curve is the fusion result, which is also a continuous rise, indicating that the gas pipeline status continues to have anomalies. Therefore, the complexity index in the proposed algorithm can effectively reflect the changes in the internal and external status of the gas pipeline.

In order to further analyze the performance of the proposed algorithm, the proposed algorithm is compared with the traditional algorithm in terms of effectiveness and stability.

Effectiveness:

Based on the results of this experimental data, both the traditional algorithm result and the proposed algorithm result are shown in Fig. 28.12.

As can be seen from Fig. 28.12, at the black dotted line, that is, at the tenth moment, the value of the traditional algorithm is 2.2833, which exceeds the upper limit of 1.1000, and the value of the proposed algorithm is 0.4420, which also exceeds the upper limit of 0.4000. Therefore, both the proposed and traditional algorithms can detect anomalies in time.

Stability:

In order to verify the stability of the proposed algorithm, the data of the last moment is changed to the data of the first moment so as to simulate the data anomaly. Both the curve of the traditional algorithm and the proposed algorithm are shown in Fig. 28.13.

As can be seen from the black dotted line in Fig. 28.13, since the traditional algorithm only considers the data at the beginning moment and the end moment, the calculated value of the traditional algorithm at the last moment becomes 0, so the

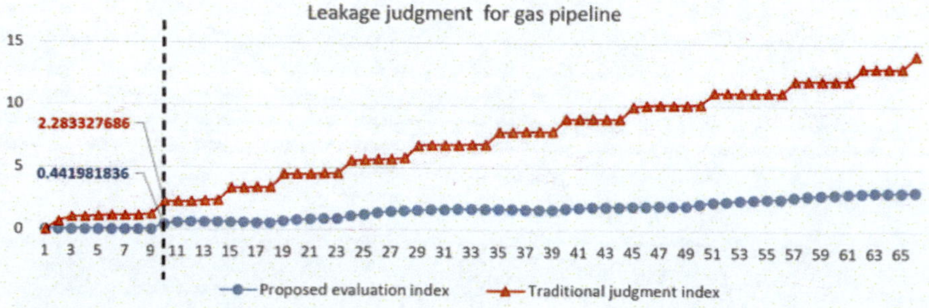

Fig. 28.12 The leakage judgment for the gas pipeline

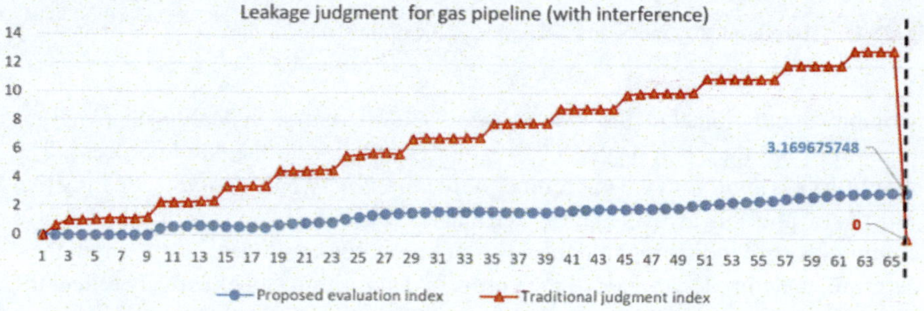

Fig. 28.13 The leakage judgment for gas pipeline with interference

final judgment result is no anomaly. However, the calculated value of the proposed algorithm at the last moment is 3.1697, which is greater than the upper limit of 0.4. Therefore, the risk is only successfully identified by the proposed algorithm, indicating that the proposed algorithm is more stable than the traditional algorithm in dealing with the abnormal status.

28.5 Conclusion

In this study, an automatic leakage inspection system for gas pipelines is proposed. By integrating the gas pressure sensor, gas temperature sensor, and atmospheric pressure sensor, a single detection device can simultaneously detect three kinds of data, and the detection data can be uploaded to the cloud in real time. The client can view the visualized detection process data and leakage warning information through the client in real-time. Therefore, compared with the traditional manual data recording detection method, this system is more accurate and can automatically complete the whole automation process from data acquisition and data processing, improving the efficiency and intelligent level of detection. Furthermore, the system's effectiveness and stability are verified by the experiment. The experimental results show that the proposed system can detect gas pressure anomalies in the pipeline in time. In the case of individual abnormal data, compared with the traditional judgment algorithm, it can correctly determine the anomalies and has higher stability.

Although this system has some advantages over traditional detection methods, there is still room for improvement. For example, when the normal fluctuation range of the detected data is large, the fixed interval setting may not be the best. Too much interval setting will significantly increase calculation cost, while too little interval setting will fail to detect anomalies in time.

Acknowledgements This work is supported by Zhejiang Provincial Market Supervision Administration (Project number: ZC2021B091) and Quzhou Science and Technology Key Project (Grant No. 2023K247 and 2023K044).

The authors thank Zhejiang Chaoyi Fire Equipment Co., LTD and Quzhou Xinao Gas Co., LTD for their support and help during the experiment.

References

1. Urban Gas Pipe Line. Journal of Failure Analysis and Prevention. 2020(3):1007–1018
2. Ma, B., Wang, H.S., Ma, X.Q., et al.: Application of information Technology In Integrity Management of Urban Gas Pipe Line. Gas & Heat **43**(2), 38–42 (2023). https://doi.org/10.13608/j.cnki.1000-4416.2023.02.017
3. Hu, J.Q., Hou, Y.C., Dong, S.H., et al.: Risk hazard identification in the emergency evacuation process of community gas pipeline accident based on STAMP. Safety and Environment Engineering **30**(2), 10–20 (2023)

4. Ahammed, M., Melchers, R. E.: Reliability of underground pipelines subject to corrosion. J. Transp. Eng. **120**(6), 989–1002 (1994) https://doi.org/10.1061/(ASCE)0733-947X(1994)120:6(989)
5. Au, S.K., Beck, J.L.: Estimation of small failure probabilities in high dimensions by subset simulation. Probab. Eng. Mech. **16**(4), 263–277 (2001). https://doi.org/10.1016/S0266-8920(01)00019-4
6. CJJ33–2005. Code for construction and acceptance of city and town gas distribution works. Bei Jing: China Architecture & Building Press (2005)
7. CJJ63–2018. Technical standard for polyethylene(PE) gaseous fuel pipeline engineering. Bei Jing: China Architecture & Building Press (2018)
8. Li, M.H., Feng, X., Han, Y., et al.: Mobile augmented reality-based visualization framework for lifecycle O&M support of urban underground pipe networks. Tunn. Undergr. Space Technol. **136**, 105069 (2023). https://doi.org/10.1016/j.tust.2023.105069
9. Zhao, L., Qi, G., Dai, Y., et al.: Integrated dynamic risk assessment of buried gas pipeline leakages in urban areas. J. Loss Prev. Process Ind. **83**, 105049 (2023). https://doi.org/10.1016/j.jlp.2023.105049
10. Wang, Z.F., Xu, J.W., He, X.Q.: Analysis of spatiotemporal influence patterns of toxic gas monitoring concentrations in an urban drainage network based on IoT and GI. Pattern Recognition Letters. **238**, 237–246 (2020) https://doi.org/10.1016/j.patrec.2020.07.022

Chapter 29
Automatic Calibration Method for High Resolution LiDAR and Fisheye Camera

Jun Hu, Zuotao Ning, Haoxiang Jie, Lifeng Liu, Hongfei Yu, and Jin Lv

Abstract With the rapid development of autonomous vehicles, researchers and engineers have brought LiDAR sensors to the academic and the industrial attention since the LiDAR sensor can not only provide accurate range information but also generate various geometrical structures. Importantly, the combination of cameras and LiDAR sensors, as a typical example, benefits the perception a lot by offering rich and precise sensing data. In this paper, we propose a novel pipeline that can estimate an accurate rigid-body transformation when calibrating the extrinsic parameter between a fisheye camera and a LiDAR sensor. This system uses a checkerboard as the target to build the correspondence between these two sensors. By using the extracted 2D feature points and 3D feature points from a camera and a LiDAR sensor, the extrinsic parameter is calculated via a reformulated approach. In addition, we demonstrate the result of the proposed method and conclude this method lastly.

J. Hu · Z. Ning · H. Jie (✉) · L. Liu · J. Lv
Neusoft Reach Automotive Technology Ltd, Shenyang, China
e-mail: jie.hx@reachauto.com

J. Hu
e-mail: hu.jun@reachauto.com

Z. Ning
e-mail: ningzt@reachauto.com

L. Liu
e-mail: liu-lf@reachauto.com

J. Lv
e-mail: lv.j@reachauto.com

H. Yu
Liaoning Petrochemical University, Fushun, China
e-mail: yuhfln@163.com

R. Kountchev et al. (eds.), *Multidimensional Signals, Augmented Reality and Information Technologies*, Smart Innovation, Systems and Technologies 374,
https://doi.org/10.1007/978-981-99-7011-7_29

29.1 Introduction

Multi-sensor system is widely applied on robots and autonomous vehicles. By using a multi-sensor system, for example, an intelligent vehicle can both collect rich color information and accurate range information from the camera and LiDAR sensor respectively. Based on this fusion system, the state-of-the-art approaches can be easily deployed to implement tasks (vehicle detection, pedestrian detection, etc.). In order to percept accurately, the camera-LiDAR fusion is utilized in navigation tasks [1], simultaneous localization and mapping tasks [2] and object detection and tracking tasks [3, 4].

Considering that an exact correspondence between different types of sensors is the base of a multi-sensor system, the extrinsic parameter calibration is essential and compulsory before setting this system. The extrinsic parameter calibration is a process to calculate the extrinsic transformation (translation and rotation) between two different types of sensors.

In this paper, we propose a LiDAR-camera extrinsic parameter calibration method. The manuscript is structured with four sections of related work, method, test and result, and conclusion. Through the above-mentioned sections, this work is reviewed, illustrated, tested and concluded respectively. The main contributions of this paper are summarised as follows:

- A target-based extrinsic parameter calibration method for a fisheye camera and a high-resolution LiDAR sensor is proposed;
- An automatic calibration system that can meet the calibration requirements at the industrial level is developed;
- A novel estimation method to calculate the extrinsic transformation is reformulated and designed.

29.2 Related Work

In terms of using a calibration target or not, the extrinsic calibration method between a camera and a LiDAR sensor can be roughly categorized into target-based method and targetless-based method respectively. For the target-based ones, it is essential to design a reference target (like checkerboards and ArUco tags) or find a corresponding feature (like boxes) to be extracted easily. Typically, a checkerboard is used as the target pattern to calculate the extrinsic calibration between a camera and a LiDAR [5, 6]. Similarly, in [7], an improved method is proposed to minimize the collecting poses of the checkerboard to one by using 3D line and plane relations. Cardboard boxes, as the reference target, are used to detect box planes to deal with the calibration task of RGB cameras and LiDAR sensors [8].

For those methods that do not require any targets, the key to calibrating is to detect and model typical patterns in the nature, which is a function solving problem on the basis of geometric constraints. In other words, the targetless methods are embedded with devices, like robots and autonomous vehicles, to calculate the extrinsic parameters by natural objects in the real world but not any certain targets in the lab. Being beneficial from deep learning, several approaches have been proposed to estimate the extrinsic parameter. RegNet [9] is the first approach to calculate the extrinsic parameter between a scanning LiDAR and a monocular by using the deep convolutional neural network (CNN). In addition, RegNet establishes a nonmanual platform that provides calibrating online. Furthermore, CalibNet [10] integrates a module that utilizes geometry information and designs a layer of 3D spatial transformer in the model. Besides the above-mentioned methods, DXQ-Net [11] brings a differentiable pose estimation module for generalization. Instead of solving the regression problem as most previous works, it builds a probabilistic model to improve the performance of calibrating.

The recent studies have achieved better performance on the LiDAR-to-Camera calibration; however, those off-the-shelf methods are mainly focusing on calibrating the extrinsic parameter between a flat camera and a LiDAR sensor. In this paper, we propose a novel method that can effectively calibrate the extrinsic parameter between a fisheye camera and LiDAR sensor. The details are described in the following sections.

29.3 Method

In this paper, we propose a method to estimate the pose between a LiDAR sensor and a camera. The overview of the calibration process is illustrated in Fig. 29.1. Firstly, the image and the point cloud are collected from the Camera and the LiDAR sensor respectively. The correspondence is built based on the 2D feature points from the image and the 3D feature points from the point cloud. To be specific, the feature points are the inner corners of the checkerboard shown in Fig. 29.2. Finally, the transformation between a camera and a LiDAR sensor is estimated by using the correspondence of the 2D feature points and 3D feature points.

29.3.1 Feature Points Extraction

In this part, we define the inner corners (shown in Fig. 29.2) of the checkerboard as the 2D feature points and 3D feature points for the image and point cloud respectively. The detail is illustrated in the following sub-sections.

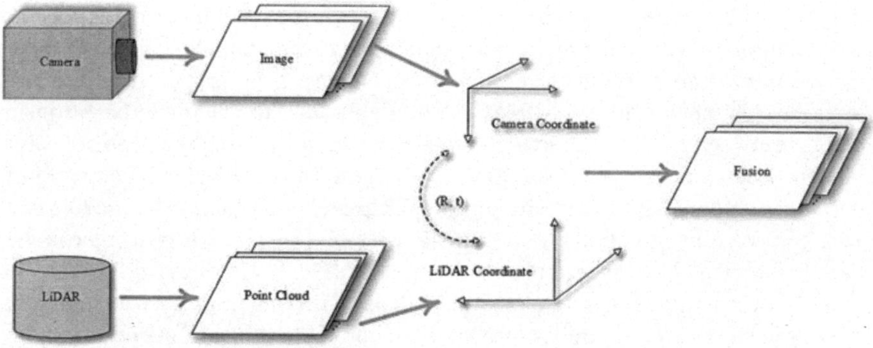

Fig. 29.1 Overview of the proposed method

Fig. 29.2 A 80 cm * 70 cm checkerboard (Left) and that with 7 * 6 inner corners (Right)

29.3.2 2D Feature Points from the Image

There are many approaches to detect the corner points of a checkerboard. To implement this task, Harris points [12] and Shi-Tomasi corners [13] are widely applied to localising the corner points in an image. The Harris corner operator is used to detect the checkerboard corner of images to be calibrated in [14]. A new method is introduced in [15] for detecting corners on checkerboards, which involves using a localised Radon transform and is implemented with large box filters to increase robustness against low contrast, image noise, and blur.

The above-mentioned methods are mainly focusing on extracting corner points via a flat camera. However, those approaches cannot meet the requirement in this work to calibrate between a camera and a fisheye camera. Therefore, we use the method introduced in [16] to detect the corners of images on fisheye cameras. Importantly, this method is designed to remedy the shortcoming on detection images with gross distortion. The main idea of this work is to introduce an improved heuristic that can be verified geometrically to work under gross distortions. And the details can be found in the cited original work.

29.3.3 3D Feature Points from the Point Cloud

In order to detect the 3D feature points, we apply the technique in [17] named ACSC(Automatic extrinsic calibration for non-repetitive scanning solid-state LiDAR and camera systems) to extract the 3D corner points from the point cloud. In terms of the 3D feature points extraction, ACSC contributes to the following aspects.

Initially, a time-domain integration pre-process is applied to make use of the character of non-repetitive scanning point clouds. Secondly, a feature refinement process is adopted to generate a noise-free point cloud measurement from the raw point cloud. Based on the above processes, a non-linear optimisation scheme is proposed to estimate 3D corner points. Specifically, a global reflectance distribution is introduced as a constrain to solve this optimisation problem. In addition, the interactively extrinsic parameter solving is used to overcome the issue of limited samples of checkerboard measurement.

Then the extrinsic parameter of a camera and a LiDAR sensor is calculated based on the above generated 2D and 3D feature points.

29.3.4 Extrinsic Parameters Estimation

Considering that transformation matrix T_{LiDAR}^{camera} cannot be directly used to optimise and solve the pose between a LiDAR sensor and a Camera, it is reasonable to think of using Lie algebraic [18–20]. Through solving this optimisation problem it can be rewritten as a structure of both a differentiable manifold and an algebraic group. The initial transformation matric is shown as in Eq. (29.1), where T_{LiDAR}^{camera} denotes the pose transformation between a LiDAR sensor and a camera, R_{LiDAR}^{camera} and t_{LiDAR}^{camera} are the rotation matric, and translation vector between those two sensors.

$$T_{LiDAR}^{camera} = \begin{bmatrix} R_{LiDAR}^{camera} & t_{LiDAR}^{camera} \\ 0 & 1 \end{bmatrix} \tag{29.1}$$

By using Lie algebraic, we define a τ as below (29.2), where the former 3-dimension (ρ) and the latter 3-dimension (φ) elements denote translation and rotation between a LiDAR sensor and a camera respectively.

$$\tau = (\rho\varphi)^T \tag{29.2}$$

Table 29.1 Sensor parameters

Sensor	Type	Field of View (FoV)	Resolution
LiDAR	Mechanal (with 128 Lines)	Horizontal 360° Vertical 40°	Horizontal 0.2° Vertical 0.1°
Camera	Fisheye	180°	1280 * 720

To solve this transformation (29.1) is reformatted as show in (29.3), where $J = \frac{\sin\theta}{\theta}\mathrm{I} + (1 - \frac{\sin\theta}{\theta})aa^T + \frac{1-\cos\theta}{\theta}a\wedge$ and $\varphi = \theta a$.

$$\mathrm{T}_{\mathrm{LiDAR}}^{\mathrm{camera}} = \mathrm{e}^{\tau} = \begin{bmatrix} \mathrm{e}^{\varphi^{\wedge}} & J\rho \\ 0 & 1 \end{bmatrix} \tag{29.3}$$

Lastly, in order to estimate the extrinsic, we define a function as the cost function shown in (29.4).

$$\mathcal{L} = \sum_{j}^{n} K \cdot e^{\tau} \cdot {}_{j}^{LiDAR}P - {}_{j}^{Camera}P \tag{29.4}$$

29.4 Tests and Results

29.4.1 Setup

In this paper, we use a mechanical LiDAR sensor and a fisheye camera to collect data. Before the LiDAR-camera calibrating process, it is necessary to calibrate the intrinsic parameter of the camera in advance. Because the method itself does not integrate the camera calibration module.

And the sensor details are described in the following Table 29.1.

29.4.2 Data Collection

Before collecting frames from sensors, it is necessary to prepare a printed checkerboard properly. As for this work, we use a 80 cm * 70 cm checkerboard divided by 8 * 7 cells as shown in Fig. 29.2(Left).

Considering the different FoV of the camera and LiDAR sensor [21, 22], we set the data collecting spots of the checkerboard into mainly 9 groups to cover near, middle and far positions by longitude and latitude distance respectively. For each group, we set 5 different poses of the checkerboard facing the sensors. Following the settings upon, we have 45 capturing frames in total for each sensor.

29.4.3 Tests and Results

The proposed method is tested in the indoor and the outdoor scenarios respectively. After collecting data as before, we have 45 frames point cloud frames and images. Based on this method, using the collected data frames, an extrinsic parameter known as T_{LiDAR}^{camera} is estimated [23–25].

Directly, we test the indoor cases as soon as we get T_{LiDAR}^{camera}. Firstly, we collect frames of point clouds and images with a checkerboard by different posed positions. Following this, for all posed point clouds, images are generated via the extrinsic parameter. In the last step, to visualise the results, we project the generated images from point clouds onto the frames of the original images. Clearly, the visualisation is shown as Fig. 29.3.

In addition, we test the outdoor cases with the same setups as the way mentioned above and collect frames of point clouds and images without a checkerboard in an outside driving scenario. In order to demonstrate the proposed method in other perspectives, we conduct a projection from images to point clouds. According to the estimated T_{LiDAR}^{camera}, it is vivid to show the calibrating results by colouring all point cloud frames by the aligned images. And the testing results of outdoor scenarios can be seen in Fig. 29.4.

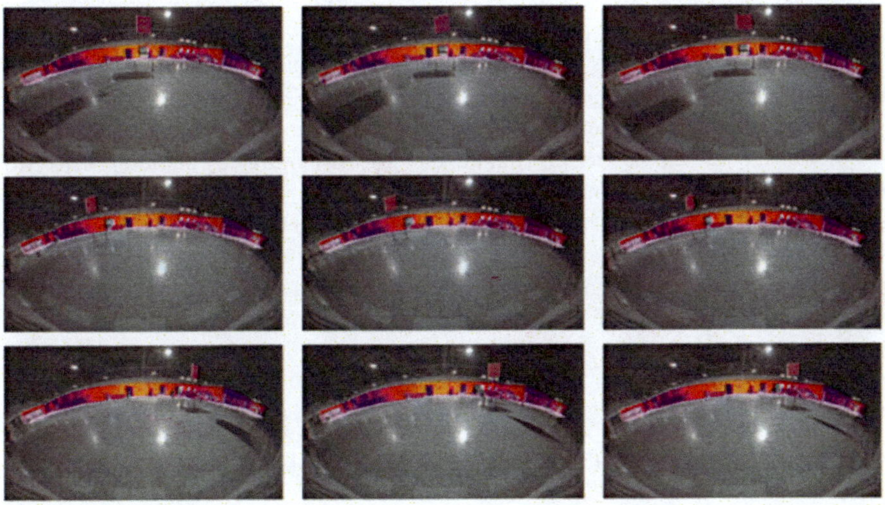

Fig. 29.3 Indoor, calibration results of point clouds projecting on cameras

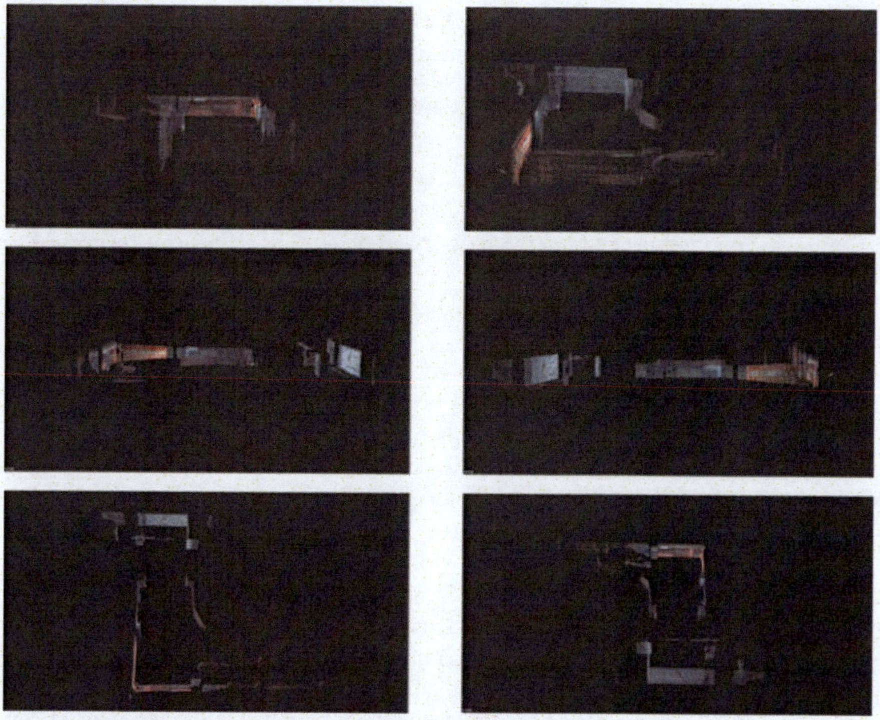

Fig. 29.4 Outdoor, calibration results of images projecting on point clouds

29.5 Conclusion

In this paper, we propose the extrinsic parameter calibration method that can meet the calibrating requirements at the industrial level. We build this pipeline to solve the extrinsic parameter calibration between a LiDAR sensor and a camera. Essentially, we introduce a new method to reformulate the transformation calculation process in order that the extrinsic parameter can be directly and correctly estimated.

In this work, we only use and test the proposed method by using a high-resolution LiDAR sensor. However, intuitionally, it may deliver a poor performance by using a LiDAR sensor with less scanning lines. Therefore, we plan to evaluate this proposed method with a less scanning line LiDAR sensor and improve this system if the mentioned issue arises.

Acknowledgements This work was supported in part by Intelligent Control Theories and Key Technologies of Heterogeneous Unmanned Electric Commercial Vehicles Formation under Grant U22A2043 and in part by Development and Application of Vehicle Cloud Collaboration Self-Evolution Platform for Advanced Automatic Driving under Grant 2022JH1/10400030.

References

1. Hussein, A., Hussein, A., Marín-Plaza, P., Martín, D., et al.: Autonomous off-road navigation using stereo-vision and laser-rangefinder fusion for outdoor obstacles detection. 2016 IEEE Intelligent Vehicles Symposium (IV), Gothenburg, Sweden. pp. 104–109 (2016) https://doi.org/10.1109/IVS.2016.7535372.
2. Zuo, X.X., Geneva, P., Lee, W., Liu, Y., Huang, G.: LIC-Fusion: LiDAR-Inertial-Camera Odometry. 2019 IEEE/RSJ International Conference on Intelligent Robots and Systems (IROS), Macau, China, pp. 5848–5854 (2019). https://doi.org/10.1109/IROS40897.2019.8967746.
3. Vora, S., Lang, A.H., Helou, B., Beijbom, O.: PointPainting: Sequential Fusion for 3D Object Detection. 2020 IEEE/CVF Conference on Computer Vision and Pattern Recognition (CVPR), Seattle, WA, USA, pp. 4603–4611 (2020). https://doi.org/10.1109/CVPR42600.2020.00466.
4. Kim, A., Ošep, A., Leal-Taixé, L.: EagerMOT: 3D Multi-Object Tracking via Sensor Fusion. 2021 IEEE International Conference on Robotics and Automation (ICRA), Xi'an, China, pp. 11315–11321 (2021). https://doi.org/10.1109/ICRA48506.2021.9562072.
5. Geiger, A., Moosmann, F., Car, Ö., Schuster, B.: Automatic camera and range sensor calibration using a single shot. 2012 IEEE International Conference on Robotics and Automation, Saint Paul, MN, USA, pp. 3936–3943(2012), https://doi.org/10.1109/ICRA.2012.6224570.
6. Zhou, L. Deng, Z.: A new algorithm for computing the projection matrix between a LIDAR and a camera based on line correspondences. 2012 IV International Congress on Ultra Modern Telecommunications and Control Systems, St. Petersburg, Russia, pp. 436–441(2012). https://doi.org/10.1109/ICUMT.2012.6459706.
7. Zhou, L., Li, Z., Kaess, M.: Automatic Extrinsic Calibration of a Camera and a 3D LiDAR Using Line and Plane Correspondences. 2018 IEEE/RSJ International Conference on Intelligent Robots and Systems (IROS), Madrid, Spain, pp. 5562–5569(2018). https://doi.org/10.1109/IROS.2018.8593660.
8. Pusztai, Z. Hajder, L.: Accurate Calibration of LiDAR-Camera Systems Using Ordinary Boxes. 2017 IEEE International Conference on Computer Vision Workshops (ICCVW), Venice, Italy, pp. 394–402(2017). https://doi.org/10.1109/ICCVW.2017.53.
9. Schneider, N., Piewak, F., Stiller, C. Franke, U.: RegNet: Multimodal sensor registration using deep neural networks. 2017 IEEE Intelligent Vehicles Symposium (IV), Los Angeles, CA, USA, pp. 1803–1810(2017). https://doi.org/10.1109/IVS.2017.7995968.
10. Iyer, G., Ram, R.K., Murthy, J.K. Krishna, K.M.: CalibNet: Geometrically supervised extrinsic calibration using 3D spatial transformer networks. In 2018 IEEE/RSJ International Conference on Intelligent Robots and Systems (IROS), IEEE, Madrid, Spain, pp. 1110–1117 (2018)
11. Jing, X., Ding, X., Xiong, R., Deng, H., Wang, Y.: XQ-Net: Differentiable LiDAR-Camera Extrinsic Calibration Using Quality-aware Flow. 2022 IEEE/RSJ International Conference on Intelligent Robots and Systems (IROS), Kyoto, Japan, pp. 6235–6241 (2022). https://doi.org/10.1109/IROS47612.2022.9981418.
12. Harris, C., Stephens, M.: A combined corner and edge detector. Alvey vision conference, Vol. 15. No. 50 (1988)
13. Shi, J. B., Tomasi: Good features to track. 1994 Proceedings of IEEE Conference on Computer Vision and Pattern Recognition, Seattle, WA, USA(1994). pp. 593–600, https://doi.org/10.1109/CVPR.1994.323794.
14. Douskos, V., Kalisperakis, I., Karras, G., et al.: Fully automatic camera calibration using regular planar patterns. Int. Arch. Photogram. Remote Sens. Spatial Inf. Sci, 37, 21–26 (2008)
15. Duda, A., Frese, U.: Accurate Detection and Localization of Checkerboard Corners for Calibration, 29th British Machine Vision Conference, pp.126 (2018)
16. Rufli, M., Scaramuzza, D., Siegwart, R.: Automatic detection of checkerboards on blurred and distorted images. 2008 IEEE/RSJ International Conference on Intelligent Robots and Systems, Nice, France, pp. 3121–3126(2008). https://doi.org/10.1109/IROS.2008.4650703.
17. Cui, J., Niu, J., Ouyang, Z., He, Y., Liu, D.: ACSC: Automatic Calibration for Non-repetitive Scanning Solid-State LiDAR and Camera Systems(2020). https://doi.org/10.48550/arXiv.2011.08516.

18. Beltran, J., Guindel, C., Escalera, A.D.L., Garcia, F.: Automatic extrinsic calibration method for LiDAR and camera sensor setups. IEEE Transactions on Intelligent Transportation Systems (2022)
19. Yu, J.Y., et al.: Automatic extrinsic self-calibration of mobile lidar systems based on planar and spherical features. Meas. Sci. Technol. **32**(6), 1–19 (2021)
20. Ma, T., Liu, Z., Yan, G., Li, Y.: CRLF: Automatic Calibration and Refinement Based on Line Feature for LIDAR and Camera in Road Scenes, arXiv e-prints (2021). https://doi.org/10.48550/arXiv.2103.04558
21. Zhang, X., Zhu, S., Guo, S., Liu, H.: Line-based automatic extrinsic calibration of lidar and camera. 2021 IEEE International Conference on Robotics and Automation (ICRA), Xi'an, China, pp. 9347–9353(2021). https://doi.org/10.1109/ICRA48506.2021.9561216.
22. Sungdae, S., Juil, S., Kiho, K.: Indirect correspondence-based robust extrinsic calibration of LIDAR and camera. Sensors, **16**(6), 933 (2016)
23. Zhou, L.: A new minimal solution for the extrinsic calibration of a 2d LIDAR and a camera using three plane-line correspondences. IEEE Sensors Journal (2014)
24. Kumar, V. R., Milz, S., Simon, M., Witt, C., Amende, K., Petzold, J., et al.: Monocular fisheye camera depth estimation using sparse lidar supervision. arXiv e-prints (2018)
25. Hausberg, J., Ishikawa, R., Roxas, M., Oishi, T.: Relative drone -- ground vehicle localization using lidar and fisheye cameras through direct and indirect observations. arXiv e-prints (2020)

Chapter 30
Design and Implementation of a Multi-point Target Position Determination Method for Industrial Robots

Qingqing Zhang, Zhiqiang Wan, and Hao Zhang

Abstract In order to solve the problem that when the target position is changed, the conventional working mode of the industrial robot is not flexible enough to pick up the target object accurately, we give an intelligent vision system and method for determining the target position of an industrial robot with multiple points A multi-point target position determination system and method are presented. The overall structure of an intelligent vision system for multi-point target location determination of industrial robots is designed, and the structural flow and working principle of the system are described in detail. The research results show that the 3D information of the surrounding environment and target objects as well as the walking path are generated in real time according to the specific working environment of the industrial robot, and thus can be flexibly adapted to different working environments.

30.1 Introduction

In industrial plants, there are often a large number of mechanical equipment and finished semi-finished workpieces. When industrial robots need to pick up the corresponding items accurately and quickly, they often need to rely on intelligent vision systems to analyze and judge the specific location, specific shape and smooth path of the items. Therefore, intelligent vision system is an important guarantee that industrial robots can work efficiently. With the rapid development of industrial intelligence, machine vision technology has become one of the key technologies indispensable for industrial intelligence [1]. In some work environments that are not suitable for manual work or the amount of work where manual vision is difficult to meet the requirements, the use of machine vision to replace manual vision can greatly improve production efficiency and automation [2].

Q. Zhang (✉) · Z. Wan · H. Zhang
Anhui Technical College Of Mechanical and Electrical Engineering, Wuhu 241002, Anhui, China
e-mail: 280066077@qq.com

R. Kountchev et al. (eds.), *Multidimensional Signals, Augmented Reality and Information Technologies*, Smart Innovation, Systems and Technologies 374,
https://doi.org/10.1007/978-981-99-7011-7_30

Nowadays, industrial robots often do not have intelligent vision systems and can only move back and forth on a pre-set route, so that the objects to be picked up can only be placed by humans at the specified location for the industrial robots to pick up, which is extremely inflexible and affects work efficiency. For example, if an obstacle appears on the path, the robot is not able to avoid it in time, resulting in collision damage, and once the target position is changed, the target object cannot be picked up accurately [3].

This paper provides an intelligent vision system and method for industrial robots to determine the target position at multiple points, in order to solve the problem that the conventional working mode of industrial robots proposed above is not sufficiently resilient, such as not being able to avoid obstacles in the walking path in time, resulting in the industrial robots being prone to collision damage, and not being able to pick up the target object accurately once the target position is changed.

30.2 Introduction of Intelligent Vision System

30.2.1 Introduction to Machine Vision Technology

The Society of Manufacturing Engineers (SME) Machine Vision Section and the Robotics Industries Association's (RIA) Vision Automation Section give the following definition [4]: "Machine vision is a device that automatically receives and processes images of real objects by means of optical devices and non-contact sensors to obtain the required information or to control robot motion".

Machine vision system is mainly through the image acquisition unit will be taken into the target converted to the image signal through the image processing unit for digital processing, according to the digital processing results such as brightness, color, area, length, number, location and other information to size, angle, shape, color, number of discrimination, etc., in order to command the actuator such as detection, measurement, positioning or identification and other corresponding controls. The actuators are commanded to perform corresponding control actions such as detection, measurement, positioning or identification. A typical machine vision system generally consists of four major functional modules: image formation, image acquisition, image analysis and control, and result output, corresponding to light source and optical imaging system, image capture card, digital image analysis software, and judgment decision and electromechanical control execution modules, respectively [5].

Typical applications of machine vision in industry are: inspection, measurement, positioning and identification.

(1) The goal of machine vision inspection is to improve the flexibility and automation of production. Machine vision inspection method is generally through the optical imaging and area image acquisition device to obtain the digital image of

the product, and then use vision algorithm software to process and analyze the image, to obtain relevant inspection information, to form a judgment decision on the product being inspected, and finally send this information to the control device mechanism, to complete the next action, such as sub-freezing and easy to not remove.

(2) The shape of the object under test often has a geometric shape, and the measured dimensions usually include multiple parameters, such as distance, angle, area, length, round hole diameter, curvature, etc., as well as product thickness, height difference, etc. The dimensional measurement method of machine vision can not only obtain the dimensional parameters, but also give timely feedback according to the measurement results to correct the processing parameters to avoid generating more defective products and reduce losses.

(3) Machine vision positioning technology can automatically determine the position of the object and output the position information according to certain communication protocols. To complete the "grasp or place" action of the industrial robot on the automatic production line, it is necessary to obtain the positioning information of the operated object.

(4) Identification is mainly the acquisition and tracking of object information. The use of machine vision for image acquisition, processing, analysis and understanding can identify a variety of different patterns of targets and objects. The most typical application in image recognition is the recognition of QR code, which stores a large amount of data information in a small QR code, through which the product is tracked and managed. Machine vision systems can read and identify 1D or 2D codes on various material surfaces, greatly improving the efficiency of modern production.

The development of machine vision technology has gone through the following stages. 1960s: The origin of machine vision technology. Roberts, an American scholar, proposed the idea of extracting three-dimensional structure from 2D images, which sparked interest in machine vision at the Massachusetts Institute of Technology (MIT) Artificial Intelligence Laboratory and other institutions, and marked the beginning of the three-dimensional machine vision.

30.2.2 Intelligent Vision System

A smart vision system is a micro-compact vision inspection system with simultaneous image acquisition, image processing and information transmission, also known as a smart camera. Smart cameras generally include image acquisition unit, image processing unit, image processing software, network communication device and other components [6]. In Fig. 30.1 the structure block diagram of the smart camera is shown.

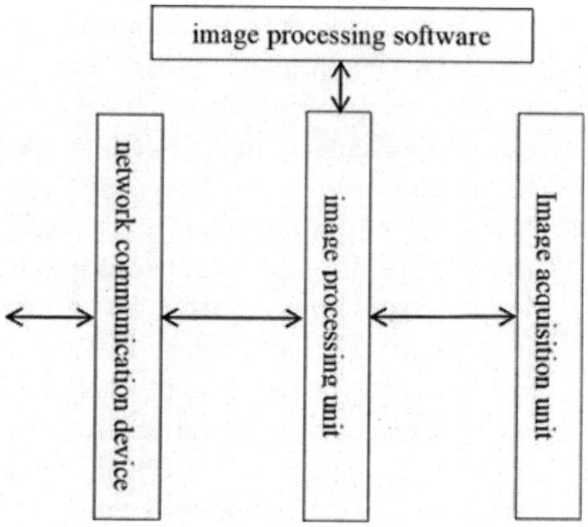

Fig. 30.1 Block diagram of the structure of the smart camera

(1) Image acquisition unit
In an intelligent camera, the image acquisition unit is equivalent to a CCD/CMOS camera and an image acquisition card in the ordinary sense. It converts optical images into analog/digital images and outputs them to the image processing unit.
(2) Image Processing Unit
The image processing unit is similar to an image capture/processing card. It stores image data from the image acquisition unit in real time and performs image processing with the support of image processing software.
(3) Image processing software
The image processing software mainly completes image processing functions with the support of the image processing unit hardware environment, such as geometric edge extraction, Blob, gray scale histogram, OCV/OVR, simple localization and search, etc. In the smart camera, the above algorithms are packaged into fixed modules, which can be applied directly by the users without programming.
(4) Network communication device
Network communication device is an important part of smart camera, which mainly performs the task of communication of control information and image data. Smart cameras generally have built-in Ethernet communication devices and support a variety of standard networks and bus protocols, thus enabling multiple smart cameras to form a larger machine vision system.

Although machine vision varies, it basically includes the following aspects [7].

- Image acquisition, the optical system acquires the image, and the image is converted to digital format and transferred to the computer memory. In a smart camera, the image acquisition unit converts the optical image into an analog/digital image and outputs it to the image processing unit.
- Image processing: The processor applies different algorithms to enhance the image pixels that are important for detection. The image processing unit stores

the image data from the image acquisition unit in real time and completes the image processing functions with the support of the image processing software.

- Holding feature extraction: The processor identifies and quantifies the key features of the image, such as position, number, area, etc. These data are then transmitted to the control program by the network communication device.
- Judgment and control: The processor's control program makes conclusions based on the data received. For example, whether the position is within specification or how the actuator moves to pick up a part.

30.3 Overall Technical Design Scheme

The intelligent vision system and method for multi-point determination of target position of an industrial robot facilitates real-time generation of 3D information of the surrounding environment and the target object as well as the travel path according to the specific working environment of the industrial robot, thus enabling flexible adaptation to different working environments and avoiding the need to travel according to a prescribed route [8]. An intelligent vision system for industrial robots with multiple points to determine the target position is shown in the following flow Fig. 30.2:

The data cloud processing platform mainly includes an industrial robot, a data transmission system and a data cloud processing platform, wherein a data acquisition system is provided in the body of the industrial robot, and the data acquisition system is connected to the data transmission system, and the data acquisition system is connected to the data cloud processing platform through the data transmission system, and the data cloud processing platform is connected to a three-dimensional construction system and a map construction system, respectively, and the map construction system is connected to a path The map construction system is connected to the path planning system, and the path planning system is interconnected with the industrial robot through the data transmission system.

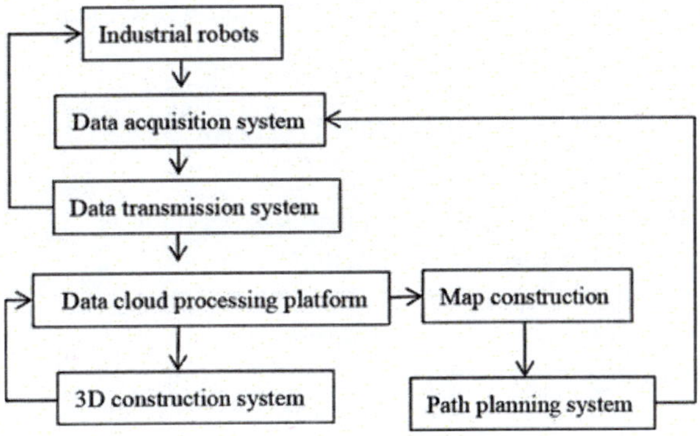

Fig. 30.2 General flow diagram

30.4 Specific Implementation Process

Specific processes for intelligent vision systems for industrial robots:

(1) Turn on the data acquisition system on the industrial robot, at which time the data acquisition system performs infrared imaging and CCD image acquisition of the external environment through two subsystems: the infrared acquisition system and the CCD image acquisition system, while the infrared distance measurement module within the infrared acquisition system determines the distance between the industrial robot and the surrounding target objects and derives detailed data;
(2) The collected image data and distance data are transmitted to the data cloud processing platform for centralized processing through the data transmission system (GPRS wireless data transmission system module);
(3) The data cloud processing platform will integrate and analyze the received data, and the data cloud processing platform will apply the 3D construction system to construct 3D images of the image data and distance data that have completed the integration and analysis, forming a 3D model of the surrounding environment centered on the industrial robot;
(4) The data cloud processing platform converts the 3D image digital model built by the 3D construction system into a live map of the industrial robot's surroundings through the map construction system, which includes elements such as the target object and the walking environment;
(5) The live map of the completed conversion is reasonably planned for the walking path of the industrial robot by means of the path planning system, and the planned path has been marked in the planned path using the obstacle avoidance system to avoid the obstacles along the way, as well as the object recognition and positioning system to mark the three-dimensional characteristics of the surrounding target objects and the three-dimensional coordinates;
(6) The planned path containing the three-dimensional characteristics of the target object and the three-dimensional coordinate markers is transmitted back to the industrial robot body through the data transmission system, and the industrial robot performs its work according to the planned path.

30.5 Data Acquisition System Working Principle

The data acquisition system in this paper includes the infrared acquisition system and CCD image acquisition system, and the infrared acquisition system and CCD image acquisition system are two independent systems working side by side at the same level, which is convenient to collect multi-point image information of the surrounding environment and objects, and avoid the situation that the collected information has been missed due to the angle problem.

Figure 30.3 shows the simple working principle of a CCD camera. 3D scenes are imaged on the CCD display through an optical device (lens), and the CCD sensor (Imager) serves to convert the received light energy into electrical charge, which is stored in the form of digital signals in the frame buffer after analog-to-digital conversion, and these digital signals can then be displayed through the display processor for image display, as well as using various machine vision algorithms for processing. The frame buffer plays a central role here, as a high-speed image memory that can store multiple images or their derivatives [9].

The infrared acquisition system includes two modules: infrared imaging and infrared ranging, and infrared imaging means image acquisition of surrounding objects under poor light conditions, and infrared ranging means the distance between the industrial robot and the surrounding target objects is measured by infrared ranging technology, which enables the industrial robot to collect object information even in the case of insufficient light, and image information is collected while the specific distance between the target object and the obstacle is measured. The distance between the target object and the obstacle can be measured at the same time. This is shown in Fig. 30.4.

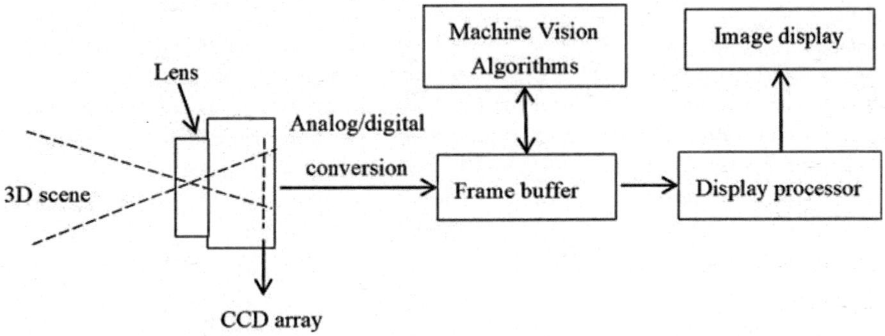

Fig. 30.3 The simple working principle of CCD camera

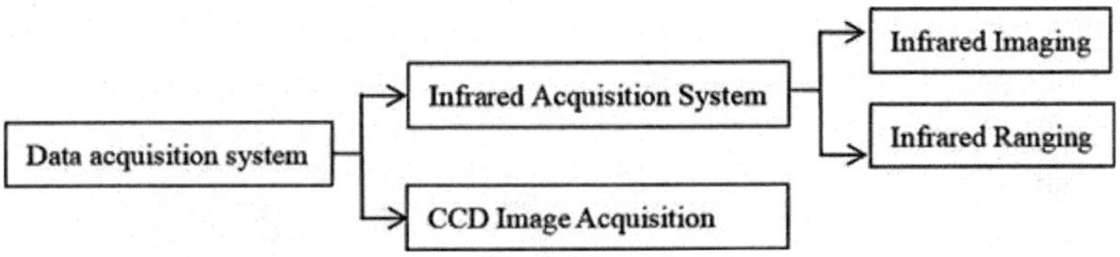

Fig. 30.4 Schematic diagram of data acquisition system structure

30.6 Data Processing System

The data cloud processing platform integrates and analyzes the received data, and the data cloud processing platform applies a three-dimensional construction system to build a three-dimensional image of the image data and distance data that have been integrated and analyzed to form a three-dimensional model of the surrounding environment centered on the industrial robot; the three-dimensional construction system uses the IBMR three-dimensional construction technology. Modeling and Rendering) is called image modeling and drawing, which is a method to generate a three-dimensional scene through certain visual processing algorithms for the collected two-dimensional images. The data cloud processing platform converts the 3D image digital model built by the 3D building system into a live map of the industrial robot's surroundings through the map building system, which includes elements such as the target object and the walking environment [10, 11].

30.7 Route Planning

The path planning system includes two subsystems: obstacle avoidance and object identification and positioning, and the two subsystems are two independent subsystems working side by side at the same level, which can avoid obstacles in the path during the travel process of picking up the target object to avoid collision and prevent the industrial robot from malfunctioning in the process;

Object recognition and positioning includes two modules: 3D features and 3D coordinates, and 3D features are the calculated 3D characteristics of the industrial robot's surroundings, and 3D coordinates are the 3D specific coordinates of the industrial robot's surroundings, which can identify the shape characteristics of the target object, and also determine the walking environment around the path, so that the target object can be picked up accurately. It is not necessary to place the target object at the specified location for the industrial robot to pick it up as shown in Fig. 30.5.

The construction of the map and localization mainly solves the problem of the robot understanding its specific location in the map. The robot continuously acquires information about its surroundings through various sensors on board, so that its radar scans the surrounding environment and digitizes it, builds a 2D map that the robot

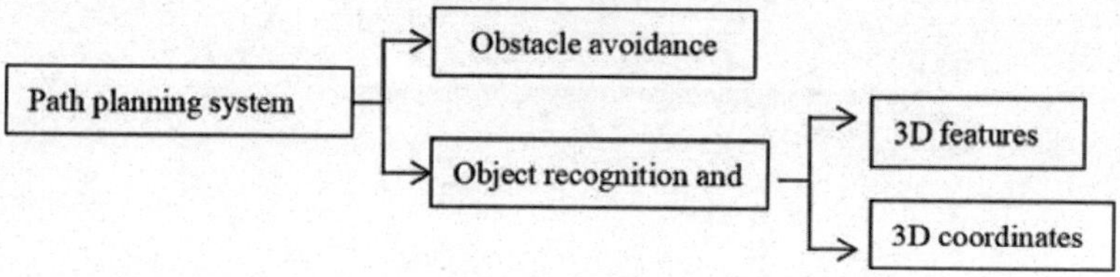

Fig. 30.5 Schematic diagram of the structure of the path planning system

can understand, and determines the specific coordinates of the robot by comparing the map with the sensor information. Next path planning, new path points are added. The path planning mainly solves the problem of how to safely reach a given target point from the current position of the robot. Under the condition that the scene map, the coordinates of the current position and the coordinates of the target point are all determined, the path planning algorithm is used to plan a safe path curve from the current point to the target point, and the local path is updated in real time to avoid collision for dynamic obstacles on the path, so that the robot can travel along the planned path. The robot travels along the planned path. Figure 30.6 shows the path planning diagram of the laboratory robot control software. Move the AGV robot so that its radar scans the surrounding environment and digitizes it, creates a 2D map, and adds new path points. After the path is added, modify the path parameters and input the location parameters x,y,yaw into the autonomous navigation AGV control software. Click "location information" to get the current AGV location parameters, click "collect path points" once for each point and click "save edit" to use the autonomous navigation The point test in the AGV control software can move the AGV to each location.

Part of the code is as follows:

```
class:Move()
def_ _init_ _(self)
self.origin_pose = (0,0,0)
self.processing_pose = (0.35,−0.05,−0.02)
self.unloading_pose = (1.04,0.8,3.1)
self.warehouse_1_pose (−1,1,0)
self.warehouse_2_pose (−1,1,0)
self.warehouse_3_pose = (−1,0,0)
```

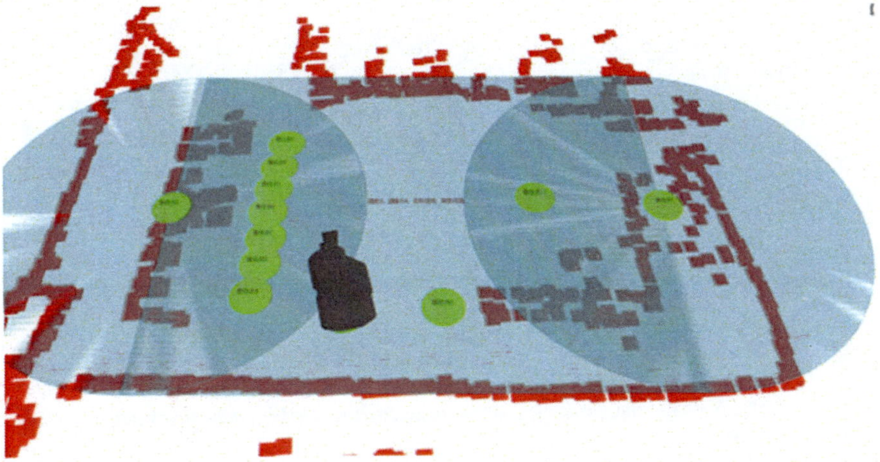

Fig. 30.6 Schematic diagram of path planning

self.warehouse_4_pose = (−0.21,1.21,−0.018)
self.warehouse_5_pose = (−0.21,0.768,−0.018)
self.warehouse_6_pose (−1,1,0)
self.warehouse_7_pose (−1,1,0)

The research results show that the 3D information of the surrounding environment and the target objects as well as the walking path are generated in real time according to the specific working environment of the industrial robot, and thus can flexibly adapt to different working environments.

30.8 Summary

The design and implementation of a multi-point target location determination method for industrial robots, on the one hand, facilitates real-time generation of three-dimensional information of the surrounding environment and the target object as well as the walking path according to the specific working environment of the industrial robot, and thus can flexibly adapt to different working environments and avoid the need to follow a prescribed route. On the other hand, there is no need to manually place the target object to the designated position for the industrial robot to pick up, which reduces the labor intensity of workers and avoids the problem of not being able to pick up the target object accurately due to the displacement of the target object.

At the same time, it prevents industrial robots from having problems with collision failures that cannot get around obstacles, and improves the flexibility of industrial robots to perform route walking, while contributing to making planned routes more reasonable.

Acknowledgements 2021 Anhui Provincial Natural Science Foundation Project on Key Technology Research on Path Optimization of Mine Rescue Robot (KJ2021A1512), 2021 Anhui Provincial Natural Science Foundation Project on Key Technology Analysis and Research on Health Condition Detection and Early Warning of CNC Machine Tools (KJ2021A1515).

References

1. Tang, X.: Machine vision inspection and application. Machinery Industry Press, Beijing (2021)
2. Zhang, M.W.: Elementary tutorial on the application of intelligent vision technology (Xinjie). Harbin Institute of Technology Press, Harbin (2020)
3. Peng, J.G., Zhang, B., et al.: Research on intelligent visual obstacle avoidance for underground mobile robots. Ind. Min. Autom. **46**(9), (2020)
4. Yu, W.Y., Shi, Y.: Automatic machine vision inspection technology. Chemical Industry Press, Beijing (2013)
5. Liu, T. et al.: Machine Vision and Its Application Technology. Beijing, Machinery Industry Press (2021)

6. Zhu, G.M.: Intelligent Vision Technology and Application. Xi'an, Xi'an University of Electronic Science and Technology Press (2021)
7. Milan, S., Vaclav, H., Roger, D.B., et al.: Image Processing, Analysis and Machine Vision Book. Beijing: Tsinghua University Press (2016)
8. Shi, Z.S.: Target location tracking methods and practices. Electronic Industry Publishing House, Beijing (2019)
9. Ding, S.H., et al.: Machine vision technology and application practice. People's Post and Telecommunications Publishing House, Beijing (2022)
10. Hua, Z., Zhang, X. et al.: Research on visualization of mine ventilation system based on IBMR technique in wavelet space. Coal Mine Mach. **27**(8), (2006)
11. Cao, T.Y., Cai, H.Y., et al.: Combined image content matching for robot visual navigation and global map construction system. Opt. Precis. Eng., **25**(8), (2017)

Chapter 31
Design of Contact Protection Intelligent Guide Robot Based on Multi Environment Simulation Detection

Zhiqiang Wan and Qingqing Zhang

Abstract Blind people are vulnerable groups without visual perception ability, and walking in urban roads with complex traffic often brings great inconvenience and danger; The detection environment of common blind guide robots is relatively simple and the detection efficiency is not high. In this paper, the detection platform, motor and anti-collision plate and other devices are designed, which can simulate the road obstacles and uneven ground environment and improve the detection efficiency. The guide robot mainly serves users with visual obstacles who are not able to move. At the same time, when the guide robot contacts with obstacles and uneven roads, it should be protected to avoid damage to the guide robot. The device can be used in many different environments and has a good market prospect.

31.1 Introduction

With the development of science and technology, the intelligent guide robot aims to provide navigation services for users with visual impairments who are not able to move. Through sensor technology and intelligent data processing technology, it can identify obstacles and path planning on normal roads, which will greatly facilitate the travel of people with visual impairments [1]. In the production and processing of intelligent guide robots, it is necessary to detect them to ensure their accurate navigation, The use of environmental simulation detection devices is an existing common means. However, the existing environmental simulation detection devices have the following problems in use:

In the normal road, there are many obstacles for the intelligent blind guide robot. The more common obstacles are the occlusion of objects on the way and the uneven road surface. In order to ensure the smooth progress of users, the intelligent blind guide robot is required to identify and circle these obstacles. The existing environment

Z. Wan (✉) · Q. Zhang
Anhui Technical College of Mechanical and Electrical Engineering, Wuhu, Anhui, China
e-mail: 307609359@qq.com

R. Kountchev et al. (eds.), *Multidimensional Signals, Augmented Reality and Information Technologies*, Smart Innovation, Systems and Technologies 374,
https://doi.org/10.1007/978-981-99-7011-7_31

simulation detection device is not convenient to switch multiple environments. The fixed position is set with multiple obstacles, which takes up space and lacks randomness. At the same time, the existing environmental simulation detection device lacks contact protection for the intelligent guide robot. If the intelligent guide robot encounters obstacles in the detection and fails to identify them in time, it is easy to collide and overturn, which is easy to damage it [2].

The purpose of this design is to provide a contact protection type intelligent guide robot multi environment simulation detection device, to solve the problem that the existing environment simulation detection device mentioned above is inconvenient to switch between multiple environments, and the lack of contact protection for the intelligent guide robot [3].

31.2 2Technical Analysis and Research

A multi-environment simulation detection device for contact protection intelligent blind guidance robot, including a detection platform, a motor, a first screw, a second screw and a collision plate, wherein a vertical plate is fixed with bolts at the top edge of the detection platform, a motor is embedded in the vertical plate, the output end of the motor is connected with a first screw through the vertical plate, the thread sleeve on the first screw is provided with an obstacle wall plate, and the inner side of the obstacle wall plate is fixed with a positioning rod. The positioning rod penetrates the positioning groove, and the positioning groove is arranged in the vertical plate. The upper end surface of the detection table is provided with an adjustment groove, and the inner shaft of the adjustment groove is connected with a second screw. The screw sleeve on the second screw is provided with an adjustment plate. The outside of the barrier wall plate is fixed with a mounting base, and the end of the mounting base is provided with an expansion groove. The expansion groove is connected with a mounting rod through a return spring, and the side end of the mounting rod is fixed with a collision plate. The embedded shaft at the outer side of the anti-collision plate is connected with a guide roller.

A multi-environment simulation detection device for contact protection intelligent blind guidance robot, as shown in Fig. 31.1, comprises of a detection table 1, a vertical plate 2, a motor 3, a first screw 4, an obstacle wall plate 5, a positioning rod 6, a positioning groove 7, a rack 8, a tooth roller 9, a belt 10, a second screw 11, a regulating plate 12, a regulating groove 13, a protective plate 14, a movable groove 15, a movable rod 16, a mounting base 17, an expansion groove 18, a return spring 19, a mounting rod 20, a collision plate 21 and a guide roller 22, The bolt at the top edge of the detection platform 1 is fixed with a vertical plate 2, the vertical plate 2 is embedded with a motor 3, the output end of the motor 3 passes through the vertical plate 2 and is connected with a first screw 4, the screw sleeve on the first screw 4 is provided with an obstacle wall plate 5, the inner side of the obstacle wall plate 5 is fixed with a positioning rod 6, the positioning rod 6 runs through the positioning groove 7, and the positioning groove 7 is arranged in the vertical plate 2, the upper

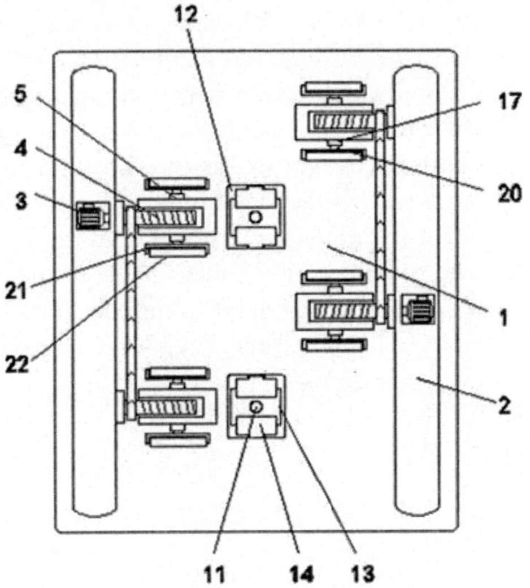

Fig. 31.1 Schematic diagram of the design structure of the guide robot. (1. Test bench; 2. Vertical plate; 3. Motor; 4. First screw; 5. Barrier wallboard; 11. Second screw; 12. Adjusting plate; 13. Adjustment groove; 14. Protective plate; 17. Mounting base; 19. Return spring; 20. Mounting rod)

end surface of the detection platform 1 is provided with an adjusting groove 13, and the inner shaft of the adjusting groove 13 is connected with a second screw 11, The screw sleeve on the second screw rod 11 is provided with an adjusting plate 12, the outside of the barrier wall plate 5 is fixed with a mounting base 17, the end of the mounting base 17 is provided with an expansion groove 18, and the inside of the expansion groove 18 is connected with a mounting rod 20 through a return spring 19, the side end of the mounting rod 20 is fixed with a collision plate 21, and the embedded shaft outside the collision plate 21 is connected with a guide roller 22.

Compared with the prior art, the design has the following advantages:

(1) Through the two sets of staggered barrier wallboards, when the motor drives the first screw to rotate, it can drive the two sets of barrier wallboards to move relative to each other, forming a barrier on the way. At the same time, the positioning rod slides in the positioning groove. When the barrier wallboard is limited, the rack on the positioning rod contacts with the tooth roller, so that the tooth roller can start the rotation of the second screw through the belt, thus driving the adjustment plate to slide vertically in the adjustment groove, When the barrier wallboard moves to block the obstacles, the adjusting plate moves upward to keep the road smooth [4]. On the contrary, when the barrier wallboard returns to its original position, the adjusting plate moves downward to make the adjusting groove and the detection table form an uneven table. By switching the barrier wallboard and the adjusting plate, different traveling environments are simulated to improve the detection efficiency;

(2) Through the protective plates set on both sides of the top of the adjustment plate, the movable slot at the bottom slides in the movable rod. When the adjustment plate moves up and down, the movable rod with spherical structure at the end slides in the movable slot, driving the protective plate to rotate at the top edge of

the adjustment slot. The gap between the adjustment block and the adjustment slot can be filled through the protective plate to avoid the blind guide robot from giving way too late, It will not overturn when reaching the sagging adjustment block;

(3) Through the anti-collision plate with arc structure set on both sides of the barrier wall panel, when the blind guide robot has no time to adjust the route, it contacts with the guide roller on the anti-collision plate, and initially buffers through the rotating guide roller. At the same time, the installation rod slides into the expansion slot under the impact force, and further buffers with the use of the return spring, effectively protecting the blind guide robot from contact [5].

31.3 3Example Given to Illustrate

In Fig. 31.2, two groups of barrier wallboards 5 are arranged on the left and right of the central axis of the detection platform 1, and two groups of barrier wallboards 5 are arranged in each group, and the two groups of barrier wallboards 5 are staggered. Through the two groups of barrier wallboards 5, when the motor 3 in the two vertical plates 2 drives the first screw 4 to rotate, the two groups of barrier wallboards 5 can be driven to slide relative to each other, forming obstacles on the detection platform 1, which is convenient for the blind guide robot to move and detect on the detection platform 1 [6];

In Fig. 31.3, the positioning rod 6 and the positioning slot 7 are connected by sticking and sliding, and the positioning rod 6 and the first screw 4 are parallel to each other. When the motor 3 drives the first screw 4 to rotate, the barrier wall panel

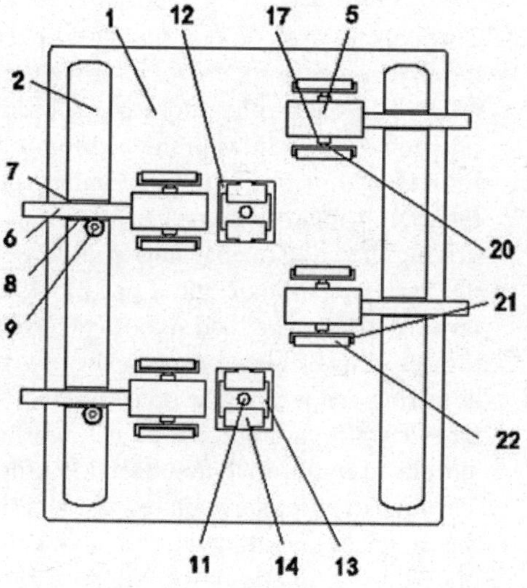

Fig. 31.2 Schematic diagram of the top section structure of the designed toothed roll. (1. Test bench; 2. Vertical plate; 5. Barrier wallboard; 6. Positioning rod; 7. Positioning groove; 8. Rack; 9. Toothed roll; 11. Second screw; 12. Adjusting plate; 13. Adjustment groove; 14. Protective plate; 17. Mounting base; 20. Mounting rod; 21. Anti collision plate; 22. Guide roller)

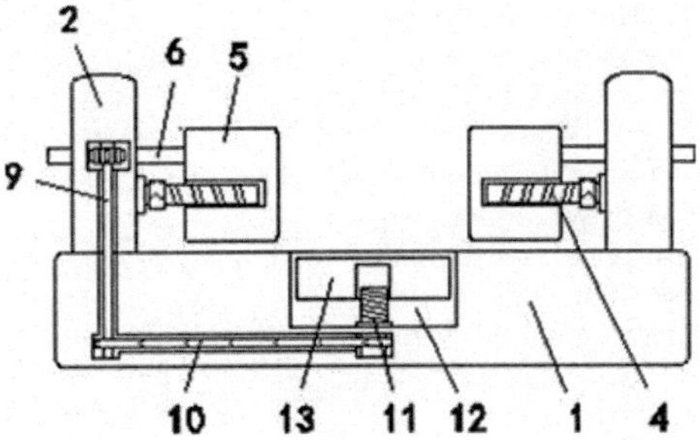

Fig. 31.3 Schematic diagram of the design orthogonal section structure. (1. Test bench; 2. Vertical plate; 4. First screw; 5. Barrier wallboard; 6. Positioning rod; 9. Toothed roll; 10. Belt; 11. Second screw; 12. Adjusting plate; 13. Adjustment groove)

5 can be moved horizontally by limiting the position of the positioning rod 6 and the positioning slot 7 [7, 8];

In Fig. 31.4, a rack 8 is fixed on the side of the positioning rod 6, and a gear roller 9 is arranged on the side of the positioning rod 6. The embedded part through the shaft of the gear roller 9 is connected to the detection table 1 and the vertical plate 2 on the left side. The gear roller 9 is located at the bottom of the detection table 1 and connected with the second screw 11 through a belt 10. The rack 8 is evenly distributed on the positioning rod 6, and the rack 8 and the gear roller 9 are meshed with each other. When the barrier wallboard 5 drives the positioning rod 6 to slide in the positioning slot 7, The rack 8 on the positioning rod 6 contacts with the tooth roller 9 to drive it to rotate [9, 10], so that the tooth roller 9 can drive the second screw 11 to rotate in the adjustment slot 13 through the belt 10;

In Fig. 31.5, the upper shaft of the inner wall at the top of the adjusting plate 12 is connected with the protective plate 14, the bottom of the protective plate 14 is provided with a movable groove 15, one end of the movable rod 16 is placed in the movable groove 15, the bottom of the movable rod 16 is fixed at the top edge of the adjusting plate 12, the protective plate 14 is symmetrically arranged with respect to the central axis of the adjusting plate 12, the adjusting plate 12 is vertically connected with the adjusting groove 13 through the second screw 11, and the top of the movable rod 16 is a spherical structure design, The top of the movable rod 16 slides with the movable groove 15. The diameter of the top of the movable rod 16 is larger than the width of the bottom of the movable groove 15. When the second screw 11 rotates in the adjustment groove 13, it can drive the adjustment plate 12 to slide up and down. At the same time, the movable rod 16 on the adjustment plate 12 slides in the movable

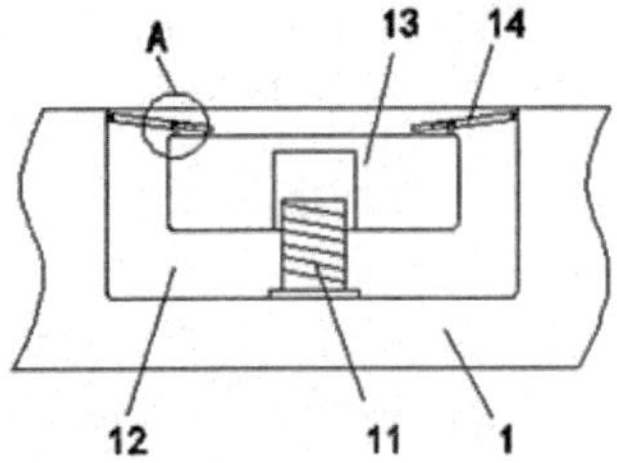

Fig. 31.4 Structural diagram of guard board. (1. Test bench; 11. Second screw; 12. Adjusting plate; 13. Adjustment groove; 14. Protective plate)

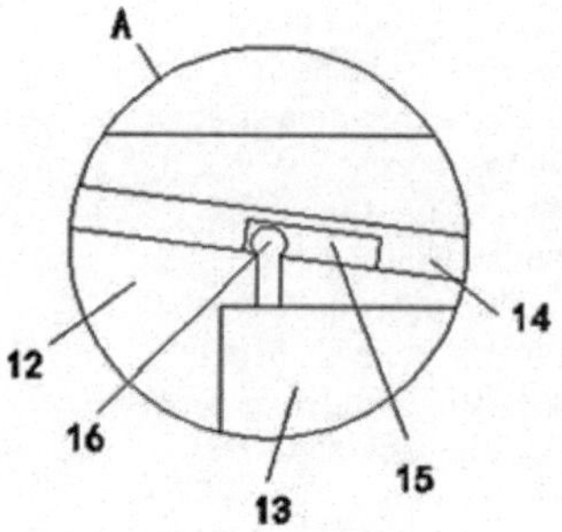

Fig. 31.5 Schematic diagram of enlarged structure at A. (12. Adjusting plate; 13. Adjustment groove; 14. Protective plate; 15. Movable trough; 16. Moving rod; 17. Mounting base; 18. Expansion slot; 19. Return spring; 20. Mounting rod; 21. Anti collision plate; 22. Guide roller)

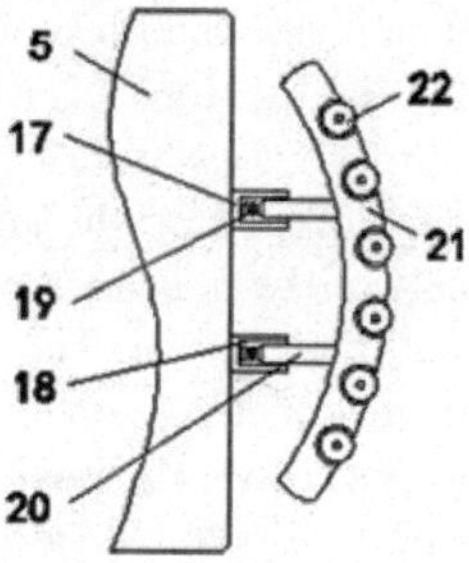

Fig. 31.6 Structural diagram of anti collision plate. (5. Barrier wallboard; 17. Mounting base; 18. Expansion slot; 19. Return spring; 20. Mounting rod; 21. Anti collision plate; 22. Guide roller.)

groove 15, which can drive the protective plate 14 to rotate in the adjustment groove 13, so that the protective plate 14 can fill the gap between the adjustment plate 12 and the adjustment groove 13, Avoid overturning when the guide robot moves into the adjustment slot 13 [11];

In Fig. 31.6, the installation rod 20 forms an elastic sliding structure between the return spring 19 and the expansion slot 18. The installation rod 20 is perpendicular to the anti-collision plate 21. The guide rollers 22 are equally spaced on the anti-collision plate 21. The anti-collision plate 21 is designed as an arc structure, and the anti-collision plate 21 is symmetrically arranged on the central axis of the barrier wall plate 5. When the blind guide robot moves to avoid the barrier wall plate 5 in time, it contacts the guide roller 22 on the anti-collision plate 21. The initial buffering is carried out by turning the guide roller 22 on the anti-collision plate 21. At the same time, the mounting rod 20 slides into the expansion slot 18 under the force. The buffering is carried out again with the use of the return spring 19 to protect the blind guide robot [12].

31.4 Conclusion

As shown in Fig. 31.7, the multi environment simulation detection device of the contact protection intelligent blind guide robot can simulate the road obstacles and uneven ground environment to improve the detection efficiency. At the same time,

Fig. 31.7 Physical picture of blind guide robot

when the blind guide robot contacts the obstacles and uneven roads, it can provide protection to prevent the blind guide robot from being damaged. The design is simple in structure and has a broad market application prospect in the field of blind guide robot.

Acknowledgements This work is supported by the scientific research project of Anhui Provincial Department of Education and funded by the University Natural Science Foundation of Anhui (KJ2021A1512 and KJ2021A1515).

References

1. Wu, Z.H., Rong, X.W., Fan, Y.: Overview of research status of guide robot. Comput. Eng. Appl. (2020)
2. Zhu, A.B., He, D.Y., Luo, W.C., Chen, W.: Research on wearable guide robot based on binocular vision method. Mech. Des. Res. (2016)
3. Chen, L.X., Qiao, Y.H., Zhang, C.Y., Wang, L., Bai, Y.L., Li, B.: Research on cup control technology of robot based on force control guidance. J. Anhui Univ. Sci. Technol. (2021)
4. Wei, Z.C., Liao, J.H., Zhang, Z., Dai, Q.F., Lv, G.H., Liang, R.S.: Intelligent voice guide based on ultrasonic distance measurement and PSD infrared distance measurement. Mod. Electron. Technol. (2013)
5. Wang, R.R., Li, X.H., Chen, T.: Research on the blind walking aid system based on tactile perception. J. Shanxi Datong Univ. (Nat. Sci. Ed.) (2018)
6. Cao, Y.Z., Liu, G., Yang, H.F.: Road zebra crossing recognition method in blind guidance system. Comput. Eng. Appl. (2008)
7. Wang, F., Yin, H.W., Jiang, F.L., Guo, Y.T.: Research and application of natural language processing for service robot. J. Anhui Univ. Sci. Technology. **32**(04), (2018)

8. Tang, Z.C., Su, L., He, C., Hu, R.J.: Research on visual recognition technology of traffic signs for blind guide robots. Comput. Technol. Dev. (2014)
9. Hu, H.M., Xi, W.Y., Li, Y., Shi, Z.F., Qi, C.H.: design and implementation of blind obstacle detection system based on google project tango. Electron. Technol. Softw. Eng. (2019)
10. Liao, Q.H., Hu, W.R., Zeng, W.Y., Hu, L., Zhu, L.: Research on guide crutches based on GPS technology and intelligent obstacle avoidance. Electron. Technol. (2016)
11. Li, S.Y., Yin, J., Chang, F.B., Wei, A.H, Zhou, D.Q.: Design of blind guidance system based on ultrasonic sensor array. China Med. Equip. (2014)
12. Li, Z.K., Xu, G.Z., Guo, M.M.: Design and research of audio-visual fusion guide robot. Prog. Laser Optoelectron. **54**(12), 121506 (2017)

Author Index

R. Kountchev et al. (eds.), *Multidimensional Signals, Augmented Reality and Information Technologies*, Smart Innovation, Systems and Technologies 374,
https://doi.org/10.1007/978-981-99-7011-7